TH
NAT·
Of
ASTROLOGY

"This is an incredible work! Others have written histories of astrology, but none of them have been histories of astrology as well as comprehensive discussions of the sociology of astrology throughout its history. This is not only a superb piece of intellectual history but also an eloquent discussion of where astrology is today and how it has gotten here. In particular Scofield has not only explained and defended astrology on philosophical as well as on other grounds, but in the course of doing so, he has also written a brilliant critique of what he calls the reductionist-mechanistic-materialist (RMM) view, which monolithically dominates modern science to the degree that any corpus of ideas that requires one to see outside of its influence is almost impossible to see. I believe he has given a brilliant critique of the RMM that is useful for any student of modern civilization, not just astrologers. This book I would say is his magnum opus."

ROBERT HAND, AMERICAN ASTROLOGER,
HISTORIAN, AUTHOR, AND SCHOLAR

"Bruce Scofield has crafted the authoritative text on natural astrology—the ancient branch of the subject in which terrestrial and celestial patterns intersect and manifest in the material world in such matters as weather and climate. Scofield artfully interweaves the history of the topic with both modern evidence and his own doctoral research on the relationship between Saturn cycles and temperature variation. The text is accessible, clear, and essential for anyone in search of a full and rounded understanding of astrology's claims and nature."

NICHOLAS CAMPION, PH.D., PRINCIPAL LECTURER
AT THE INSTITUTE OF EDUCATION AND HUMANITIES
AND ASSOCIATE PROFESSOR OF COSMOLOGY AND CULTURE
AT THE UNIVERSITY OF WALES TRINITY SAINT DAVID

"This is a book that astrologers have been awaiting for decades. Drawing on a baker's dozen of cutting-edge sciences, Bruce Scofield levels a potent challenge at pseudoskeptical critics of astrology by setting out a solid basis in reason and evidence for the ancient science of the stars."

JOHN MICHAEL GREER, AUTHOR OF *THE TWILIGHT OF PLUTO*

"*The Nature of Astrology* is a valuable and timely contribution to the field and a necessary examination of the ongoing stigma against this complex and greatly misunderstood subject. Drawing upon historical and contemporary scientific research as well as his own investigations, Scofield methodically reveals how the Earth and the life upon it are influenced by the greater cosmic environment. He also presents an in-depth and rich history of astrology, including new and fascinating insights on astrology's decline, and provides possible avenues for its renewal. Scientists, academics, astrologers, and skeptics will all benefit from reading this captivating and edifying work."

MARLENE SEVEN BREMNER, AUTHOR OF
HERMETIC PHILOSOPHY AND CREATIVE ALCHEMY

"This scholarly tour de force deserves a place on the bookshelf of everyone seriously interested in the widest and deepest terrain of astrology. A richly rewarding read, it fulfills its promise of the history, philosophy, and science of astrology. It is actually a complete university education on the subject. Scofield does an excellent job of answering the perennial question: 'How does astrology work?' Not only answering many of astrology's critics, he plots out an elegant future for this largely misunderstood and underappreciated branch of knowledge."

FREDERICK HAMILTON BAKER, AUTHOR OF
ALCHEMICAL TANTRIC ASTROLOGY

"Scofield's well-researched arguments qualify him to assert that astrology *is* a science. He points out that it has an empirical body of knowledge and relies on the repeatable practices of brilliant ancient astronomers. These rules, procedures, and methodologies, perfected thousands of years ago but still understandable to this day, award that status. In this one book, a giant step in human understanding of nature's solar system and its ultimate, supreme influence has been taken. Without doubt, it describes the genesis of all spiritual understanding and religious symbolism."

ALISON CHESTER-LAMBERT, AUTHOR OF *ASTROLOGY READING CARDS*

"Bruce Scofield's book lives up to its title, *The Nature of Astrology*. Building on his Ph.D. work in the geosciences (with a dissertation titled *A History and Test of Planetary Weather Forecasting*) at the University of Massachusetts (Amherst), Scofield discusses the scope, history, science, sociology, and philosophy of astrology. An important aspect is the place of astrology within a broader cultural and scientific context, which raises fundamental issues regarding the nature of science and scientific evidence, including alternatives to the 'reductionist-mechanistic-materialistic' (to quote Scofield) trend in modern science. Scofield favors systems thinking that goes beyond reductionism; systems can exhibit emergent properties and self-organization. Astrology at its best can be considered a form of systems thinking that has been practiced for millennia. This is a fascinating book that anyone with a serious interest in the intellectual development of humanity should have in their library."

ROBERT M. SCHOCH, PH.D., AUTHOR OF *FORGOTTEN CIVILIZATION:*
NEW DISCOVERIES ON THE SOLAR-INDUCED DARK AGE

THE
NATURE
OF
ASTROLOGY

History, Philosophy,
and the Science of
Self-Organizing Systems

BRUCE SCOFIELD

Inner Traditions
Rochester, Vermont

Inner Traditions
One Park Street
Rochester, Vermont 05767
www.InnerTraditions.com

Text stock is SFI certified

Cataloging-in-Publication Data for this title is available from the Library of Congress

ISBN 978-1-64411-617-3 (print)
ISBN 978-1-64411-620-3 (ebook)

Printed and bound in the United States by Lake Book Manufacturing, LLC. The text stock is SFI certified. The Sustainable Forestry Initiative® program promotes sustainable forest management.

10 9 8 7 6 5 4 3 2 1

Text design by Kenleigh Manseau and layout by Priscilla Baker
This book was typeset in Garamond Premier Pro with Democratica and Gill Sans used as display typefaces

To send correspondence to the author of this book, mail a first-class letter to the author c/o Inner Traditions • Bear & Company, One Park Street, Rochester, VT 05767, and we will forward the communication, or contact the author directly at **naturalastrology.com**.

CONTENTS

Acknowledgments

While hiking with friends many years ago, I was offered a job as a teaching assistant at the University of Massachusetts. The offer came from Lynn Margulis, a well-known scientist, and she added I would require a union card, that is a doctorate, which she would supervise. Lynn said that for my thesis I could test astrology using scientific methodologies as it relates to nature (not people) and that this could be the start of a book that explained a subject she knew nothing about. I took her up on the offer and focused my thesis on testing a single astrological claim in the field of astrometeorology, that Saturn's position relative to other solar system bodies correlates with terrestrial weather in certain ways. This correlation is the kernel of *The Nature of Astrology,* a starting point for a broad look at how the cosmic environment shapes nature and how astrology is a way of studying this phenomenon. Without the initial encouragement and support of Lynn, however, this ambitious project would likely have never happened.

Not many others have traveled down the paths I take in this book, and it began and ended as a solo project. With only a few exceptions, my friends in science and those in the astrological community (I have been active in it for decades) were not very interested in my ideas. I do want to thank Jim MacAllister for reading the manuscript and making comments, and Barry Orr for critiquing content and spotting some grammatical mistakes in Chapter 13. Most important to me has been the support and tolerance given to me by my wife, Nancy. I thank her for putting up with the relentless obsessions with tidbits of science, history, and philosophy that were required to accomplish this project.

PREFACE

There are certain things about humans that make getting to the facts not so easy. One is that this upright mammal will believe almost anything—and this has been true since the beginning of recorded history right through to the present. Examples include beliefs in the power of supernatural gods and demons, confidence of an upcoming apocalypse, and trust placed in countless bizarre medical remedies. We've come a long way since the origins of settled life, yet even today a sizable number of Americans believe evidence-free political propaganda and conspiracy theories, one of which states that a secret cabal of Satan-worshiping, cannibalistic pedophiles, composed of Democrats, is running a global child sex-trafficking ring. All these beliefs were, or are, considered normal by the community that holds them. It appears that allegiance to irrational beliefs has survival value probably because it serves to create tribal solidarity, and this unification, by itself, can be enough to negate obvious facts that disturb the status quo. Another trait of the species is that people automatically form hierarchies and establish rank in social situations. From a sports team to a nation state, this is accomplished using whatever differences are available between individuals, whether they be visual cues, genetics, preferences, abilities, money, connections, or knowledge. These two traits—tribal beliefs and social rankings—have served to bind humans into larger groups that are better suited to survive the perils of the environment than just a few individuals. They are ancient and instinctive traits, and they continue to have a powerful influence on what we do or don't do, including the hard work of getting to the facts and determining what is real and what isn't.

In this book I am attempting to loosen the grip that tribal beliefs and the power of high-ranking individuals have on a controversial subject, one that is almost always labeled as fantastic, irrational, and surely wrongheaded by the culture's sanctioned thought police. This subject is astrology. A balanced understanding of the subject is very hard to come by and there is nothing in the educational systems of our culture that encourages anything other than learning how to suppress the subject further. At the same time, many of the practitioners of astrology project a vision of the subject that doesn't hold up to scrutiny and discourages investigation. So, when it comes to assessing the facts of what is a truly complex matter, it comes down to finding sources of legitimate information. If you don't look in the right places, you don't know, and that's where things have been for centuries, and they still are. In addressing what I also perceive to be an injustice, a wide range of information, some of it technical, needs to be explored to make up for centuries of ignorance, misunderstanding, and knee-jerk dismissal. My hope is that some readers will come to the conclusion that many who claim to know exactly what this subject is (including both critics and some who call themselves astrologers) don't know what they are talking about. Real learning takes place when a deep understanding of the issues, not motivational bias (or what we want to believe), precedes judgment. Supplying information for that deeper understanding is one of my goals in this book.

A general overview of the varied territory ahead may be helpful. Part 1 of the book begins with a review of modern biological studies that have deepened our understanding of how Earth's rotation and the rhythms of light, gravity, magnetism, and solar radiation have been used as navigational cues and activity triggers by life. These signals, and life's utilization of them, are evidence that our solar system neighbors, particularly the Sun and Moon, do affect and shape life on our planet in profound ways. Following that, scientific studies that have uncovered links between the solar system and the terrestrial weather and climate systems on much larger time scales are considered in some depth. One thing that is apparent from the first two chapters is how doing science on systems, such as organisms or weather, is not easy. After this factual introduction of astronomy in nature comes a history of the group effort in ancient times to understand observed linkages between astronomical motions and events on the ground. With the Greeks, this project became known as astrology, that is the body of knowledge that has attempted to bring astronomy to Earth by mapping and interpreting observed geocosmic connections in both people and in their environments, both social and natural. Then, an important branch of astrology becomes the focus of the next two chapters. This is astrometeorology, a method of weather forecasting that has been studied and employed from ancient times through the Renaissance and that looks to the solar system for correspondences with weather and other environmental processes. My own research program of testing this methodology forms a chapter on its own.

It is hoped that Part 1 will have adequately prepared the reader for a series of discussions on the historical, philosophical, and scientific issues that astrology, in all its forms, has raised. The central portion of the book, Part 2, focuses on the history of science and the seventeenth-century revolution in scientific thinking, a complex change of world-view that here is not limited to the usual focus on Copernicus, Galileo, and Newton. It is also an account of how both religious and secular institutions, and sometimes just a few individuals, control definitions of reality and shape history. In a matter of a few centuries, astrology changed from being a respected, though controversial, practice and natural science to a lower-class activity. The stage is then set for Part 3, a critique of the scientific evidence and theoretical explanations for astrology. Finally, some observations on the contemporary condition of this much maligned subject and a summary of what astrology actually is are offered to the (by now presumably well-informed) reader.

It is important that this book also be seen as a strong defense of science as a method of producing potentially useful and democratically-derived knowledge. To be clear, the general philosophical stance behind my thinking in this book is one of methodological naturalism with a tolerance for ambiguity, and some agnosticism when appropriate. This is not a metaphysical doctrine I hold, it is an approach to the study of nature, including people, that uses scientific methodologies and critical thinking, but does not subscribe to supernatural explanations for phenomena under investigation or engage in beliefs. It is a realistic position because it accepts that knowledge is never complete and it distinguishes between different kinds of knowledge. For example, answers to questions like *What is the meaning of life?* and *What happens when we die?* remain unsolvable by science, but answers have been made in logical arguments, books written by true believers, and the revelations of certain charismatic individuals, the latter being, more or less, how religion handles these mysteries. Many, if not most, people accept such answers as a kind of knowledge and use it as a way of filtering facts and making meaning out of the events and issues that they experience in their lives. But this is mostly somebody else's subjective knowledge, and it requires belief on the part of the adopter.

Practical knowledge, on the other hand, is acquired through agreed-upon experiences. If everyone who eats a certain fruit gets very sick, then knowledge of this fact spreads and it becomes part of a consensus understanding. Most people call this common sense, and skilled practical knowledge is called know-how. Science is like common sense but is more systematic, formalized, and socially organized. Scientific knowledge is always based on verifiable evidence and uses rules of logic and abstractions, such as mathematics and models, to move the understanding of nature forward. Over time it builds a set of facts, distinctions, and explanations that are open to public scrutiny, hence it is democratic. But the knowledge science produces is never absolute fact because science is a work in progress. When it comes to certain subjects such as astrology, where the science hasn't been done properly, or at all, it is important that clarifications be made between what we don't know and what we want to believe (i.e., motivational bias). The lack of clarity inherent in a complex situation like this one can generate anxiety in some people and cause them to double down on positions they are more comfortable with. So, in an effort to alleviate cognitive dissonance over the mere mention of the word astrology, this book aims, at the very least, to clarify what is known and what is not known about the subject.

There is another theme central to this book that concerns science, perhaps the most important one. Since the seventeenth century, the preferred methods of science have become mechanistic and reductionist in the context of a materialist metaphysics (I refer to this in the book as reductionist-mechanistic-materialist, or RMM, in my view a useful generalization of the dominant trend in science over the last ~350 years). In this version of science, which includes methodologies of falsification and checks of data against hypotheses, the parts of something being studied are isolated, tested, measured, and explained in terms of a mechanism. This particularist approach, which works so well in physics and chemistry, now also informs molecular biology, behavioral psychology, parts of the social sciences, and modern medicine in general. In contrast to this is systems thinking, a way of doing science that approaches phenomena like weather and climate systems, organisms, mind, personality, consciousness, and society from a broader perspective. Here, the dynamic properties of a system in its entirety—or, in other words, the sum of connections, patterns, feedback loops, and emergent properties (which resist reduction)—are the starting point in a scientific investigation. This more general and process-oriented approach to knowledge is multidisciplinary and is practiced in some branches of engineering and psychology, and also ecology and organismic and evolutionary biology.

Systems are complicated and not static, which is why reductionist science is not always the right tool for investigation. Systems also exhibit many unusual properties and processes that turn out to be extremely resistant to reductionism, including the phenomena of emergence and self-organization, which are central in understanding topics like the origin of life, mind, and self-consciousness. Another important quality of systems is that they often display strong responses to very weak signals. The field of astrology, as I will argue in this book, has been studying systems for the past three millennia, at least. It is not fortunetelling, pseudoscience, nor bunk. The field of astrology studies a type of environmental signaling, and it is actually a kind of system science, or at least a set of techniques for mapping (including temporal trajectories) and analyzing self-organizing systems. In this sense, it has the potential to contribute toward a more inclusive scientific program.

PART I

From Natural Science to Natural Astrology

I

Life Internalizes the Sky

If we understand why cycles are so ubiquitous in the world, I believe, we can use this knowledge to develop more sustainable behaviors and relationships with the [rest of the] natural world.

Phil Loring, Anthropologist, Storyteller, Futurist, author of *Finding Our Niche*

Indeed, since the organism is such an apparently delicately poised physico-chemical entity as to reflect small fluctuations in both identified and probably still unidentified, subtle geophysical factors, the question arises as to how essential to life itself are the various components of the natural subtle geophysical complex of the earth's atmosphere.

Frank A. Brown, "Subtle Factor and the Clock Problem"

The biosphere of our planet, a thin layer that extends only a few kilometers below the surface and not much more into the atmosphere, is where the action is. Under a ceaseless rain of charged particles and rhythmic variations of light driven by astronomical motions, chemical elements undergo transformations that control the composition of the atmosphere and hydrosphere. In turn, these fluid reservoirs force changes in the hard, spatial shapes of the lithosphere. Solar radiation bombards this zone in a constant daily cycle creating temperature and pressure gradients that move winds and ocean currents on local and global scales. Lunar tidal forces and sun-driven winds stir both land and sea in periodicities short and long. Magnetic fields generated from both Earth's core and our local star, varying in strength on hourly and multi-year timescales, are yet another kind of environment. At times huge rocks from space crash into the planet bringing instant destruction and change. All of this activity and flux is the geocosmic environment within which Earth and its biosphere exist. By evolving sensory systems that measure and manage this relentless cascade of information, life has learned to navigate both its spatial and temporal worlds—and has survived by doing so for at least 3.5 billion years.

The natural environment is the sum of the physical realities against which all life evolves and, as Darwinian evolution states, is a selector of organisms. The environment we function within is not limited to the conditions of our immediate spatial location. It is not just food and water supply, competitors, mountains, canyons, deserts, and ocean barriers. We also live in a temporal flux of signals generated by astronomical factors: cycles of light, gravitational tugs, streams of solar radiation, cosmic rays, magnetic field variations, and possibly undiscovered forces. These organized pulses of light and charged particles are as real as hills and rivers and have contours, amplitude, and periodicities. Life has evolved in this temporal matrix as much as it has in the more obvious spatial environment, though evidence of this is more subtle than biological examples, such as the development of the eye for visual navigation, or a long neck to feed from the tops of trees. Life uses the natural rhythms of the environment, most of them being driven by geophysical or astronomical phenomena, as a framework onto which the various processes of the physical body are built. Most rhythms are fundamentally solar; others are lunar or variations of the two. Of the many rhythms that have been used by life as a kind of scaffolding or grid, the rotation of the Earth relative to the Sun, meaning the solar day of twenty-four hours, is the primary geophysical periodicity to which nearly all organisms are known to respond and adapt.[1] Our bodies and those of other animals and plants know this day/night cycle well. Most obvious is the wake/sleep cycle, but hundreds of other daily rhythms exist in our bodies and those of other organisms that are essential to well-being, including feeding behaviors, metabolism, body temperature, hormone production, reproduction, blood levels, brain activity, and cell regeneration.

In addition to the alternation of light and dark that makes up the solar day (a periodicity), there is the steady rhythm of light intensity produced by Earth's tilt and the solar year of 365.24 days, which marks one orbit of Earth around the Sun. During the solar year the ratio of light and darkness varies in a cycle (a sequence of stages or phases), creating distinct seasons that are most pronounced in higher latitudes. Plants have used this cycle, called annual photoperiod, as a basis for flowering, the crucial stage in plant reproduction. Many animals, especially insects, have reproductive processes that are coupled to plants, and so they also live in sync with annual photoperiod. These two rhythms of light, those of the day and the year, are the most fundamental cycles of light on our planet, and many organisms in the five kingdoms of life on our planet—bacteria, protoctists, animals, fungi, and plants—utilize both of them for the regulation of internal functions and external behaviors.

Biological rhythms as a property of living things were recognized in prehistoric times by fishermen and hunters who learned the habits of their prey. Some organisms are diurnal, others nocturnal. Animal behavior changes during the transition periods between night and day, something that close observers of wildlife know quite well. Many animals have evolved to be crepuscular, that is, most active during dawn and dusk, as a means of avoiding predators that are entirely diurnal or nocturnal. This adaptation is, of course, often compromised by a simi-

---

1. The solar day is the time from one meridian transit (noon) of the Sun to the next one at a specific location. A sidereal day is four minutes shorter because the marker for the meridian transit in this definition is a star, a fixed point in space. Each day, as Earth moves along its orbit, the Sun (from Earth) will appear to have moved ahead by 1 degree in the zodiac, and so, when a star is used as a marker, another four minutes of Earth axis rotation will be needed to catch up to it.

lar adaptation by predators, but in either case, it is light or lack of it that establishes a temporal environmental selector that is as real as geographic isolation on mountain summits or low oxygen levels in caves.

Chimpanzees are known to hunt as a coordinated group using their knowledge of the behavior of their prey (macaques), and they also know how to use specific plants for healing purposes. This implies learning and accumulation of knowledge, and by extension, we can assume the ancestors of humans were also studying their world for millions of years. With the advent of *Homo sapiens* some two to three hundred thousand years ago, we might say that some of these early humans were natural historians, and having roughly the same brain power we have today, they probably knew volumes about the features of their environment. But nothing remains of this knowledge except as may be found on some paleolithic artifacts. Alexander Marshack investigated tiny markings on ancient bones, ice age artifacts of the Upper Paleolithic, and concluded that some of these were counts of twenty-nine-day lunar cycles. He suggested that these notations carved in bone, a kind of calendar counting, may have been the precursors to numbers and written language. The cave paintings in Europe, the best-known being those of Lascaux, France, are dated to about seventeen thousand years ago and contain rows of dots under some animal images. Bernie Taylor, in his book *Biological Timing,* has proposed that these dots placed under specific animals in the cave paintings may actually be soli-lunar guides to the migration of herds, information that is still used by some hunting and gathering cultures (Marshack 1972; Taylor 2004). He suggests that some Paleolithic cave paintings are actually time-maps that denote when during the year certain herd animals are most likely to be found in the vicinity of the cave.

If herd animals follow the changes of light from the Sun and Moon, hunters studying their prey would learn the patterns and adjust. The same goes for fishing. Fish are known to migrate by the Moon as it relates to the solar year. Taylor describes how Native Americans along the Pacific coast have long known when the salmon would be running upriver—but this period of just a few days is not the same from year to year. Apparently, the fish are entrained to the changing light of the Moon during the monthly cycle and use this information for feeding and predator avoidance. In addition, they use the length of day (photoperiod) as a reference to time the annual migration upriver to spawn. For early humans to eat and to obtain other resources needed for survival purposes, such as furry skins for clothing and hard bones for tools, knowledge of the patterns of the temporal environment was crucial.

The first written records to show an organized attempt to correlate the rhythms of the sky with weather and life on Earth are the astronomical and astrological records made on cuneiform tablets in Mesopotamia. As we will see in Chapter 3, not only were the rhythms of day and night and the cycle of the year mapped out and correlated with weather, vegetation (agriculture), and animal and human behaviors, correlations with the five visible planets were also noted. The methodical recording of weather, biological, and human events that began in Mesopotamia all correlated with astronomy, has persisted over three millennia and is today known as astrology. Astrology and astronomy were inseparable for most of this time, and in regards to data collection and pattern recognition, the Mesopotamian tradition was at the forefront of establishing what we call science today.

At least one thousand years after the earliest geocosmic correlations were first recorded in Mesopotamia, a new kind of thinker, one not content with explanations involving gods, came to prominence in the Greek world. They were the first natural philosophers and proto-scientists, and they began to question everything. These thinkers attempted to understand the world and human existence by using one tool—their logical minds. In regard to nature, they interpreted the world, in a general sense, to be alive, and they saw seasonal biological rhythms as a kind of evidence of this. Thales (624–546 BCE) of Miletus (western Turkey) was the first in a line of philosophers called the pre-Socratics who primarily studied nature. Roughly two centuries later Aristotle (384–322 BCE) noted that Thales had made a fortune in the oil business, a reference he used as evidence that philosophers, who mostly ignored money-making activities because they had more important things to think about, were not just airheads. The story goes that Thales bought up olive presses in the year before there was a bumper crop of olives and then rented them to the farmers who had to immediately process the crop when it was ripe. Exactly how he did this is not stated, but Aristotle says "he knew by his skill in the stars while it was yet winter that there would be a great harvest of olives in the coming year."

Perhaps he was noting correlations between bumper crops and some kind of astronomical data that was used at the time for predicting weather (Aristotle 2017). Since he was also credited with predicting an eclipse, he probably knew of the nineteen-year Metonic cycle, where the Sun and Moon will be in the same phase on the same day every nineteen years with an error of only a few hours. We don't know to what extent Thales may have been influenced by Babylonian ideas, or if he was tracking weather cycles, which the Greeks also regarded as being related to astronomical factors. We don't even know for sure if this story is true, but it at least shows an awareness among the earliest Greek philosophers of correspondences between Earth and sky and the practical use of this knowledge.

BIOLOGICAL CLOCKS

Only a few references to biological rhythms survive from the time of the Greeks and Romans. The physicians Hippocrates (~460–370 BCE) and Galen (130–300 CE) noted a rhythm in the progress of an illness that was attributed to the Moon as it formed quarters relative to its position at the onset of the problem. What they called "critical days" were spaced at these seven-day intervals. Another example is Aristotle who, among his other titles might be called the first great biologist, left some notes of the lunar rhythms found in a marine organism, the Mediterranean sea urchin. In his description, Aristotle mentioned that the size of its ovaries varies according to the lunar cycle, being largest at full moon, though he attributes this cycle to the "heat" of the full moon, something thought to be needed by a cold-blooded animal (Aristotle 1961, vol. V, 329). Not much else about biological rhythms has survived from the Greek scientific tradition; for the next two millennia, the detailed study of astronomical cycles and their correlations with life and the environment was left to the astrologers who mostly focused on weather prediction, medicine, and the fates of individuals.

The study of biological rhythms was re-started when French

astronomer Jean Jacques d'Ortous de Mairan (1678–1771) experimented with *Mimosa pudica,* a plant of the pea family that is known for its touch-sensitive leaf movements (Klarsfeld 2013). It is a popular household plant today usually referred to as the sensitive plant. In addition to closing when touched, *Mimosa pudica* leaflets have a diurnal cycle in that they open by day and close by night. De Mairan's 1729 experiment was to place the plant in continuous darkness and observe its leaf movements in the absence of light. They did move in the darkness, though in a less pronounced way, which suggested to him that the plant was somehow still able to sense the Sun. It was commonly assumed that the movements were a passive response to light, but the evidence suggested to de Mairan that something internal to the plant was the cause. Over a century later Darwin (1809–1882) took up the subject of plant movements and published his experiment-informed observations in an 1880 book, *The Power of Movement in Plants.* Darwin was primarily concerned with gravity and light as external stimuli and the actual details of plant response. Circumnutation, cycles of leaf movement, he thought were central to understanding plant movements, and he reported that most plants, excepting insectivorous species, display a periodicity in which leaves move upward at dusk and evening and then drop at morning (Darwin and Darwin 1881).

A contemporary of de Mairan was Carl von Linné, or Carl Linnaeus (1701–1778), of Sweden. He is one of the great names in biology and the person who created the modern system of classification for plants and animals, in which each organism has a genus name and a species name. In his 1751 book *Philosophia Botanica,* Linnaeus published the idea of a *Horologium Florae,* or floral clock, that allowed one to tell the time of day (Gardiner 1987). From his careful observations of diurnal blooming patterns he came to know at what hour a particular plant would flower. With knowledge of the daily rhythms of a variety of flowering plants, a clock-like circle of flowers could be planted. (It's not known if he actually executed his plan, however, but others did and still do.) Linnaeus's idea was an application of knowledge of cycles particular to each plant over the course of one day, that is, the circadian cycle. During the early twentieth century, American government research on efficient tobacco propagation led to the discovery that plant flowering at specific times in the year was actually influenced by the day/night ratio that varies within the annual cycle. This property was named photoperiodism, and in the 1930s German scientist Erwin Bünning (1906–1990) published a hypothesis concerning biological rhythms in which he stated that the timing mechanisms of photoperiodism were the same mechanisms behind daily leaf movements (Gardner and Allard 1920; Bünning 1958, 1960).

Bünning, whose life work was dedicated to exploring biological rhythms, thought they were natural oscillations in organisms that had evolved to be synchronized with Earth's rotation, and that this feature improved adaptation to the environment. After decades of experiments on plants (especially bean plants) and insects (fruit flies of the genus *Drosophlia*) he arrived at some conclusions that have formed much of the central subject matter of an entirely new branch of biology, now called chronobiology. He found that the daily rhythm was not exactly twenty-four hours; rather, as with Linnaeus's floral clock, it was circadian (*circa* meaning "about" and *dian* meaning "day"), which was apparent when organisms were placed in total light or darkness or subjected to abnormal alterations of light and dark. Under experimental conditions of constant light or constant dark the organisms were allowed to "free-run," and it was found they would settle into a periodicity approximating, but not exactly equal to, twenty-four hours. He also found the free-run rhythm was unique to the organism and was inheritable. Another finding was that the circadian cycle in plants is linked to the timing of flowering during the year, or the photoperiod. Bünning argued strongly for a twenty-four-hour endogenous, or internal, rhythm that could read signals of light or other cues from the environment, but was not at all dependent on them. This rhythm was thought of as a biological clock, and this term became part of the standard terminology for the study of these natural rhythms.

The nature of the actual mechanism behind biological rhythms was debated for many years. In opposition to Bünning, an argument was made for the "clock" in the organism being more of a passive sensory receptor to any number of environmental signals. A few studies suggested the clock was driven by an external *zietgeber* (meaning "time giver"), or several of them. Frank A. Brown, Jr. (1908–1983) of Northwestern University was the champion of this exogenous hypothesis, and he conducted a number of studies that seemed to confirm his view. One was a series of experiments using the fiddler crab of the genus *Uca* that showed it maintained an exact twenty-four-hour rhythm (not a circadian rhythm) over a wide range of temperatures and that it had another rhythm that was synchronized with the lunar day of 24.8 hours, which is the tidal rhythm. It turned out, however, that these crabs were an exception and most organisms aren't so precise in their twenty-four-hour rhythm. Colin Pittendrigh (1918–1996) of Princeton, Brown's primary opponent in the endogenous/exogenous debate, argued that the lack of a precise rhythm found in other organisms was a logical result of selection pressure during the evolution of the clock. The daily resetting (phase adjustments) would correct for any errors in the clock, making the organism more adaptable and therefore better fit for survival. Brown didn't dispute the circadian qualities of the clock, which he named autophasing, but suggested that it is the laboratory's constant conditions that produce it and that true daily rhythms found in nature were far more closely in sync with the solar and lunar periods.

In another experiment, Brown moved oysters from New Haven, Connecticut, to Evanston, Illinois, and observed their tidal behaviors in constant light, specifically the opening and closing of their shells, which had remained synchronized with the tides at New Haven. After two weeks, however, the oysters shifted their tidal movements to three hours later in the day. Brown noted that this shift corresponded to the time of the upper and lower meridian transits of the Moon, which would move a tide, if there were one, in Evanston, Illinois. In regard to mechanism, his methodology and data ruled out light and barometric pressure, which led him to consider that more subtle factors such as geo-magnetic and geo-electrostatic fields may be the forces the oysters were responding to.[2] One later study tested this result: mussels collected in Massachusetts that were taken to California maintained their east coast tidal rhythm and did not adjust to west coast tidal rhythms—until they were exposed to these tides (Sweeney 1987).

---

2. Earth has an electrostatic field as do organisms. An electrostatic field is one of charge without movement, or in other words, an electric field at rest (Brown, Hastings, and Palmer 1970).

Brown conducted many sophisticated and creative experiments with marine organisms, hamsters, chicken embryos, and even potatoes and carrots (he measured their daily cycles of respiration) that showed not only the expected circadian cycle, but over longer periods of time, rhythms in the amplitude of the daily rhythm that he thought might be a response to solar variations. His experiments placed organisms in spaces completely blocked to light, temperature, and pressure, yet there were still clearly discernable rhythms in metabolism that tracked the day and the seasons, and also lunar cycles. His experimental data suggested to him that living organisms, in addition to responding to the daily cycle, demonstrated sensitivities to very subtle signals during the solar or lunar day, including magnetic and electrical fields and even fluctuations of cosmic radiation (cosmic ray flux is inversely related to magnetic field strength). These findings make the scientific ideal of interference-free laboratory conditions far more difficult to create, if not impossible in some cases, and were probably very disturbing to other researchers.

Science is a communal process that seeks consensus, which, while democratic, guarantees the dominance of the norm and the exclusion of the fringes. This can be seen throughout the history of science and accounts for long lag times between the discovery and general acceptance of strange findings like Brown's. In regard to circadian rhythms (which are explained in more detail in the next section), it was a landmark conference on biological rhythms at Cold Spring Harbor in 1960 that set the direction for research. Scientists like Bünning, Pittendrigh, and Jürgen Aschoff, who are considered to be the founders of modern biological rhythm studies (chronobiology), were prominent figures at that conference, and they focused on the endogenous and bio-chemical model of the circadian clock. Their research program came to dominate the field, and their approach has led to testable models of the circadian clock with an emphasis on the properties of phase shifting. When geneticists became engaged with this problem during the 1990s they discovered clock genes that led to a molecular model of the clock. Relative to most of the other chronobiologists who attended the 1960 conference, Brown was something of a maverick and many of his ideas on exogenous forces were criticized or ignored, though he defended his findings and opinions at the meeting that year and afterward until his death. To be fair, Brown said his studies didn't disprove an endogenous timer, and he added that the existence of both external and internal timing mechanisms would be an example of redundancy, something common in nature. He envisioned the biological clock as being a duality in which an internal responder to subtle signals from the environment is overlain by an endogenous timing mechanism.

During the 1960s Brown's ideas and experiments were ignored or even rejected as they ran against the officially sanctioned search for an endogenous clock mechanism. He continued to debate others in the field, but eventually he lost much research funding. The last nail in the coffin was probably a paper published in *Science* that, without naming Brown, mocked his experiments and statistical methodology by using randomly generated data to find a cycle in "the unicorn." Eight years later, however, a physiologist and mathematician named A. Heusner published a paper demonstrating that the author's debunking method was deeply flawed—but the damage had been done (Cole 1957; Heusner 1965). Still, Brown didn't give up and he continued to study rhythms in a series of creative, comprehensive, and low-budget experiments that were described in published papers into the 1970s. Much of his research was on correlations between rhythms and magnetic fields, gamma rays, and other subtle signals in the natural environment. He was actually ahead of his time as the study of magnetic fields and life, now called magnetoreception, was picked up a few decades later by a number of scientists, including geologist Joe Kirschvink (born in 1953, and also sometimes called a maverick), who has investigated both the geomagnetic field itself and magnetic sense in organisms.

For the majority of researchers, the endogenous/exogenous debate in biological rhythms was over following the Cold Springs Harbor meeting. An increasing number of experiments during the 1960s (using hamsters, fruit flies, and bread mold) supported the idea of an endogenous oscillator that is self-sustaining without any obvious environmental cues. One of these studies took place at the south pole where the research organisms were placed on a rotating table that canceled out Earth's rotation. In the 1980s, a study with a bread mold from the genus *Neurospora* was conducted on a space station that orbited the Earth every few hours. The results were the same as other studies: there was a roughly, but not exact, twenty-four-hour rhythm that would establish itself regardless of light, temperature (within a range), or gravity, but it was weaker and there was some arithmicity (Sulzman et al. 1984). These findings were taken to have settled the exogenous/endogenous debate, though anomalies remained. It is now accepted doctrine in chronobiology that biological rhythms are, for the most part, environmentally independent and driven by complex internal molecular processes, not by external signals or forces. These rhythms are entrained by environmental cues, however; thus external signals are essential for their synchronization with nature, which presumably gives them survival value, though studies that support this are difficult to do and consequently there are few of them. All of this tells us that life has evolved in such ways as to reproduce internally a major part of its temporal environment and needs only occasional cues to establish accurate synchronicity with the external world.

CIRCADIAN RHYTHMS

The alternation of the light/dark cycle, the fundamental geophysical period caused by the rotation of our planet, is certainly the most important environmental signal for organisms that live near the Earth's surface. It is essentially a digital signal—a repeating pattern of light and dark that occurs as the Earth faces toward or away from the Sun during the course of a solar day. This signal is as much an environment for life as is the spatial surface of the Earth, and it should be no surprise that biological rhythms of approximately twenty-four hours are now known to be ubiquitous in living things on our planet. They are a general feature of the physiological organization of a wide-range of organisms including prokaryotes, protoctista, plants, fungi, and animals. Since the 1950s these rhythms have been studied extensively and are referred to as circadian rhythms, the term coined by another scientific maverick, Franz Halberg (1919–2013), who also came up with the name of the scientific study of such things, chronobiology. More on Halberg later.

Chronobiologists today define circadian rhythms as having three main characteristics. First is the persistence of a period close to twenty-four hours in constant light or dark conditions. Persistence of the rhythm and deviance from exactly twenty-four hours, the period, is species specific, and there are also differences between individuals. Some organisms have rhythms that will continue without external signals for only several complete cycles; others can free-run for weeks or months. Secondly, circadian rhythms will persist throughout a wide range of temperatures allowing an organism to maintain a stable rhythm regardless of temperature variations throughout the day or year. Third, circadian rhythms can be entrained by light/dark cues, by sharply changing temperature cues of more than 10 degrees Celsius, and in some cases by other environmental signals including food availability, social cues, electromagnetic field strength, and atmospheric pressure.

Circadian rhythms in an organism will establish a phase relationship between subjective, or internal, day and external day (or the reverse in nocturnal organisms) by responding to an external environmental signal. While chronobiologists tend to favor and focus on the mechanics of the endogenous rhythm, synchronization doesn't work without an exogenous trigger. Some organisms establish the phase of their circadian period by the onset of light, others by the onset of darkness. Studies have shown that circadian rhythms can be modulated, or phase shifted, by the presence of environmental signals such as light pulses at points in the cycle other than the normal onset of light or darkness. It is this ability to phase-shift that accounts for the ongoing adaptation to the changing light/dark ratio of the seasons, and it is this process that unifies the daily circadian and yearly photoperiod rhythms.

A number of models and mechanisms for biological rhythms have been proposed. One model described in the 1970s considered electrical charge gradients and the rhythmic flux of ions in the cell membrane.[3]

---

3. Membranes in cells are boundaries not only between the inner cell material and what is outside of it, but they also wrap around structures within the cell (e.g., organelles). Because membranes regulate what passes through them, including ions (atoms or molecules that have a net electrical charge due to the possession of extra or missing electrons) organized so as to establish a gradient that can drive flow, this regulation may be one that occurs in a consistent timeframe and functions like a clock (Koukkari and Sothern 2006, 158–67).

A more specific mechanism behind circadian rhythms was a matter of speculation until the 1990s, when geneticists entered the world of chronobiology and developed a molecular model for the circadian oscillator (Hardin, Hall, and Rosbash 1990). The laboratory organisms from which this model was originally developed were from the genera *Neurospora* (bread molds) and *Drosophila* (fruit flies). In its standard form, called the transcription-translation oscillator model (or transcription translation feedback loop), a photo-receptor composed of specialized cells, such as those in the eye, recognizes alterations and variations of light and dark. This information is then fed to an oscillator that is essentially a negative feedback loop in protein production within the cell. The model posits that specific clock genes are transcribed in the nucleus of the cell, taken to the cytoplasm, and then translated into proteins that accumulate until a threshold is reached. At this point the process shuts down until there is too little of the protein, which then triggers a signal calling for more. This thermostat-like cycling built on negative feedback is the roughly twenty-four-hour endogenous clock. The period and phasing of this endogenous time-making process is capable of being reset by the photo-receptor, should photic information reach it at critical points in its cycling, and the reception of such sensory information will then link the internal molecular cycle with the external cycle or, in other words, day and night. From the clock cells in which the rhythm is produced, chemical messengers relay timing information to other parts of the organism, which keeps the living system properly regulated. This molecular "hourglass" model, highly simplified here, has become a widely accepted explanation for circadian rhythms, although both old and new findings point to its limitations. The model is classically reductionist in that it is a scientific approach that builds an understanding on parts, as in a clock or machine.

CIRCADIAN RHYTHMS IN THE FIVE KINGDOMS

Here is an important point to emphasize: circadian rhythms have evolved in every kind of organism. Life on Earth has been classified in various ways, as in animal, vegetable, mineral. Today the three-domain scheme (archea, bacteria, and eukaryota) is taught but is probably not as useful, certainly not in educating non-scientists, as the more intuitive and practical five kingdom scheme of Robert Whittaker (Whittaker 1969). To begin to understand this scheme, it makes more sense to first distinguish between the two primary kinds of life on our planet: prokaryotic and eukaryotic. Prokaryotes, including archea and bacteria, are basically strands of DNA floating inside a lipid membrane. While these organisms have devised a variety of metabolisms, they reproduce by fission when conditions, such as availability of food, permit it. Eukaryotes are all products of ancient symbioses; they are mergers of different types of archea and bacteria or possibly another kind of microbe that went extinct leaving little or no fossil evidence. These organisms, which include the kingdoms of plants, fungi, animals, and also protists and algae (collectively named Protoctista), are far more complex than prokaryotes in their life history and modes of reproduction. Until the late 1980s it was assumed that circadian rhythms existed only in eukaryotes.[4] The assumption was that, since circadian systems appear to control the cell-division cycle, then prokaryotes, which can divide in periods of less than one day, could not have circadian cycles because cell division in periods of less than twenty-four hours would cause an uncoupling of any circadian clock. In the 1990s, however, this assumption became obsolete when several circadian rhythms were found in the marine cyanobacterium of the genus *Synechococcus,* a ubiquitous photosynthetic marine coccoid bacterium that contributes greatly to the supply of oxygen on our planet.

Cyanobacteria (often incorrectly referred to as blue-green algae) are one of the most ancient life forms on Earth. It was this organism that first evolved photosynthesis, the making of glucose (food) from carbon dioxide and water using solar energy, with oxygen as a byproduct. Organisms also need nitrogen to make amino acids, proteins, and DNA. The problem is that oxygen interferes with the process of nitrogen fixation (making compounds that include nitrogen); so, there is a need in photosynthesizing organisms to separate nitrogen fixation from

---

4. Evidence for rhythms in the bacterium *Escherichia coli* had been reported earlier (Halberg et al. 2003).

photosynthesis. Cyanobacteria do it in two ways, depending on the species. Filamentous cyanobacteria get around this problem of isolating oxygen and nitrogen by a primitive form of multi-cellularity. Specialized cells called heterocysts are formed that separate photosynthesis and nitrogen fixation spatially. The other way is through a circadian rhythm that turns on nitrogen fixation for part of the day, then shuts it down and turns on photosynthesis. This bacterial trick was discovered by measuring daily oscillations in nitrogenase, the enzyme responsible for fixing nitrogen, relative to daily photosynthetic activity. Strains of bacteria from the genus *Synechococcus* also exhibit other circadian rhythms including nitrogenase mRNA abundance changes, amino-acid uptake, protein synthesis, light/dark entrainment, and the cell division cycle itself. That's a lot of organismic regulation, apparently easy for a bacterium, all running off the internalization of an external natural rhythm.

Protoctista is a category based on exclusion; it is an assemblage of eukaryotes that are not plants, animals, nor fungi. Some of the best known of this grouping are the amebae, slime molds, diatoms, species of *Paramecium,* and most algae including species of *Gonyaulax, Acetabularia, Euglena, Chlamydomonas,* and many others. One of the very first eukaryotic organisms to be studied for its circadian rhythms was *Gonyaulax polyedra,* a dinoflagellate single-celled alga known for its ability to glow in the dark and also for causing red tides. It was found to have a circadian rhythm of bioluminescence that persists in constant light as well as circadian cycles of photosynthesis and cell division. All the rhythms of *Gonyaulax* species are entrainable by either the natural light/dark sequence or by single short light pulses; the latter are capable of phase-shifting the rhythm if applied at the proper time. Cell-division appears to be keyed to the ending of the dark, or night, phase and light cues presented then will reset the clock. The circadian rhythm of photosynthesis, on the other hand, seems to be keyed to maximum light at midday with entrainment sensitivity at the beginning of the light phase (dawn). *Gonyaulax* species also have at least two separate oscillators, each receptive to different wavelengths of light that regulate the cycle of swimming behavior or aggregation. In this single-celled organism we see how circadian rhythms are sensitive and responsive to phase signaling, light levels, and wavelength, which is a lot more than just the alternation of light and dark (Sweeney 1987).

Acetabularia is a genus of giant single-celled algae that are five centimeters high with a "cup" that is about one centimeter in diameter. It is generally found in shallow, sheltered waters attached to rocks and other shallow substrates in the vicinity of tropical coral reefs, and is used as an aquarium plant and as a treatment for gallstones. It is of interest here because it illustrates the location problem of the circadian oscillator, something one would expect to be a simple task in a one-celled organism. Experiments involving species of *Acetabularia* targeted the cell nucleus, and it was found that removing it did not terminate circadian rhythm; instead, the cell material, membranes, and cytoplasm were all that was needed to keep the beat going. It was also found that a replaced nucleus picked up the rhythm that was apparently established by the cytoplasm of the cell. So, with no nucleus transcription-translation oscillator mechanism found in this case, the problem of locating the circadian mechanism is still not completely resolved by the current molecular model, at least in this genus of algae, and also in a few others like yeast and blood cells (Sweeney and Haxo 1961; Woolum 1991). Blood cells have no nucleus to begin with, so they really do present a problem for the model. Studies along these lines have shown that biochemical reactions in the cytoplasm, which may be linked to membrane processes, seem to be a kind of foundational clock on which other clocks rest, including the molecular model. It appears that genetic material is not the ultimate organic mechanism many reductionists hoped—or assumed—it would be and that biological rhythms involve other processes in the organism.

The transcription-translation oscillator mechanism of the circadian system has been studied extensively in a number of model laboratory organisms. For example, the cyanobacterial oscillator is similar in general principle to those in eukaryotic systems like species of *Drosophila* (fruit flies), *Neurospora* (bread molds), *Arabadopsis* (mustard plants), and *Mesocricetus* (hamsters), but the clock genes and proteins are completely different, which suggests an independent evolution. The anatomical location of the circadian oscillator itself varies widely among animals. In gastropods it is found in the eyes, in some crustaceans it is located in the eyestalks, in insects in the brain, and in other organisms in the brains and abdominal ganglia. In reptiles and birds it is located in the pineal gland, and in mammals in the hypothalamus. The genetic mechanisms for the circadian system in animals are fundamentally the same; even mammals and insects share at least some clock parts. This does suggest a common ancestral clock, but maybe one not so ancient as it may be that the real ancestral clock is in the cytoplasm or works off ion gradients in the membrane.

The circadian system in vertebrates consists of photoreceptors (usually eyes), the pineal organ, and the suprachiasmatic nucleus (SCN), which is located in the hypothalamus. It is thought that this particular circadian system is highly conserved and was established with the origins of vertebrates roughly five hundred million years ago. The SCN, the central pacemaker located in the hypothalamus, produces circadian rhythms by gene product negative feedback loops (the transcription-translation oscillator model) in specialized cells, which then relay cycle information via neural and endocrine pathways to the rest of the organism, including other peripheral clocks. The SCN also controls the release of melatonin (a factor in jet lag and seasonal affective disorder) into the bloodstream and to the rest of the body. The processing of light by vertebrates for the circadian oscillator differs significantly from the processing of light for vision and appears to involve brightness receptors (for photon counting) that are separate from rods and cones in the eye. For example, the mammalian eye has parallel pathways for vision and brightness; the latter pathway has dedicated photo receptor cells that comprise about 1 percent of the total ganglion cells in the retina.

To bring this all together, here are a few take-home points from this overview on circadian cycles that are worth considering. First, it is clear that life on this planet has somehow internalized Earth's "digital" twenty-four-hour spin-cycle of light and dark. The fact that circadian rhythms are found even in prokaryotes, the first life, point to very, very early origins for this function, and the fact that they are found throughout all the kingdoms of life points to these rhythms having a selective advantage, though testing that assumption hasn't been so easy to prove in a lab. The existence of multiple rhythms in organisms, and the ability, through phase-shifting, to account for photoperiod, is really quite astounding; it suggests how important attunement to the natural environment is to the regulation of a self-organizing system, whether single cell or multicellular. Then, there was the rush in the 1990s to find a genetic mechanism for this clock, a development that took place around the time of the human genome project. The model that was produced turns out to be incomplete, however. If the research had elucidated a simple molecular machine inside the cell that explained everything, biology might have moved closer to the precision of physics. But life doesn't lend itself that easily to mechanical explanations. The insights of biologists with organismic perspectives (a more holistic view that includes evolutionary relationships, adaptations, and ecology) can also claim some success in understanding the problem. Finally, there is the case of Frank A. Brown Jr., a very competent scientist whose work pointed to the influences of a far more complex cosmic environment than simply light and dark.

PHOTOPERIODISM

The diurnal cycle expressed as circadian rhythms is one thing, but photoperiod is another. The annual progression of day length is explained by the astronomy of the seasons: the Earth's tilt and its orbit around the Sun. Day length is modulated over the course of the year by the 23.45-degree tilt of the Earth's axis (relative to the plane of its orbit around the Sun), which is held steady as it orbits the Sun. When Earth's axis leans toward the Sun during its orbit, in either the northern or southern hemisphere, it will be summer. The reverse, when the axis is leaning away from the Sun, of course produces winter. Tilt and orbit produce a day of approximately twelve hours of sunlight and twelve hours of night at the equator, but up to twenty-four hours of either in latitudes over 66 degrees.[5] The annual cycle of day-length challenges the ability of organisms to adapt to a continuously changing light/dark ratio, particularly in the higher

---

5. Another way to express the astronomy behind photoperiod is to see it as the annual cycle of solar declination. Declination is distance measured north or south of the celestial equator.

latitudes where the effect is so extreme. To solve the problems of seasonal changes, organisms have evolved an adaptive timing mechanism to what are called circannual rhythms. Again, a wide range of organisms display circannual cycles as a response to photoperiodism, more evidence that life has adapted to geocosmic phenomena.

Circannual rhythms allow organisms to locate in time opportunities for flowering, feeding, reproduction, growth, molt, migration, and hibernation. Circannual rhythms can also provide information on distance and direction that are crucial to migratory behaviors and, in mid to high latitudes, important timing information for animals that den and hibernate for the winter. In general, photoperiodism is more robust and precise in species living in mid to high latitude, and also in long-distance migrating species, than in tropical species where the annual light cues are minimal. However, species living near the equator have been found to respond to subtle light intensity levels and also display a photoperiodic cycle. For decades chronobiologists asked if circannual rhythms require their own clock, or if they piggyback off the circadian system. The current general understanding is that these rhythms are managed by the circadian system. It is the continuous entrainment of the circadian oscillator by constant phase resetting, which happens as light levels shift during the seasons, that photoperiod adjustments are accomplished.

The study of circannual rhythms challenges researchers who have to invest years of study (and years of requests for funding) before reaching conclusions. In spite of these problems studies have been done and it has been found that the persistence of endogenous rhythms varies widely in organisms, and in some cases, there is a greater need for exogenous cues. True circannual rhythms have been found in sheep, deer, bats, and starlings (a long-living bird), among others. A second type of seasonal rhythm is common in shorter-lived species such as mice, hamsters, and many birds and reptiles. Here the photoperiod response is not free-running, and light cues are needed to keep the endogenous rhythm synchronized with local seasonality. Plants, where photoperiodism was first scientifically studied, use photoperiod timing for growth, adjustments to seasonal changes, the induction of flowering, and also germination time for seeds at the appropriate time of year. Plants can measure day length very accurately. Vascular plants sense the seasonal changes by discriminating day from night, measuring passage of the night interval, and then integrating this information with changes of growth, germination, and flowering. Phase shifts are how the circadian clock of the plant reads the day length changes. When light occurs earlier in the circadian cycle, during the night portion, the plant adjusts for the changing ratio of light and dark (it phase shifts), and this then signals other functions in the plant. Lab studies have found that free-running periods longer than twenty-four hours enhance the ability of the plant to track dawn. This allows for more rapid phase-shifting, that is adjusting to changing day length, and therefore better seasonal adaptation and consequently general fitness (Todd et al. 2003).

Photoperiodism is a reflection of the annual cycle of the seasons during which each individual day varies a small amount from others in two ways. One is the changing light/dark ratio, the other is in the change of total solar radiation received, this due to the elevation of the Sun relative to the local horizon. Interestingly, a number of studies have reported on correlations found between season of birth and individual characteristics. For example, bison born early in the year tend to be more socially dominant than those born later in the year, which makes sense as those individuals have a longer time to mature as they grow into synchronization with the seasonal cycle. Studies on mice have discovered correlations between winter births, depression, and the cycling of neurons in the master biological clock, the SCN. Humans also display individual differences that are linked to the part of the year they were born in, this information coming from an emerging subfield called seasonal biology. Findings include a negative correlation between females with a novelty-seeking personality and births in winter, higher reward dependence for men born during autumn, and a higher frequency of bipolar depression and schizophrenia for people in general born in winter. Various temperament scales have been used in these studies and strong correlations have been found between cyclothymic (manic) behavior and summer birth, depressive temperament and winter birth, hyperthymic (upbeat) and spring and autumn birth, and a negative correlation between irritable temperament and winter birth.

Studies have been done on people from infancy through adulthood, and it does appear that the biological clock, and its regulation of the various neurotransmitters that regulate emotions, arousal, and other behavioral responses, is attuned to seasonality (Chotai, Lundberg, and Adolfsson 2003; Ciarleglio et al. 2010; Gonda et al. 2014).

TIDAL AND LUNAR RHYTHMS

The periodicities of the day and the year, both based on Earth-Sun relationships, are relatively simple. Lunar cycles are far more complex, inexact, and multi-faceted. The most identifiable of several lunar cycles is the synodic cycle of 29.5 days. This period is defined as the number of days between successive alignments (syzygy) of the Moon and the Sun, such as the new Moon (Sun-Moon-Earth) or full Moon (Sun-Earth-Moon).[6] The combined gravitational forces of the Sun and Moon at the times of full and new Moons generate high tides, the Sun's gravitational contribution being about 45 percent. The distance between the Moon and the Earth also varies by about 13 percent over a cycle of 27.5 days, this period being called the anomalistic month. At perigee, the closest distance, the Moon's gravitational force is strongest and the coincidence of perigee with a full or new Moon produces the highest tides. At present, the Moon orbits the Earth at a mean distance of about 385,000 kilometers (or about 239,000 miles). This 27.5-day period and the orbital distance have increased during Earth's history. Since the Moon's orbit recedes at a rate of about 4 centimeters a year, we know that millions of years ago tides were stronger because of a closer Moon, and they were also more frequent due to the faster rotation of Earth.

The Moon is a major feature of Earth's rhythmic cosmic environment. During the course of one twenty-four-hour rotation of the Earth, the Moon advances in its orbit by roughly 13 degrees of celestial longitude on average, requiring an additional 0.8 hours of Earth rotation to catch up with it. This means that the gravitational pull of the Moon produces a daily cycle of two tides spaced about 12.4 hours apart;

---

6. The parentheticals refer to the order of alignment in each type of syzygy.

twice this figure is the tidal or lunar day of 24.8 hours. This daily tidal rhythm is seen in the behaviors and internal functions of many marine organisms. The lunar period, commonly found in both land and marine organisms, is the synodic cycle, or the cycle from new Moon to new Moon, that averages 29.53 days. Half of this, 14.75 days, which spans the period from new to full Moon, is called the semilunar period. High tide occurs on the side of the Earth facing the Moon where its gravitational force is strongest, but a high tide also occurs simultaneously on the opposite side of the Earth. This is generally explained by the inertia of the water as Earth and the Moon pull toward each other while they orbit their common center of mass (barycenter), this point being located within Earth, about a quarter of the way to the center. Tides are complicated by the shape of coastlines and the location of land masses, which cause local daily tide schedules to be highly irregular. The selection pressure for a species to evolve a tidal clock, allowing it to anticipate the substantial changes in the coastal environment as the tide changes, is high, but local conditions require unique adjustments. Perhaps for this reason free-running tidal rhythms in many marine organisms are generally not as precise as circadian rhythms.

Tidal and lunar periodicities are prominent features of many marine organisms and a few, including the fiddler crab of the genus *Uca,* have been studied extensively (Brown, Hastings, and Palmer 1970; Palmer 1995). These crabs live in burrows, emerging and feeding as the tide ebbs, and so are subjected to repetitive fluctuations of light, temperature, and tidal submergence. As was discussed earlier in this book, Frank A. Brown Jr. studied the genus *Uca* in depth and found that they have a precise 24-hour circadian cycle of shell color that is darker at dawn and lighter at dusk. However, their running activity retains a tidal periodicity of approximately 24.8 hours, 50 minutes later each day. Further, the reproductive rhythm is semilunar as their larvae are released at the new or full Moon at the hour of high tide. It appears that more than one clock is operating in this peculiar organism.

Tidal rhythms in marine organisms such as those of the genus *Uca* are known to vary widely among individuals; data analysis can be complicated when the same organism is studied in groups. One view on how tidal rhythms are maintained suggests that a circadian clock with a long free-running period is able to adjust itself, or phase shift, by entrainment to changes in the environment, including such factors as hydrostatic pressure, temperature, water agitation, and others. This view is supported by the rapid shifting of certain organisms (e.g., diatoms in the genus *Euglena*) in the lab from a 24.8-hour tidal rhythm to a 24-hour light/dark cycle. A second view, called the circa lunidian clock hypothesis, proposes that two coupled clocks run simultaneously within the organism, each one tracking a separate 12.4-hour period (Palmer 1995). Evidence for multiple clocks include the finding that the two tidal periods of each day appear to be scanned at different rates.

Lunar rhythms are even found in marine protoctista. Many species of single-celled planktonic foraminifera have a reproductive cycle that is characterized by the alteration of two generations, haploid and diploid, with different modes of reproduction that in some species may take two years to complete.[7] Reproduction in the vastness of the ocean requires a large number of gametes, about three to four hundred thousand, and the gametes that are released benefit from consolidation in time and space. To accomplish this requires precise synchronization in order to secure gametic fusions (reproduction) and the continuation of the species. At least some of these single-celled species appear to coordinate these processes with the lunar synodic cycle.

In one study, samples of three species of foraminifera were collected every two days for forty-seven days, at the same time and depth each day (Bijma, Erez, and Hemleben 1990). It was found that the number of all species was at its minimum in the period of three to five days

---

7. Haploid cells, which are usually sex cells (like sperm or egg cells in humans), have only one set of chromosomes. Diploid cells, which in animals and plants comprise the rest of the organism, have two sets. The cyclic reproductive process of certain species of foraminifera is complicated but runs as follows: The diploid form of the organism undergoes meiosis (the splitting of the diploid cell), producing haploid agametes with one set of chromosomes. These become the sexually reproducing gamont generation of the organism. The gamont, when fully grown, undergoes mitosis (duplication), producing hundreds of thousands of flagellated, free-swimming haploid gametes. These gametes, still with one set of chromosomes, then leave the test or shell of the foraminifera, meet with others of their kind, and fuse, becoming the diploid form.

after the full Moon. It was also found that the numbers of two of them, *Globigerinoides sasculifer* and *G. ruber,* were at maximum numbers nine days before the full Moon, but those of *Globigerinella siphonifera* were at maximum three days before full Moon. Apparently, each species has its preferred phase of surface activity relative to the full Moon, a light intensity environmental factor. So, it appears that the cycle of moonlight as a zeitgeber is a possibility, but light as a cue would be dependent on weather conditions. In the laboratory the three species of foraminifera did not synchronize, and reproductive timing was found to be affected by other conditions (i.e., food availability, light intensity, population size, etc.). However, the lunar cycle did persist in the lab for a related species, *Hastigerina pelagica.* With no moonlight reaching the lab, there was 85 percent gametogenesis at three to seven days after the full Moon. How the organism picked up the lunar signal was not clear. Other possible lunar signals that the organism may be receptive to include Moon-induced fluctuations of the magnetic field and changes in lunar gravitational strength (the anomalistic month). These studies point out how very difficult it is to study lunar rhythms in both field and lab.

An outstanding example of a response to the lunar synodic cycle is the reproductive behavior of the Palolo worms of the south Pacific. As with the release of haploid gametes by planktonic foraminifera into the sea, the Palolo worms require precise daily timing, but also timing relative to the seasons. The terminal body segments of this marine polychaete are genetic capsules that are released at dawn on a specific day during the spring in the southern hemisphere. This date, which changes each year, coincides with the October or November third quarter Moon, the lunar phase that produces tides (neap tides) that aren't as agitating as those at the new or full Moon. This timing serves to increase the number of reproductive events. Records of spawning show that the worms, by following the third quarter Moons in either October or November, appear to be tracking the eighteen-year Metonic cycle, where a specific phase of the Moon occurs on the same date after exactly eighteen years. Even more remarkable is the lunar-synchronized reproduction of at least 107 species of coral along five hundred kilometers of the Great Barrier Reef in the western Pacific. Nearly all release gametes three to six days following the full Moon in October or November, and about four hours after sunset (Endres and Schad 1997).

At present there is no known mechanism by which lunar influence is entrained in genetic processes. Light is certainly a factor in many organisms. Studies of juvenile salmon show grouping (protection in numbers) occurring at full Moon and movement/migration at new Moon (invisibility), these being strategies that limit predation. The vertical migration (in the water column) of zooplankton is influenced by different degrees of lunar illumination. Higher levels of moonlight, which encourage predation, increase the vulnerability of zooplankton populations when they are at crucial developmental stages. Since the phase of the Moon is not the same at a constant point in the seasonal cycle, predation will eventually select for organisms, in this case the zooplankton, that can entrain to the lunar cycle. Fish, feeding on these smaller organisms, will then be entraining on the cycle as well. And animals, including humans that feed on the fish, will also adjust to this rhythm, a bottom-up entrainment.

Many fishermen know that fish are more likely to hit a lure or grab bait at dawn or dusk, transition points in the day-night cycle. In 1926, John Alden Knight (1890–1966) formulated a product that he promoted as a guide to the best times for catching fish, a method based on his knowledge of folklore and the habits of game fish. He had observed that fish are more active at dawn and dusk, but they were also biting at other times of the day, which he found to be when the Moon was rising or setting. Knight did not think light was the external factor driving behavior. He reasoned that it was the Moon's tidal effect that was stirring up aquatic insects and even microscopic life, which, in turn, agitated the fish who feed on this part of the food chain. Knight never conducted a formal study, instead he created tables showing the best times of the day for fishing, based on the solar and lunar diurnal cycles, and sold these as a kind of almanac that he called Solunar tables. The popular outdoor sporting magazine *Field and Stream* included his graphic tables of the Sun and Moon in each issue for many years (Knight 1972).

The Moon's influence on the environment is not limited to rising and falling tides and variations in nocturnal illumination. Other possible aquatic lunar-cycle triggers include temperature changes, turbulence, water pressure, and salinity changes. The atmosphere also displays lunar periodicities as the upper layers are affected by the Moon's gravity, which then modulates air pressure that is detectable on the surface. Lunar gravity also has an effect on the electrically conductive conditions of the upper atmosphere, a subtle but real effect called the lunar magnetic variation. One study found that the frequency of natural (not induced) human births correlates with the actual gravitational force of the Moon, that is the distance from the center of the Moon to the center of Earth, but not with the phase of the Moon. Another study found human sleep to be synchronized by lunar phase over a wide range of conditions, including urban and rural environments. The authors noted that this finding could be explained by available light, but also by lunar gravity as gravitational forces at the new and full Moon occur at different points in the solar day (Wake et al. 2010; Casiraghi et al. 2021). Magnetic field data does show correlations to the Moon's diurnal cycle and its phases, so this signal also exists, and many organisms are known to be sensitive to magnetic fields. Observed throughout human history is the correlation between human female menstruation and the lunar cycle. Studies show that the match is quite close and that for women who don't take birth control pills, or become biologically synchronized with other women, menstruation tends to occur near the new Moon.[8] At present, few studies have been done that compare the responses of organisms to the various influences of the Moon mentioned above and the subject is still, in many respects, in its infancy.

THE GEOMAGNETIC FIELD AND LIFE

The Earth is shielded from a steady rain of charged particles and cosmic rays by its own self-generated magnetic field. This field, called the magnetosphere, is comet shaped as a result of the constant pressure of the solar wind, a stream of hot electrons and protons emanating from

---

8. This topic has been controversial, but data supports a linkage between human fertility cycles and the Moon, one that has also been found in apes and monkeys (Koukkari and Sothern 2006, 227, 265–71).

the Sun. The sunward side of the field, the magnetopause, is compressed to a depth of about eleven Earth radii. The magnetotail, on the opposite side away from the Sun, has the form of an elongated tail that may extend as much as one thousand Earth radii. Viewed from space, with its magnetic field made visible, Earth looks like a comet headed toward the Sun. The magnetic field not only establishes an Earth-encircling electro-magnetic dipole framework, like a bar magnet, but it also exhibits fluctuations and periodicities. Geomagnetic micropulsations (caused by lightning storms, particle bombardment, etc.) that last less than one second to a few minutes occur constantly. Longer periodicities reflected in magnetic field variations include responses to the Earth's rotation (24 hours) and the lunar day (24.8 hours), the synodic month (29.5 days), solar rotation (~27 days), the solar year (365 days), variations in the solar wind (1.3 years), and the sunspot cycle (11 and 22 years on average). A lot of information is contained in Earth's magnetic field variations, although recording and displaying it as data presents many problems. It is known, however, that the Earth's magnetic field is utilized by certain organisms for its directional and temporal information content. This sensing of the field is accomplished by either direct contact to it or as a response to weak electric fields induced by the geomagnetic field. The ability to "read" the geomagnetic field requires remarkable discrimination, something one would think only possible in "higher" organisms like animals, but it is found in certain bacteria that derive selective advantages from this response.

Magnetotactic bacteria are a diverse group of motile, mainly aquatic, anaerobic, or microaerobic gram-negative prokaryotes that were first recognized by Salvatore Bellini of the University of Pavia in 1963. Twelve years later, Richard Blakemore, a graduate student in microbiology at the University of Massachusetts, rediscovered them when he noticed that the bacteria he was looking at under the microscope were following a magnet he was fiddling with in his hand. He published a paper on this finding and called the behavior magnetotaxis, which is the term now used. Magnetotactic bacteria use Earth's magnetic field for navigation purposes to seek their favored environment. They have a negative response to oxygen in the water and so seek anoxic regions in the sea floor, which they locate by using the north-sloping direction of the Earth's magnetic field (in temperate latitudes) as a guide to vertical mobility. Their response and alignment to magnetic fields is possible due to a chain or chains of small, biologically-produced crystals of magnetite inside the body of the bacteria. These particles are called magnetosomes (Blakemore 1975).

Magnetosomes are intracellular, single crystals of magnetite (Fe_3O_4), the mineral that makes up lodestone. Magnetite can form inorganically, but only at high temperatures and pressures, yet it is synthesized by these bacteria from materials found in their environment. Magnetosomes are an example of biologically-controlled mineralization (e.g., bone and tooth formation), an organic matrix-mediated process that begins in the magnetosome membrane. This specialized membrane then regulates the deposition of the particle and controls its position in the cell relative to the other particles. In terms of size, all magnetosomes fall within the thirty-five to one hundred nanometer range, just big enough for internal polarization but small enough to avoid multiple regions of polarization. Magnetosomes in magnetotactic bacteria are arranged in one or more chains of about twenty particles that behave as a single magnetic dipole. These bacteria are very widespread in the oceans, and some believe they once existed on Mars! Very tiny pieces of magnetite found in the Martian meteorite ALH84001 were proposed to be a signature of bacterial life on Mars, though subsequent analysis has substantially weakened the case. It is instructive, and relevant to the subject matter of this book, to consider the complexity of magnetic field reception and the structures that have evolved to do this—in this case in a bacterium.

In addition to bacteria, other organisms have been found to contain magnetite crystals in their bodies. These include pigeons, which are able to navigate using the Sun, topography, or the magnetic field. Honeybees have magnetite in their abdomens. Other animals with magnetite include tuna, trout, blue marlins, green turtles, whales, and dolphins. The marine mollusc *Tritonia diomedea,* a sea slug, uses the magnetic field for direction, but this response is correlated with the lunar cycle (as is also the case with pigeons, drosophila, and other organisms). This may be a case of two separate sensory systems working together, or the slug is sensing both Earth's magnetic field and the lunar magnetic field

Figure 1. A magnetotactic bacterium with internal chain of
biomineralized magnetite crystals.

variation over the course of the lunar month. In either case, the slug's direction of movement shifts at phases in the lunar cycle, causing it to move in a circle or spiral, which may aid in foraging or mate-seeking. The existence of magnetite in the human brain, along with the ability to sense direction, has been controversial, but magnetosomes have been found in human tissues. Magnetoreception in mammals may be a basic feature. Whether dogs have magnetite in their brains is unknown, but they apparently align their body axis to the magnetic field when eliminating, specifically the north/south axis. It has been suggested that magnetic sensory systems evolved very early in life history, are completely separate from other sensory mechanisms, and have increased in sensitivity over time (Lohmann and Willows 1987; Kirschvink et al. 1992; Hart et al. 2013; Kirschvink, Walker, and Diebel 2001). The actual mechanisms involved in magnetoreception are not fully understood, and possible explanations are sought in subdisciplines like bioelectromagnetism, biofield physiology, and quantum biology, which includes in its research program quantum entanglement in avian eyes and the electron transfers involved in photosynthesis.

SOLAR CYCLES AND LIFE

Sunspots, caused by a convergence of magnetic field lines on the Sun's surface, are dark regions of a lower temperature than the surrounding solar surface. They move with the Sun's rotation and are visible evidence of the cycle of solar magnetism that affects Earth's magnetic field. Solar activity varies over time, the numbers of sunspots intensifying at solar maxima and decreasing at solar minima. The numbers and groups of sunspots also change daily, and their positions on the Sun's surface change over time. Sunspot counts are a way to track the solar cycles, the most recognized of several is the Schwabe cycle of about eleven years, during which the number of sunspots increases and then decreases. The Hale cycle of approximately twenty-two years is two Schwabe cycles, which accounts for the magnetic reversals that occur between them.

Many correlations have been reported that link biological and solar activity. Well-documented rhythms of about ten years, close to the Schwabe cycle, include crop yields, fish catches, and boreal forest mammal population changes (Hoyt and Schatten 1997). Insect populations are regarded as sensitive climate monitors because they show close correlations with the Schwabe cycle. Tent caterpillar populations peak predictably about two years before sunspot maxima and other insect populations appear to do the same. It has been suggested that these population changes are due to the increased warmth and ultraviolet radiation that follows the solar cycle. At least one unicellular organism appears to show a response to a solar cycle. Species of the genus *Acetabularia,* its model-busting circadian cycle mentioned above, display rhythms found in measures of geomagnetic activity, and also in the solar magnetic field. In a database covering fourteen years of *Acetabularia* circadian cycle research done by chronobiologist Franz Halberg, the prominence of a 1.3-year cycle that is found in the rhythms of the solar wind cycle was reported (Halberg 2004). It is thought that entrainment to climate cycles that are driven by the solar cycle occurs in some species, more so in regions where such climate fluctuations are more pronounced. These adjustments, which may correlate with increasing and then decreasing food supplies, will consequently influence other species' population levels. It may be that entrainment to a natural rhythm in one species low on the food chain is all that is needed to produce cycles in the behavior of many other organisms.

BIOLOGICAL RHYTHMS AS FUNDAMENTAL TO LIFE

The study of biological rhythms is a study of how life has adapted to and internalized its temporal environment. Using the natural geophysical and astronomical periods as frameworks, life has organized many complex processes, including the general stabilization of multiple internal systems, feeding, navigating, and reproduction. In 1960, at the important Cold Springs Harbor meeting on biological clocks, it was stated that "circadian rhythms are inherent in and pervade the living system to an extent that they are fundamental features of its organization" (Pittendrigh 1960). Biological clocks that match environmental periods are a nearly ubiquitous characteristic of life, but there are wide variations between species and even between individuals. In some species, one clock, or one clock system, controls all life functions. In other species multiple clocks may be coupled, and in others are found master clocks, like the SCN discussed earlier, that drive slave oscillators. The fact that the circadian rhythm persists in the cytoplasm of the cell in certain organisms after the nucleus is removed, which contradicts the transcription-translation oscillating model, means these rhythms are not completely understood.

For organisms in the lab, biological clocks are not essential to staying alive. The lack or malfunctioning of a clock in nature is another matter. It is thought that a defect of rhythm will cause an organism to be active at the wrong times of day and that errors would be made in feeding and in mating timing, ultimately resulting in fewer offspring due to predation. One study tested the fitness of several cyanobacteria strains each having different periods. It was found those strains with rhythms that matched the light/dark cycle outcompeted the others in terms of reproductive fitness (Ouyang et al. 1998). Another study involved the day-active antelope squirrel. The SCN of a number of these animals were removed, which induced arrhythmia and caused them to be active at night. During the study, a feral cat managed to enter the research enclosure and proceeded to remove 60 percent of the lesioned animals, but only 29 percent of those with an intact SCN. A second study using chipmunks was conducted in a completely natural setting. Again, the SCN was lesioned in a number of chipmunks that didn't seem to create any serious problems for two years of stable food conditions. However, after two years of abundant acorns, the population of chipmunks increased and so did their predators. By the end of the summer season more SCN lesioned animals were killed by weasels than those without (DeCoursey, Walker, and Smith 2000).

The endogenous versus exogenous debate over the circadian rhythm appears to be less a case of one or the other and more a case of complementation. The current molecular model of the circadian oscillator has established a foundation for the endogenous source of circadian rhythm. (One exception to this model, yet to be fully understood, is a connection to the cytoplasm rhythm (Zivkovic 2011).) The endogenous oscillator is a system separate from the environment, and studies have shown that the free-running period of an organism is almost never exactly twenty-four hours, but usually very close to it. Again, this apparently allows for better tracking of environmental signals, more accurate adjustments and phase-shifting, and hence better fitness. However, the endogenous oscillator requires sensory input and external triggers to set the phase. Any number of sensory mechanisms may be involved in phase setting, although visible light is the primary trigger for the circadian system in most organisms. Models for lunar rhythms have yet to be described with a similar degree of confidence as those for the circadian rhythm. It may be that magnetic field variations and the solar wind are also used as triggers for rhythms, suggesting that life has deep links to a broader definition of the environment.

While the mainstream of chronobiological research has been focused on the reductionist molecular model and its usefulness in understanding human behaviors and health requirements, some scientists have taken other paths. One of the most remarkable of the biological clock mavericks was Franz Halberg, who we met briefly earlier in the book. Originally from Austria, he was a biophysicist who worked at the University of Minnesota and studied a wide range of biological cycles, many involving humans. He took a bio-medical approach to his subject matter and, among numerous other interests, studied the effects of medications for cancer and other disorders given at different times of day—that is, different points in the circadian cycle. His findings during the second half of the twentieth century are at the foundations of what is now called chronopharmacology. Halberg was a tireless worker who wrote or contributed to an enormous number of papers, socialized with others in the field at the international level, received many awards for his work, and was even nominated for a Nobel Prize a few times. He founded the chronobiology laboratory at the University of Minnesota and, in addition to coining the terms circadian and chronobiology and researching both photic (visible light) and non-photic (i.e., solar wind, electromagnetism, cosmic rays, and gravity) periodicities, he developed a unique framework and vocabulary for his findings.

Here I'll attempt a short summary of a few of the many cycles Halberg found and studied in a wide range of organisms and natural phenomena (Halberg et al. 2003; Halberg 2004). First, there is the simple and common single-day circadian cycle found in life. Then there is the roughly seven-day circaseptan cycle he thought is linked to the fourth harmonic of the rotation of the sun (one-quarter of its twenty-seven-day cycle). He found examples of this biological "week" of about 6.75 days in the genus *Acetabularia,* in the geomagnetic Kp index, and in the amplitude of melatonin in mammals. Half-year cycles, he found in geomagnetic activity, the human brain (vasopressin levels), length of infants at birth, and human longevity. He was very interested in what he called the transyear (~1.3 years), which is a cycle of the speed of the solar wind, and found correlations to adult human blood pressure and heart rate, and also growth patterns in bacterial colonies. He also found the transyear cycle in chloroplast movement in that odd algae *Acetabularia mediterranea.* A long, roughly ten-year cycle, which he called circadecennian and correlated with the approximately eleven-year Schwabe solar cycle, was found to match cycles of heart rate variability, deaths from myocardial infarctions, religious proselytism, human productivity, and vascular mechanisms underlying mood.

Halberg had a predecessor, the controversial Russian biophysicist Alexander Chizhevsky (1897–1964), who had influenced his thinking. Chizhevsky was the founder of heliobiology, that is, solar-Earth research, and studied the effects of the Sun's activity on terrestrial phenomena. One of the correlations he made was that high solar activity, mostly at sunspot maximum, produces mass human excitability and often coincides with wars. Halberg's data, collected from extensive university medical records that he had access to, showed that the cardiovascular system is sensitive to geomagnetic disturbances that are more frequent when the Sun is active, an observation that has been confirmed by other studies (Alabdulgader et al. 2018). This knowledge of the coincidence of individual and environmental variables he believed was of critical importance to society. Unfortunately, contemporary society seems to have its own agendas and his suggestions have not yet been implemented, or even considered.

Halberg's study of the concordance of biological and environmental rhythms he called chronomics, which is the study of chronomes (time structures), defined as transiently self-sustaining organized structures in time and space. Chronomes (which I interpret to have properties fundamental to self-organizing systems) consist of chaotic changes and rhythms undergoing trends (internal feedback loops) for a sufficient duration to possibly reproduce themselves and, thereby, to evolve. In chronomics, photic and non-photic cycles of the physical environment are seen to affect both the biosphere (realm of life) and the noosphere (realm of mind). In Halberg's schemata, which was depicted in complex diagrams jam-packed with information, were also cycles of the socio-psycho-physiological realm, what he called the ethosphere, that he thought may offer a key to the health and diseases of society. At this level, chronomics was a perspective that might be tried as a scientific way to identify the underlying mechanisms of the diseases of civilization, which he believed was a necessary step toward creating a better human world. Yet another term he coined to describe this last challenge was chronobioethics. Halberg's ideas on society are not well-known but he was pioneer in the field and was developing his own comprehensive time-systems model that gathered together rhythms of all lengths, from less than a day to millions of years, all linked to photic and non-photic geophysical and astronomical signals. In this sense, Halberg was a visionary, and his work undoubtedly contains insights that may one day be commonplace.

ב

THE EARTH CYCLES

Earth's orbit and axis orientation are constantly changing because they are being deformed by the gravitational attractions of other [planetary] bodies. These changes affect the distribution of sunlight hitting our surface, which in turn affects climate, and the kinds of sediments that are deposited. That gives us the geological record of solar system behavior.

PAUL OLSEN, IN "SCIENTISTS TRACK
DEEP HISTORY OF PLANETS' MOTIONS
AND EFFECTS ON EARTH'S CLIMATE"

. . . [W]e've clarified how the planets can affect the magnetism of the Sun by amplifying tidal resonance. We've also generalized the theory by extending the application of magneto-tidal resonance to Earth's magnetosphere. We're saying that some of the planets can have a direct influence on Earth's magnetic field.

PERCY SEYMOUR, "THE MAGUS OF MAGNETISM"

Single and multi-cellular organisms are self-organizing systems that are open to exchange with their environment and also tightly coupled to it. They are composed of many parts linked by feedback loops out of which emerge forms and behaviors that the parts alone do not predict. This emergent phenomenon is what we call life, and it has not yet been made in the lab, nor has its origin been explained definitively by modern reductionist science. One central property of an open system, that is a system that can exchange matter, energy, or information with its external environment, is that it resists the inevitability of entropy, the movement toward disorder expressed in the second law of thermodynamics. Systems that somehow manage to slow entropy can also be abiotic or a combination of biotic and abiotic. Consider that a hurricane, an organized entity that feeds off rising water vapor and gives off wind and rain, is an inorganic open system that "lives" for a time until its energy sources fail. An ecosystem is a dynamic entity, a result of multiple interactions between living and non-living components in a mostly contained area or region. Ecosystems have resilience; they can survive disturbances and, like a thermostat, return to an optimal state.

The largest terrestrial self-organizing system, what James Lovelock has called Gaia, is spread over Earth's surface, plus or minus a few miles below and above it. Here organisms and their environment are coupled in such a way as to somehow maintain conditions favorable for life. The flux of matter and energy between the atmosphere, hydrosphere, lithosphere, and biosphere in combination and acting as a single unit is called by geoscientists the Earth system, while climatologists working with the same components (including the cryosphere as a subset of the hydrosphere) refer to the climate system. In either case our inorganic environment is hardly static or dead, it is interactive with the biosphere, and over long periods of time all of it can be seen as a self-organizing open system, one that is affected by solar radiation that fuels the biosphere, and thermal activity from the core, which drives plate tectonics. The Earth system is also sensitive to external information in the form of tidal forces and magnetism, and host of extraterrestrial radiation, including cosmic rays.

Extraterrestrial influences on Earth's atmosphere, oceans, and continents exist in good measure though these do not fall under a single heading at this time because the subject matter is scattered about in the fields of climatology, meteorology, geology, oceanography, astronomy, and astrophysics.[1] One category of extraterrestrial influence shared by astronomy and the geosciences are the periodic impacts of extraterrestrial bodies, the big one that ended the dinosaur age being the best known. Whether or not these occur in cycles of millions of years is a topic full of hypotheses that will be briefly touched on. More tangible are orbital cycles, which are driven by variations of the Earth's orbit and extend beyond the span of recorded history. These were first proposed in the nineteenth century and now occupy an important, even integral, position in the geosciences, especially paleoclimatology. Orbital cycles, being controlled by the gravitational tugs of the other planets on Earth, are ultimately solar system–influenced cycles. Today they are accepted drivers of climate, ice ages, and geological processes that go as far back as late Triassic (about two hundred million years ago) sedimentation rates preserved in the ancient lakes of New Jersey (Olsen and Kent 1996).

Cycles of solar activity are also a known source of climate modulation, though on much shorter timescales, and are studied in the fields of climatology, meteorology, and astrophysics. It is not known definitively how solar cycles originate but, as will be discussed below, the orbits of the planets may play a role. There is a relation between solar cycles and atmosphere-penetrating cosmic rays that some scientists see as a driver of global temperatures. On a much larger scale, the orbit of the solar system around the center of the galaxy causes it to pass periodically through regions that raise the level of cosmic rays (cosmic ray flux) entering Earth's atmosphere on long timescales. It is thought by some astrophysicists and climate scientists that such variations influence climate and related geological and biological processes over periods of millions of years. The fourth category are tides. Tidal forcing from the Moon and the Sun affects both the oceans and the atmosphere and studies showing correlations with climate, and even earthquake activity, do exist. Tidal resonance as a force, discussed later, drives the evolution of planetary orbits and may even modulate the solar cycle, which in

---

1. A major point I want to make in this book is that extraterrestrial influences might be productively organized as a single topic, and that to a large extent such a classification exists, that it has a long history, and that it has been marginalized mostly by religious factors.

turn modulates Earth's climate. Now outlined, we look at each of these topics individually.

EXTRATERRESTRIAL IMPACT EVENTS

While astronomers use the term *bolide* to describe an extremely bright meteor, geologists use it for impactors that may be asteroids or comets. Large bolide impacts disrupt the Earth system and can even trigger a mass extinction. Once the Cretaceous–Paleogene (K–Pg) extinction event sixty-six million years ago was linked to a bolide impact, an idea proposed by physicist Luis Alvarez and his geologist son Walter in 1980, the search for evidence of other such impacts in the fossil record began. Possible correlations of those found to other extinction events have been suggested, though none have been shown to be so overwhelming as the K-Pg event that wiped out the dinosaurs. However, there is some evidence that impacts may follow a cycle. In the 1980s a number of papers on extinction cycles were published, many of them based on a dataset of fossil marine life that was assembled by paleontologist John Sepkoski. The fossil data, which covered 250 million years and included several thousand vertebrates, invertebrates, and protozoans, clearly showed five major mass extinctions of the past half billion years when the data was organized at the family level (under family comes genus and then species), along with other minor ones. Sepkoski collaborated with another paleontologist, David Raup, and the two published a paper in 1982 that described a pattern in the data they proposed was evidence of an extinction rhythm of about twenty-six million years. Given the regularity, they suggested it might be driven by extraterrestrial factors. Physicist Richard Muller got involved and proposed that Earth has a companion star, possibly a red dwarf, that orbits the Sun in a twenty-six-million-year orbit. He reasoned that when this star, which he called Nemesis, comes close to the Oort cloud (which is the name for the belt of objects, including comets, at the outer edge of the solar system), its gravity disrupts orbits, sending asteroids and comets toward Earth.

The solar system orbits around the galactic center in a period of about 220 to 250 million years. At times during its orbit the solar system will pass through a spiral arm, a slower moving, denser feature of the galaxy. The shock of entering a spiral arm destabilizes some stars, causing them to supernova, events that produce cosmic ray storms and magnetic turbulence. These passages have been proposed as the ultimate cause of a number of terrestrial changes, the proximate cause being the perturbation of the Oort cloud (Napier and Clube 1979). This, called the theory of terrestrial catastrophism, proposes that spiral arm passages, where the solar system runs into a density wave, disrupt the orbital stability of comets in the Oort cloud, sending some into the inner solar system. This would then lead to more frequent bolide impacts that may also trigger large igneous province formation (broad surface volcanic eruptions), which is known to play a role in mass extinctions (Filipovic et al. 2013). Another connection has also been made between Oort cloud disturbance and the orbit of the solar system around the galaxy, but this one doesn't involve density arm passages. The solar system crosses the plane of the galaxy (oscillates vertically relative to the galactic equator) about every thirty-two million years, this figure being close to what was found in the fossil record. Because clouds of gas and dust are concentrated along the galactic equator, passage through this equatorial dust may disturb the Oort cloud and unleash comets.

Another related idea, from geologist Michael Rampino and astrophysicist Richard Stothers, is that the solar system's passage through the clouds of dust concentrated near the galactic equator every thirty million years or so heats Earth's core, triggering volcanic eruptions. These geological events (large igneous provinces) then cause extinctions by adding massive amounts of carbon dioxide to the atmosphere that trap heat and acidify the seas. A more recent variation on this idea, from physicist Lisa Randall, is that galactic equator crossings bring the solar system into regions full of a new kind of dark matter that exists along the galactic equatorial plane. When the solar system crosses this line, its density disrupts the Oort cloud, once again sending rogue comets toward Earth. Others suggest that more cosmic rays exist north of the galactic equator and, when the solar system passes through it, these (of course) disturb the Oort cloud. While most of these studies are multidisciplinary, with astrophysicists modeling what geologists have found, they are generally considered to be speculative science and not definitive (Raup and Sepkoski 1982; Raup and Sepkoski 1984; Davis, Hut, and Muller 1984; Rampino and Stothers 1984; Randall and Reece 2014). Still, the fact that the solar system does pass through differing regions of the galaxy during its orbit around the galactic center does suggest the possibility of very long terrestrial cycles and impact events that punctuate the geological record, a process that has been named for the destroyer god in the Hindu trinity—the "Shiva Hypothesis" (Rampino and Haggerty 1996).

ORBITAL FORCING

In middle school science we learn that the annual orbit of Earth around the Sun drives the seasonal changes. As was discussed earlier, it is the 23.45-degree tilt of Earth's axis, maintained in one position relative to the background stars as Earth moves through its orbit, that over the course of a year shifts the proportions of solar radiation reaching the hemispheres of our planet. For part of each year the northern hemisphere more directly faces the sun, which heats it, and this correlates with northern hemisphere summer. Simultaneously, the southern hemisphere experiences winter. Six months later it is the southern hemisphere that faces the Sun and is heated as the northern hemisphere shifts into winter. It is the tilt of the Earth that not only creates the seasons, but also establishes the photoperiod (i.e., the day/night, light/dark ratio) that is so important in studies of biological rhythms. But the tilt of the Earth's axis is not constant, and as we will see, this and two other variations, the dimensions of Earth's elliptical orbit and the direction of the axial tilt relative to the Sun, will change the amount of solar radiation reaching Earth over very long periods of time. These long cycles that modulate terrestrial climate in profound ways are relatively recent discoveries; only in the past few decades have they been included in science textbooks.

Variations of Earth's orbit around the Sun were first correlated with global climate history in the nineteenth century. Sir John Herschel, the son of astronomer and composer William Herschel, proposed in 1830 that the slight ellipticity of the Earth's orbit should correspond to climate changes. But his calculations convinced him that this variation was too weak to cause ice ages, so he abandoned the idea. A few years later the French mathematician Joseph Alphonse Adhemar took the subject up in his 1842 book *Revolutions of the Sea*. Citing the ellipticity of the Earth's orbit and the resulting unequal distribution of sunlight on the hemispheres over the course of a year (northern hemisphere winter is about a week shorter than southern hemisphere winter), he argued that the southern hemisphere was cooling and that this would change over the course of the twenty-two-thousand-year precessional cycle (the change in direction of the tilt). These two orbital variations, Adhemar thought, accounted for the ice ages, and he was the first to make that connection. Then there was the Scottish carpenter, tea merchant, insurance salesman, janitor, and self-taught scientist James Croll who presented his own theory of ice ages that was also based on orbital variations. In his 1875 book *Climate and Time,* which was well-received and cemented his status as a world-class scientist, Croll introduced the idea of snow accumulation and its consequences: increase in reflectance (higher albedo) of solar radiation. This reflecting back into space of sunlight would then act as an amplifier of cold during periods of lower sunlight. His theory predicted ice ages would occur in cycles and that these were driven by variations in the ellipticity of the Earth's orbit, which at times amplified the effects of precession. He was also aware of the axial tilt/obliquity cycle but lacked the necessary data to properly integrate it into his theory.

Croll had taken the idea of astronomical cycles and ice ages as far as he could in his time. It was left to the Serbian mathematician, scientist, and engineer Milutin Milankovitch (1879–1958) to finish the project in the twentieth century. After studying the effects of solar radiation on weather and climate, and having perfected the math such that he also calculated the surface and atmosphere temperatures of Mars, Venus, and the Moon, Milankovitch published his arguments in two major works: *Mathematical Theory of Heat Phenomena Produced by Solar Radiation* in 1920 and *Canon of Insolation of the Earth and its Application to the*

Problem of Ice Ages in 1941. He demonstrated mathematically that the three key orbital cycles—precession, obliquity, and eccentricity—modulate the amount of insolation (originally from the Latin but now explained as *in*coming *sol*ar radi*ation*) Earth receives over long periods of time. These three orbital cycles, also called Milankovitch cycles, have become an integral component of modern paleoclimate science and are basic to the climate computer modeling that is used in making long-range forecasts for the effects of anthropogenic climate change.

The shortest of the three orbital cycles is precession, first officially noted by Hipparchus about 127 BCE. This is the cycle of the very gradual change in the direction of the Earth's polar axis. The gravitational torques exerted by the Sun and Moon on the bulge of the Earth at its equator (Earth is a slightly flattened sphere due to rotation) cause the axis to slowly wobble like a top. This motion is called axial precession, and it has a cycle of about twenty-six thousand years, during which time the north pole traces a circle in space pointing to different stars. Today the star Polaris is near the imaginary extension of Earth's pole out into space, but about five thousand years ago it was near the star Thuban and in the ninety-first century it will point toward Deneb. Precession can also be observed over time by the movement of the vernal equinox against the constellations (not signs) of the zodiac, which is why the motion was traditionally referred to as the precession of the equinoxes.

There are some distinctions about precession that need to be made, however. Axial precession, whether observed by either pole or equinox movement, is measured against the background of the relatively "fixed" stars. This roughly twenty-six-thousand-year cycle, the precession of the equinoxes, was known in ancient times.[2] A complicating factor is the fact that Earth's elliptical orbit is rotating slowly. This can be visualized by drawing a line passing through the ellipse's two foci, one of which is where Earth is located. Where the extension of this line cuts Earth's

---

2. Precession of the equinoxes, as this cycle is generally called, was known in Hellenistic times (Hipparchus) and probably before by the Egyptians. It is most likely also the basis of the Maya Long Count, which is one-fifth of the cycle. The so-called astrological ages, such as the Age of Aquarius, popularized in the nineteenth century, are also based on this motion.

orbit, which are also the points furthest and closest to the Sun, are the points called the apsides. So, it is this line, or axis, that rotates, creating the cycle. Apsidal precession takes into account this very slow rotation of the elliptical orbit of Earth around one focus of the ellipse, a rotation that takes about 112,000 years to complete. The effects of orbital rotation results in changes in the location of the Earth relative to the Sun at the seasonal markers, therefore this kind of precession relates to the seasons. (All of this is a bit difficult to visualize, even from diagrams, but instructive animations may be found on the internet.) Apsidal precession is even more complicated because torques from other planets drive variations in the cycle that amount to periods of about nineteen, twenty-two, and twenty-four thousand years, all of which are used in insolation calculations. One can see that understanding the distinctions between the two kinds of precession is not so simple, but apsidal precession has been shown to be a powerful driver of climate change.

Obliquity is what the cycle of Earth's axial tilt is called by climatologists, and it's a much simpler motion than precession. At present Earth is tilted from the vertical to the plane of its orbit by 23.45 degrees. Over time this tilt varies from a minimum of about 21.8 degrees to a maximum of 24.4 degrees within a full cycle of approximately 41,000 years. This tilt, kept more or less stable by the Moon, has prevented episodes of extreme climates—a very good thing for life on Earth. Since it is the tilt of the polar axis that produces the seasons, it follows that when the angle of the axis is low, the contrast between the seasons will be less, especially at the higher latitudes. The effects of changing obliquity are therefore strongest at the poles and weakest at the equator, but insolation in both northern and southern hemispheres is modulated equally. As noted, variations in Earth's obliquity are stabilized and kept in check by the gravitational force of the Moon. In comparison, Mars, whose two moons are minuscule, has an obliquity cycle far more extreme that can produce very wide climate swings over time—not so good for life, which prefers stable conditions.

Now we come to the longest of the three orbital cycles, eccentricity, based on the actual orbit of the Earth around the Sun, which ranges from nearly circular to slightly elliptical. The length of the cycle of ellipticity varies but is generally given as roughly 100,000 years, though there is also a second order eccentricity cycle of about 413,000 years. Ellipticity amplifies the variations between the seasons when Earth's orbit is at its most elliptical. When ellipticity is minimal, and consequently the distance from Earth to Sun over the course of the year varies little, the contrast between seasons is not so extreme. When ellipticity is greatest, Earth will be another six million miles farther from the Sun when at the most distant part of its orbit, the point called aphelion (perihelion is the point closest to the Sun), enough to amplify the seasonal differences.

We then have three orbital cycles that over very long periods of time modulate the amount of insolation that reaches Earth. But with these three constantly changing relative to each other, what is the connection to ice ages? Milankovitch theorized that the total summer radiation received in the northern latitudes, near 65 degrees north where ice sheets have formed previously, is the trigger for the development of an ice age. The fact that there is more land mass (which reacts to temperature changes far more rapidly than open ocean) in the northern hemisphere is the reason. Precession effects, which Milankovitch calculated should have the strongest effect on insolation because they regulate where the equinoxes are relative to the annual aphelion and perihelion passages (eccentricity factors) of Earth, will then accentuate the seasons. According to Milankovitch's model, when (1) summer in the northern hemisphere coincides with aphelion (furthest from the Sun), (2) eccentricity is high (greater distance from the Sun, at least half the year), and (3) obliquity is at minimum (the least contrast between seasons), insolation will be very low and conditions most conducive to glaciation. The triggering of an ice age is thus thought to occur during periods of cool summers that allow for an accumulation of ice and snow from year to year in the higher latitudes. Eventually this retained snow and ice cover builds into an ice sheet that reflects solar radiation back into space preventing heating. And so a positive feedback loop is established and the cooling of Earth accelerates. Climate change driven by orbital cycles is constant but slow, very slow when compared with anthropogenic greenhouse gas forcing, the scientific name for human-driven global warming. Skeptics and uninformed politicians have confused and distorted these two processes and have sold out our future by doing so.

It is important to emphasize that Milankovitch's modeling of insolation that drives glaciations is based primarily on precession, but does include a weaker obliquity signal and a minuscule eccentricity factor. This is logical as the calculated variations in insolation for the three cycles are 8 to 13 percent for precession, 5 percent for obliquity, and 0.2 percent for eccentricity. You can see that, of the three cycles, eccentricity produces by far the least amount of change in insolation, and it was believed to have the least effect on climate (remember John Herschel gave up on it). But this math doesn't match the climate record, in part because eccentricity is not a simple one-hundred-thousand-year cycle. That figure is just a rough estimate of a complex orbital modulation driven by perturbations on Earth's orbit from Jupiter and Venus that actually consists of three periods of approximately 95,000, 125,000, and 400,000 years. Because eccentricity does correlate with major changes in the climate record, we have here a good example of how a very weak signal can have powerful effects on a self-organizing system (climate), in this case one operating over vast amounts of time.

Evidence of the effects of ellipticity can be seen in oxygen isotope variations found in climate records. Oxygen comes in three stable forms that vary in the number of neutrons packed into the nucleus, these variations being called isotopes. Nearly all oxygen atoms have eight neutrons and eight protons, this being standard oxygen or ^{16}O. Heavier oxygen atoms with nine or ten neutrons also exist, most of these being the latter, or ^{18}O. When water (H_2O) is evaporated, more molecules having the lighter ^{16}O bonded to the two hydrogen atoms will be taken up as water vapor to be blown by the winds and deposited as rain or ice somewhere else. If this somewhere else is land, and the transfer is in the form of ice that stays in place, the ratio of oxygen isotopes in ocean water will change and can serve as an indication of how much water has been removed from the seas and, from that, what sea level was like at the time.

It is the microorganisms found in marine sediment cores that record shifting oxygen isotope ratios. In the seas, oxygen is used by foraminifera (single-celled protists) to build shells, which accumulate on the ocean floor and become a record of past ocean conditions. So, if you know the standard ratio of ^{16}O to ^{18}O, and the tiny shells in a section of a marine core show that they contain more ^{18}O than the standard ratio, you can use this information as an indication of the climate at the time the organism lived. For example, a section of a marine core that has higher than normal ^{18}O in the microscopic foraminifera shells suggests that the missing light oxygen was carried away to ice sheets and that this was the time of an ice age and lower sea levels. The location of this section in the length of the core tells us how long ago this happened. Both marine cores from the ocean floor and ice cores from glaciers are records of Earth history and contain information (called proxy data) that yields a chronology of past climates, including ice ages, sea level changes, large volcanic eruptions, and more.

In both marine and ice cores, it is the ^{18}O values that often show the strongest correlations to the basic roughly one-hundred-thousand-year cycle of eccentricity, with obliquity and precession signals being weaker. This is the reverse of what would be expected, but it has only been the case for roughly the previous eight hundred thousand years. Prior to this time, the forty-one-thousand-year obliquity cycle predominated, which tells us that there is apparently no consistently dominant cycle of ice ages. In spite of these anomalies, and there are others, most geoscientists and climatologists have accepted orbital forcing and the weak one-hundred-thousand-year cycle as a solution to the mystery of ice ages during the past approximately eight hundred thousand years. From a longer perspective, it appears that climate becomes entrained for a time to one signal or another, or a combination, depending on what the conditions are at the surface, these being regulated by other factors such as land mass location and ocean currents. Orbital forcing, or how solar system astronomy is crucial to understanding Earth history, has answered many questions and is now a valuable and integrating tool in the geosciences.

Orbital cycles are complicated by the fact that the Earth's axis and its orbit are perturbed by the gravitational forces of the other planets, noted earlier in regard to eccentricity. These perturbations are not one-dimensional. When the solar system is viewed on its side, assuming the plane of the Earth's orbit as a horizontal reference plane, the orbits of the other planets will intersect this plane at low angles. As planets follow their orbits and move above and below this plane, they will exert slightly different gravitational forces on Earth due to changes in geometry, and these will modulate Earth's orbit and, consequently, climatic frequencies. There are so many gravitational forces from the planets affecting Earth that it is difficult to produce accurate isolation curves for more than about twenty million years into the past or future. At two hundred million years there can be as much as a 40 percent error. One exception is the roughly four-hundred-thousand-year ellipticity cycle that appears to be very stable and has been found in geological data from the Newark Triassic-Jurassic basin discussed below. These geological records now serve as a means of adjusting and focusing the celestial mechanics of orbital variables in the remote past—this being a situation where geology informs astrophysics, which then informs climatology.

Exactly how the eccentricity cycle modulates climate has not been easy to explain. In the 1950s, Milankovitch's theory on orbital forcing was rejected by most geologists due to the advent of radiocarbon dating and the results it produced in regard to the geological epoch named the Pleistocene (~2.5 million to ~12,000 years ago). It was thought, based on the data available at the time, that there had been warm climate intervals as recent as twenty-five thousand years ago and other ice fluctuations over eighty thousand years in the past that were at variance with Milankovitch cycles. In the 1960s, however, the analysis of deep-sea cores produced evidence that sea level fluctuations did appear to occur in a twenty-one thousand-year precession cycle, but the one-hundred-thousand-year cycle of ellipticity, theoretically the weakest one, was far more prominent than had been expected. A one-hundred-thousand-year cycle of solar activity was proposed, and it was suggested that orbital forcing by the technically stronger nineteen- and twenty-three-thousand-year precessional frequencies were what was timing the terminations of glacial episodes; when they coincided with ellipticity at its highest, a case of wave reinforcement. Another idea used to explain the marine core data was that ice-volume fluctuations, or the ice ages, were just modulated, not driven, by orbital forcing in a highly complex system. Then there is the issue of carbon dioxide levels over time.

Absorption of CO_2 by microorganisms in the oceans has been found to have a one-hundred-thousand-year cycle, something that would affect greenhouse gas levels in the atmosphere and could account for a cyclic cooling on land. But why the oceans would draw down this gas on a regular cycle is unknown. So, this matter of a tiny force, ellipticity, having a big effect is still not as yet settled and serves as a good example of not only how complicated the study of a system can be, but how little is known about these things, even with substantial grant money available to well-trained geoscientists.

Orbital cycles have been found in sedimentation patterns in the Mediterranean Sea floor. The Messianian Salinity Crisis, which began about six million years ago in the late Miocene, was a period when the Mediterranean alternately evaporated and filled at the precessional rhythm (which explains the vast salt beds under the sea floor). As many as fifty-five precession-induced sedimentary cycles have been found. Another example, in the late Triassic (two hundred million years ago) strata of the Newark rift basin in New Jersey, are the lake level cycles that were found to correlate with orbital cycles (Olsen and Kent 1999). The Newark Basin Coring Project, funded by the National Science Foundation, retrieved rock cores covering a period of about thirty-two million years in the late Triassic and Jurassic, a time when the region was located in the tropics and the era of dinosaurs was just beginning. The cores revealed regular cycles of precipitation and evaporation that were controlled by orbital cycles, specifically precession and ellipticity. The time-scale produced from these records was then used as a template for magnetic field polarity reversals, essentially an astronomical tuning of the geomagnetic time scale. The analysis of the data suggests that the quadruple ellipticity cycle of approximately four hundred thousand years, which is modulated by the gravitational forces of Venus and Jupiter, is stable over long periods. The Mars-Earth orbital relationship over the past two hundred million years also shows up as a signal found in the data. Apparently, the other bodies in the solar system, not just the Sun and Moon, exert a significant influence on Earth's climate and geology over long periods of time through the rhythmic modulation of its orbital configurations.

THE SUN, SOLAR CYCLES, AND EARTH

By an enormous margin the Sun is the most important astronomical body in our near cosmic environment. Its light, or lack of it, is the basis of the circadian cycle and, due to photosynthesis, is the driver of nearly all life in general. Photosynthesis, a chemical process that releases energizing oxygen as a byproduct, allows complex multicellular life to thrive and also to get big. It first evolved in bacteria that were later assimilated (subsequently becoming chloroplasts) by other organisms that evolved into modern algae and plants. These Sun-dependent organisms, cyanobacteria, algae, and plants, are autotrophs, the self-feeders on which just about the rest of the biosphere is sustained. They make the air breathable with their oxygen waste and their bodies are the food for animals and other kinds of life. There is another kind of metabolism, however, that uses chemical reactions to produce food, called chemotrophy, but it is too slow a process to sustain much more than a bacterium. These are found in inhospitable places, like those where the Sun doesn't shine, and may exist on other planets or moons.

The Sun is about 4.6 billion years old and categorized as a second-generation star in the main sequence, the most common type of star. Like other stars, the Sun generates energy as a result of a chain of nuclear reactions in its core. It is the nuclear fusion of hydrogen atoms into helium, driven by immense gravitational pressures within the Sun, that produces solar energy. The temperature of the Sun's core is an unimaginable 15 million degrees Kelvin, while the visible outer layer of the Sun, the photosphere, has a temperature of only 5,700 degrees Kelvin. The heat produced in the Sun's core causes photons to be radiated from the photosphere into space, a small portion of which are encountered by the Earth in its orbit around the Sun. About 70 percent of this solar radiation, which includes the wavelengths of infrared, visible, and ultraviolet light, is absorbed by the Earth and fuels the biosphere; about 30 percent is reflected back into space.

Earth is also bathed in solar wind, which is a stream of charged particles, mostly electrons and protons, that are lost from the Sun. Because the Sun rotates, the solar wind radiates out from the Sun like streams of water from a rotating lawn sprinkler. The number of sunspots on the Sun at any given time is a good indication of the power of the solar wind, as sunspots indicate higher solar activity. The solar wind also carries with it magnetic fields that spiral out into the solar system, creating what is called the Interplanetary Magnetic Field. If Earth or another planet happens to be in the way of one of these spiraling waves of magnetically charged solar wind, it gets blasted. When the charged particles of the solar wind contact Earth's magnetic field, they push against the field like water being displaced by the bow of a ship in motion. When the solar wind is particularly active, such as after a solar flare occurs, its collision with Earth's magnetic field will generate aurorae, shimmering curtains of electrical activity called the northern or southern lights, near the poles. The total omni-directional energy output of the Sun is called its solar luminosity. The amount of solar radiation that the Earth intercepts (assuming a surface perpendicular to the solar beam at the top of the atmosphere) is called total solar irradiance (TSI). The mean of solar irradiance, at all wavelengths, is called the solar constant, and it has been calculated to be about 1,366 watts per meter squared. This figure, thought to be constant within a narrow range of roughly 0.1 percent, is the total radiant power of the Sun at one astronomical unit, which is the average distance of the Earth from the Sun. There is some fluctuation of this energy reaching the Earth, however, and it is related to the eleven-year sunspot cycle.

To understand what sunspots are and how they may form, consider the Sun as a dynamic system operating in a fluid state. Its fluid nature is demonstrated by its axial rotation: twenty-five days at its equator and thirty days at its poles, with an average rotation of twenty-seven days. This difference in motion causes magnetic complications. The Sun's magnetic field, which extends far into space, is a polar dipole magnetic field like that of Earth that has north and south magnetic poles. With the solar equator rotating every twenty-five days and the poles every thirty, the north to south magnetic field lines on the Sun will twist and coil, creating a constantly evolving magnetic field—what's called the solar dynamo. In the process of twisting, magnetically charged sunspots form on its surface, these being large regions that are about 1,500 degrees Kelvin cooler than the surrounding surface, and they reduce the radiated energy from the region they occupy. Somehow sunspots grow and decay in a more or less regular cycle, and they do account for changes in total solar irradiance, which raises two issues: How exactly is the solar cycle generated and what effects does it have on Earth's climate and biosphere? Before these are addressed, we need to consider the scientific history and the physical nature of the solar cycle.

The roughly eleven-year solar cycle is evident in the numbers of sunspots counted on the Sun over a period of time, as was discussed earlier. Heinrich Schwabe (1789–1875) is generally credited with discovering this cycle, which bears his name. Schwabe was an amateur astronomer who systematically observed the Sun through a telescope from 1826 to 1868 in hopes of spotting a solar transit of Vulcan, a hypothetical inter-Mercurial planet. He never found Vulcan, but he did record a succession of sunspots and thus created a database. In 1843 Schwabe announced that he had discovered a cycle of ten years during which time sunspots increased and then decreased in number. Rudolf Wolf (1816–1893) refined this discovery. He noticed that sunspot cycles varied from as short as 8 years to as long as 17 years, though the average was about 11.1 years. Wolf worked in Zurich, Switzerland, and established an index of sunspots that is now called either the Wolf or Zurich index. Wolf also examined older records of sunspots and pushed the index back as far as 1700, effectively creating the first multi-century solar activity database.

Another important nineteenth-century figure in the history of solar science is Richard Carrington (1826–1875). In addition to his observation of a massive solar flare in 1859 and noting connections to magnetic field disturbances on the Earth, he also discovered that the Sun has differential rotation and that sunspots form at high solar latitudes and move toward the equator as the cycle progresses. This phenomenon was graphically illustrated by E. Walter Maunder (1851–1928) in 1904, and it has since become known as the Maunder butterfly diagram because of its resemblance to butterfly wings (see diagram below). In the early twentieth century George Ellery Hale (1868–1938) recognized that sunspots were actually gigantic magnetic storms on the surface of the Sun, and not dark clouds. He also noticed that the magnetic polarity of sunspots shifts from positive to negative at each eleven-year cycle and a complete pattern of reversal takes twenty-two years, which is thought to be the proper length of the solar cycle. This cycle is now commonly referred to as the Hale cycle. A number of other solar cycles have been discovered or hypothesized. The Gleissburg cycle, given as about eighty-seven years, is a cycle of the amplitude of the Schwabe cycle, which can vary from as few as eight years to as many as seventeen. Evidence of it is found in beryllium and carbon-14 isotope counts in ice cores and tree rings. A roughly two-hundred-year cycle, called the Seuss or de Vries cycle, is found in climate records but not in sunspot counts. Other longer cycles are proposed, though few show strong correspondences with climate data.

Exactly what causes the Sun to produce a solar cycle of sunspot numbers and magnetic field changes is not completely understood. Most textbooks will explain it entirely in terms of differential rotation. This model, the Babcock solar magnetic dynamo model, assumes that the faster equatorial rotation of the Sun, which distorts or "winds up" its magnetic fields,

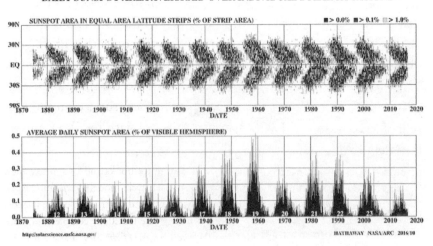

Figure 2. The solar cycle seen as sunspot numbers and formation over time. The upper chart shows time on the x-axis and solar latitude on the y-axis. Sunspot density moves toward the solar equator over the course of each cycle. The lower chart, with matched x-axis, shows the percentage of sunspot cover on the Sun's surface since 1875.

is the ultimate cause of the solar cycle (Babcock 1961). At the start of the cycle the Sun's magnetic lines of force are like longitudinal lines running through the solar surface from pole to pole, called a poloidal field. Because the solar equator rotates faster than the polar regions, these magnetic field lines twist and distort, eventually becoming parallel to the equator and establishing what is called a toroidal field. From these windings it is hypothesized that subsurface bands develop polarized kinks and coils that emerge from below the photosphere to become pairs of sunspots. Eventually, these spots migrate toward the solar equator and dissipate, setting the stage for a new poloidal field with reversed polarity. The periodicity of sunspots is then something that apparently, and somewhat mysteriously, emerges out of the Sun's rotation. Other versions of this dynamo model have been proposed, but at present none concur with all observations.

The alternate model to explain the cause of sunspot cycles, not often mentioned in solar scientific literature, involves planetary gravitational forcing. Paul D. Jose from the Office of Aerospace Research analyzed the movements of the center of mass of the solar system (it's not the Sun, it's the barycenter) and found that its eccentric paths in and out of the body of the Sun followed a 178.7-year period (which some suspect is the basis of the Seuss cycle) during which 8 Hale cycles of 22.37 years occur.[3] He argued that this cycle is driven by the angular momentum of the solar system in total, and that the larger planets—Jupiter, Saturn, Uranus, and Neptune—contribute greatly to the effect. Jose plotted the movement of the solar system's center of mass relative to the Sun's surface, and also the Sun's motion around it, and correlated this data with the Wolf sunspot

---

3. A look at two major introductory astronomy college textbooks and two books about the Sun are evidence of the information deficit in solar cycle models. *Explorations: An Introduction to Astronomy* (Arny 1998) offers only a simplified description of the Babcock dynamo model and *Astronomy: The Evolving Universe* (Zeilik 2002) completely ignores the issue of solar cycle causes. *The Cambridge Encyclopedia of the Sun* (Lang 2001) is a moderately technical book dedicated entirely to the Sun, and it offers only one page on this matter. The *Guide to the Sun* (Phillips 1992, 67–72) goes into some detail in regard to the dynamo model but also doesn't mention the planetary hypothesis. In contrast, *Dark Matters* (Seymour 2008, 114–29) discusses the dynamo model and its weaknesses, and also the planetary model in some detail. More discussion can also be found in the article "Sun's Motion and Sunspots" in *The Astronomical Journal* (Jose 1965).

index. He concluded that the solar cycle is influenced by the gravitational torque of the planets on the Sun, which agitates the solar atmosphere. Planetary tidal forces, and not differential rotation alone, then play a role in moving magnetic fields in the Sun that leads to sunspot formation. It follows that, because the orbits of the planets are regular, this will keep the solar cycle regular; Jose drew attention to the 178.7-year cycle as fundamental to understanding the process. Other observations pertinent to this planetary model of the solar cycle are that Jupiter's roughly 12-year orbital period is very close to the approximately 11-year Schwabe cycle; the roughly 20-year synodic cycle of Jupiter and Saturn is close to the approximately 22-year Hale cycle; and that 6 Saturn cycles equals 15 Jupiter cycles, which equals about 178 years.

$$(6 \times 29.5 = 15 \times 11.9 = \sim 178)$$

Critics of this planetary forcing of the solar cycle point to the fact that there were few sunspots during the Maunder Minimum, a period of low solar activity and cold climate during the late sixteenth and early seventeenth centuries, but the planets were presumably still orbiting the Sun in those years. This critique hasn't stopped investigation into the phenomena, however.

Geologist Rhodes Fairbridge (1914–2006), and also climatological mavericks Timo Niroma (?–2009) and Theodore Landscheidt (1927–2004), are among those who have incorporated planetary forcing in their solar and climate studies. Niroma argued that the orbit of the gas giant Jupiter has a primary role in the solar cycle and, using the Wolf index, pointed out that sunspots increase when Jupiter is at aphelion, but decrease when the planet is at perihelion. Landscheidt developed a unique method of forecasting solar events (flares, coronal mass ejections, and the intensity of individual solar cycles) by analyzing Jupiter's gravitational torque on the Sun, combined with that of Earth, and the rotational velocity of the Sun around the barycenter of the solar system. Like Niroma he was a global warming critic and he made many climate predictions including one that forecast increasing twenty-first-century cold for several decades based on his predicted low in the solar cycle. This forecast continues to be cited by global warming skeptics (Mackey

2007; Landscheidt 1989; Niroma 2009). There have been fewer sunspots during the second decade of the twenty-first century, but global temperatures are up, so the burning of fossil fuel as the cause of the heating is by far the best explanation.

A paper on planetary forcing of the solar cycle suggests there is a spin-orbit coupling between the Sun's rotation and the Jovian planets (Wilson, Carter, and Waite 2008). Changes in the Sun's equatorial rotation rate are argued to be affected by shifts in the center of mass of the solar system (the barycenter) and that this connection is modulated by the approximately twenty-year synodic period of Jupiter and Saturn. One finding reported is that alignments (syzygy) of Jupiter and Saturn occurring after solar maximum establish a trend where the Wolf sunspot number remains above eighty per year. This pattern is then sustained for a period that is roughly the length of the Gleissburg cycle (about eighty-seven years) after which there is a collapse of the pattern when the syzygy occurs before maximum. The authors of the paper argue that it is a resonance lock between the planets and the Sun that is broken when the cycles drift apart after about ninety years, or roughly four Hale cycles, which is the approximate length of the Gleissburg cycle. Obviously, this topic is complex and not reducible to a single correlation, and it raises questions that require understanding on the systems level, where multiple causes drive emergent phenomena.

In the 1990s, British astronomer Percy Seymour proposed a specific correlation between peaks of solar activity and the peaks of the combined tidal effects on the Sun generated by Earth, Venus, and Jupiter. The mechanism behind his proposal is the phenomena of orbital resonance, where matched periodicities are able to selectively amplify and transfer energy. Seymour's theory of magneto-tidal resonance states that the gravitation of two or more planets moving together around the Sun pull on the plasma of the Sun (just as the Moon pulls the oceans of the Earth, creating the ocean tides) and that their combined effect can be much greater than the simple sum of their gravitational influence. This amplification of tidal forces is then capable of generating certain emergent properties on the Sun. Almost three decades later a paper was published that proposed a similar planet-driven explanation for the solar dynamo. In this model the gravitational forces of Venus, Jupiter, and Earth modulate the solar magnetic field and together produce the 11.1-year Schwabe cycle, and they may also affect the stratification of plasma within the Sun (Seymour, Willmott, and Turner 1992; Stefani, Giesecke, and Weier 2018). In spite of all these studies and proposals, the theory of planetary modulation of the Sun as a cause of the solar cycle has not interested the majority of solar cycle researchers (possibly because it smells of astrology), but it remains a viable alternative because it is difficult to debunk. Scientific papers on the influence of planets and the behavior of the solar system's barycenter over time continue to be published, and hopefully, more will be learned about how the Sun produces spots on a more or less regular schedule.

SOLAR IRRADIANCE AND ITS VARIABILITY

Having looked at sunspot cycles, we return to the crucial issue of determining the extent to which the Sun plays a role in weather or climate change, and how it may affect the biosphere and the Earth system in general. One would think this is a no-brainer based on experience, like why birds and people go south for the winter—or even why there is life on Earth. The Sun dominates the sky, is obviously hot, and it heats the planet. A search on the internet for images of "energy budget" will bring up countless diagrams providing information on how much solar radiation makes it through the atmosphere, how much gets reflected back into space, and how much actually heats the planet. This is basic information, but it is only a snapshot, a brief moment in time that says nothing about any variations or cycles. It also says nothing about solar variability and to what extent the Sun, which is generally assumed to be a constant force, undergoes changes that can move global temperatures in discernable patterns. We know for sure that the Sun keeps the bulk of Earth's biosphere going, but as far as having behaviors that definitively drive climate, that turns out to be not so simple and is another example of the kind of uncertainty in science that is often found in deep investigations of self-organizing systems.

This issue of solar effects on climate has come to be quite polarized, with some climate scientists discounting solar effects and others, especially climate change skeptics, touting it. Data collection is crucial to sorting this out, and much effort has been spent over the past century on precisely measuring solar energy reaching the Earth. One of the more productive projects was carried out by Charles Greeley Abbot (1872–1973), who ran the Smithsonian Astrophysical Observatory from 1902 to 1957. During this time, he carefully measured solar brightness using many delicate instruments, and his records remain to this date a useful dataset of solar variability. Greeley found that sunspots will, on the short term, decrease solar irradiance (called sunspot blocking), but that on the long term, the Sun at sunspot maximum is actually brighter, a counter-intuitive observation. In the 1970s and 1980s accurate data on solar variability was gathered by orbiting solar observatories. It was found that total solar irradiance did vary from what's called the solar constant (~1366 watts per square meter) by only 0.1 percent over the course of the approximately eleven-year cycle. On a daily basis, however, the change could be as much as 0.5 percent. Another finding was that faculae, very bright portions of the Sun surrounding sunspots, more than compensate for sunspot blocking. So, Greeley was correct, the Sun at sunspot maxima does put out more energy. But solar radiation is not so simple. It is known that the solar spectral peak, which is the center of the visible spectrum, is at a wavelength of 0.5 micron, but the Sun also radiates energy in other wavelengths. For example, short wavelengths (ultraviolet) show significantly greater variation, up to 10 percent, over the course of the solar cycle, though critics of a solar influence on climate point out that short wavelengths less than 0.3 micron don't even reach the Earth's surface; they are absorbed high in the atmosphere. However, this radiation does heat the stratosphere, which may indirectly affect climate.

During sunspot maxima, sunspots decrease the solar output at all wavelengths. But along with more sunspots come more faculae, the bright photosphere regions surrounding sunspots. These faculae radiate strongly in the blue and ultraviolet (UV) portions of the spectrum and account for a net increase of UV radiation reaching the Earth's atmosphere during times of solar activity. UV radiation accounts for

20 percent of the total solar energy reaching the Earth. Since faculae are mostly located in the central regions of the Sun, their radiation is highest in the direction of the plane of the solar system (the Sun has a low axial tilt) and will therefore intercept the Earth's orbit. All of this amounts to saying that UV radiation increases are substantially higher than other wavelengths over the course of the solar cycle, and that these may have some affect on the upper atmosphere, which, in turn, affects the lower atmosphere and consequently climate. This chain of events and amplifying mechanisms—processes like those that operate in non-linear systems—need to be considered in determining to what extent the Sun drives climate.

Taking a broader perspective, we can ask if solar variability is found in other stars, and if so, at what amplitude? Stars other than the Sun are now known to display cycles of variability in energy output. In fact, most stars display variability to a much greater extent than does the Sun, leading some scientists to speculate that our Sun is atypical in this regard—or it is currently in a resting phase. Younger stars are more active than older stars, and the cycles of variability observed typically range from seven to twenty years. Solar-type stars also display received irradiance variations of up to 0.6 percent. With instrument data extending only a century or two into the past, and human historic records extending not more than a few thousand years, it is difficult to say whether or not the Sun displays over time the range of variability now known to exist in other stars.

In summary, a case can be made for solar irradiance being variable, but not with a particularly high amplitude except in the UV range. Many solar variability cycles have been described or proposed and several have been correlated with climate data. One is the twenty-seven-day rotation of the Sun, which is thought to produce a variability when one hemisphere of the Sun contains sunspots and the other doesn't. With two different sides of the Sun, each with a different radiative value, specifically in the higher wavelengths, solar rotation would then produce something like a digital signal. But this period is so close to that of the 27.3 lunar sidereal cycle that terrestrial data displaying a 27-day periodicity is difficult to ascribe to a specific mechanism, solar or lunar. The best known of the solar cycles, the Schwabe 11.1-year (average) cycle of sunspots, also has a variable cycle itself that ranges from as few as 8 years to as much as 17 years, with some cycles expressing more sunspot blocking and others expressing more facular emission. The Hale 22-year double sunspot cycle includes magnetic field change, which amounts to one full solar reversal cycle. It is thought by some that these magnetic field reversals in the solar cycle affect Earth's magnetic field, which in turn influences the upper atmosphere in terms of electrical charge—something that can affect thunderstorm frequency and weather in an 11-year cycle. These linkages are an example of how small forces become amplified.

Other, longer solar cycles are recognized. A roughly eighty-three-year cycle was first noticed by Wolf in 1853 but was then rediscovered decades later when, in the 1930s, Wolfgang Gleissberg published a number of papers on a cycle of about eighty-seven years that groups the lengthening-shortening trend of Schwabe cycles together. What is now called the Gleissberg cycle is considered to be one of the more significant solar cycles because it does show up in climate records of carbon, beryllium, and oxygen isotopes. More of a slow wave than a pulse, it also shows that the Schwabe solar cycle is not a true periodicity. Another long cycle in the range of 180 to 210 years, now called the Seuss-de Vries cycle, is also found in ice core isotope data and sunspot counts. It has been linked to the barycenter cycle, specifically the 179-year recurrence cycle of Jupiter and Saturn (15 Jupiter orbits are equal to 6 Saturn orbits). A number of other solar cycles have been proposed, including the approximately 2,400-year Bray or Hallstatt cycle, which has been argued to be driven by the four gas giant planets (Scafetta et al. 2016). So there seem to be at least some solar effects on climate that are recorded in proxy data, but how do they affect surface temperatures, if they do?

The science behind a correlation between solar activity and environmental conditions on Earth over short or long intervals (weather and climate) turns out to be very complicated. (From what has been presented so far in regards to geocosmic connections this should be no surprise.) Besides funding, this work requires accurate and long datasets (steadily improving due to satellite data) and computer modeling methodologies that account for disruptions, such as volcanic eruptions that send dust into the atmosphere cooling the planet, anthropogenic greenhouse gas forcing, El Niño events, and changes in ocean currents that affect weather. Because this research involves different disciplines, explanations that are not commonplace, and often limited data, the results of studies are often debated for decades. All of this illustrates just how excruciatingly difficult it is to fully understand the multiple processes in an operating system and actually put numbers on it.

SOLAR ACTIVITY AND CLIMATE

In this next section I will attempt to describe the history and results of decades of solar activity research that, I think, makes the case that pinning down correlations between solar system activity and the Earth system is not for amateurs, or those with a low tolerance for ambiguity. This is an important issue to resolve in the climate change debate, however, and it requires considerable knowledge, mathematical skills, and vast institutional support to do the science properly. There are no unequivocal answers as yet, and this is a weakness that climate change skeptics exploit to push their agenda. Their success is a testimony to human discomfort with ambiguity and the need for authoritative answers—cognitive closure reduces stress. More importantly to the themes of this book, it also illustrates how a chain of causes is often required to explain a correlation and also how relatively weak forces can be amplified and have big effects.

In 1976 astronomer John A. Eddy published a landmark paper that drew attention to a period of low solar activity in the late seventeenth century and early eighteenth century (Eddy 1976). During this period historical records of aurorae and observations of the Sun's corona during eclipses were absent and sunspots were rarely seen. When they were sighted by the astronomers of the day, they were regarded as special discoveries and scientific papers were written about them. For the same period ^{14}C (radiocarbon) records from tree rings were high, this being an indication of a weak solar wind that allowed more cosmic rays (which collide with nitrogen to form ^{14}C) to enter the atmosphere. From this data Eddy reconstructed past solar activity and found it to have been extremely low between 1645 and 1715. He found that it correlated with a period of cold climate, at least for Europe, and he named it the Maunder Minimum. Eddy also found declines in solar activity not only then, but also between 1460 and 1550, a period he named the Sporer Minimum (named for Gustave Sporer, another sunspot science pioneer). These periods mark low points of solar activity in a larger block of time that has been named the Little Ice Age. In contrast, very low ^{14}C levels are found between 1100 and 1250, a warm period that Eddy referred to as the Grand Maximum (also known as the Medieval Maximum), the time when Vikings were living comfortably in Greenland. Eddy's paper made a strong case that solar activity has a lot to do with surface temperature, at least in some parts of the world. This all sounds straightforward but, unfortunately, the situation is far more complicated and unsettled, as will be seen over the next several paragraphs.

Correlations between the various solar cycles and climate have been made since the early days of sunspot observation, but proponents of a correlation have differed as to whether or not sunspot maxima coincided with higher or lower temperatures. By the 1960s, this long history of a lack of consensus led to the entire field of solar-climate research falling into disrepute. The major problem seems to have been that sunspot cycle maxima did correspond to higher temperatures between 1720 and 1800, but it correlated with lower temperatures between 1600 and 1720 and 1880 to 1920. The correlation reversal that occurred in the 1920s was one during which a number of Sun-climate relationships changed or broke down completely, these reversals of correspondence being called by climatologists "sign changes."

Inconsistent correlations are often reported in climate studies. In 1991 a paper correlated a 130-year record with the approximately 11.1-year solar cycle, but found that the temperature record actually led the solar cycle, negating a cause-and-effect relation. The authors then looked at the solar cycle lengths (eight to seventeen years), which, when tweaked to fit into the eighty- to ninety-year Gleissberg cycle framework, removed the solar lag and corresponded neatly to the temperature records. The study turned out to be flawed, however, by some data inconsistency; an update reported that the correlation, which existed in the past, failed in more recent decades. A correlation between globally averaged sea surface temperatures, which are more evened out than land temperatures (because it takes water longer to heat than air), and Wolf's long-term solar activity record, was reported in 1987. The study found that the approximately eighty-seven-year Gleissberg period appears as the dominant cycle of sea surface temperatures and also that the necessary change in the solar constant to produce this long-term effect was under 1 percent, within the range of what's been measured from the Sun. In 2010 a study once again reported a small correlation between sea surface temperatures and the eleven-year solar cycle. Using a dataset from 1854–2007, the authors calculated that about a quarter of the warming over the past one hundred and fifty years was probably due to the Sun, the rest to human fossil fuel burning (Friis-Christensen and Lassen 1991; Reid 1987; Zhou and Tung 2010).

The solar cycle appears to correlate with more than just temperatures. Correlative evidence for a link between the solar cycle and precipitation has been found, but it has also proven inconclusive. Again, there is a long history of studies of this nature, but as soon as one apparent correlation has been found and a paper has been published, the next study disproves it. The central problem has to do with the fact that rainfall varies spatially, and any given region's annual rainfall

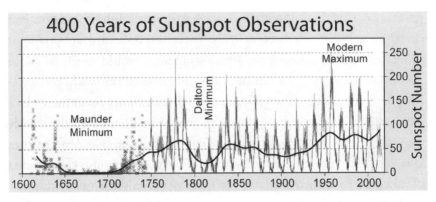

Figure 3. Sunspot numbers and estimated temperatures (solid line) over four centuries. Before 1740, sunspot number figures are based on ^{14}C records.

may not accurately reflect the larger pattern. Another problem is that cyclic variations in records for precipitation show both an 11- and an 18.5-year cycle. The latter figure is presumed to represent a response to the lunar 18.6-year lunar declination cycle; this periodicity is shown by the precession of the lunar node (the intersection of the Moon's orbital plane with the plane of Earth's orbit around the Sun). It is thought that variations in tidal strength due to the changing elevation of the Moon (declination) has an effect on the surface area of the oceans, the amount of surface area available for heating. A roughly 20-year drought cycle, which some believe to be a reflection of the 22-year Hale cycle, has been found in tree rings. Others have argued that this could also be a reflection of the lunar 18.6-year cycle—or both.

A mechanism to explain cyclic patterns in rainfall driven by solar activity involves the expansion and contraction of the Hadley cell, a major latitudinal circulation feature of the atmosphere. In the Hadley cell, hot, humid air at the equator rises, moves toward the poles losing its moisture along the way as rain, and then, depleted of water, sinks to the ground at about 30 degrees latitude north and south, the regions where most deserts and arid lands are located. The Intertropical Convergence Zone, a belt of clouds over the equator where trade winds converge and where the highest levels of incoming solar radiation are received, would, theoretically, expand and intensify along with an increase in total solar irradiance. This expansion would force the moisture-depleted descending limb of the Hadley cell further north and south of the equator, moving the desert regions poleward. It is thought that a 0.15 percent increase in solar irradiance would result in about the same percentage increase in precipitation. Since the world precipitation average is 100 centimeters per year, this amounts to an increase of 0.15 centimeters per year, though this would be distributed differently over the globe. But this is just a hypothesis, and the proposed mechanism has yet to be demonstrated conclusively. The research is also complicated by the burning of fossil fuels (anthropogenic greenhouse gas forcing).

Storm frequencies are another weather variable that appear to reflect the solar cycle. The thought behind this connection is that higher levels of solar irradiance produce higher sea surface temperatures and that this leads to more evaporation and the production of more cyclones. Both higher and lower frequencies of storms at solar maxima have been noted, and apparently both the phase and the actual location of the study is again crucial. The annual number of cyclones does appear to follow an 11.3-year pattern, but there is also a peak at 51.2 years, the latter thought to correlate with a solar activity peak that occurs at 52.7 years (yet another cyclicity—and also one found to correlate with the economy). Thunderstorm frequencies appear to follow the solar cycle, and a mechanism has been proposed to explain this. Low levels of solar activity will result in a thinner and weaker solar wind, allowing more cosmic rays to enter the Earth's atmosphere. Increased cosmic ray bombardment causes changes in the electrical charge of the ionosphere and consequently variations in the global electrical circuit, thus affecting thunderstorms with electrical discharges.

Further evidence of possible solar effects on climate may be seen in population cycles of various organisms. Studies have shown an eleven-year cycle in insect populations, often high just before sunspot maximum. Reasons given for this include increased heating or precipitation (depending on species and region) and also increased solar UV radiation, which is thought to stimulate the general activity of organisms. Locusts, a plague species, have been correlated with the sunspot cycle, with populations waxing between solar minimum to maximum and declining between solar maximum to minimum. Tent caterpillar populations in New Jersey also show a roughly eleven-year cycle that peaks about two years ahead of sunspot maximum. Interestingly, recent satellite data has shown that the Sun brightens at about this point in the solar cycle. Tree rings, which are a highly regarded climate proxy of both temperature and precipitation, reveal a weak eleven-year cycle, but its appearance is also inconsistent, leading to a general distrust of this connection (Hoyt and Schatten 1997).

Correlations between the approximately eleven-year solar cycle and economics have been proposed. The astronomer (discoverer of Uranus) and composer William Herschel (1738–1822) followed the solar cycle for forty years and observed that periods of few sunspots correlated with the price of wheat. This correlation is still debated, and papers both pro and con continue to be published. The economist and logician William Stanley Jevons (1835–1882), one of the founders of economics as a science, took Herschel's ideas further. Not satisfied with the notion that "commercial moods" caused the more or less regular fluctuations of the business cycle, he looked into what might be causing them. Jevons came to the conclusion that the solar cycle drives weather that affects agricultural harvests, which, in turn, affects investment and speculation and consequently the larger economy (Peart 1991). This is a multi-link theory that Jevons saw as suggesting two things. One is how a weather cycle, as it filters through human needs and commercial infrastructure, is observable in economic data, and the other is the notion that a small signal can have a broad effect. Jevons published several scientific papers on this subject but, as with Herschel's, his ideas have had a mixed reception and have been challenged by critics but also applauded and applied by speculators.

It should be obvious at this point that, for a long time now, the solar-climate relationship has not been a clear-cut situation. Much depends on how questions are formulated and the actual tolerances of the climate system. Exactly what is being sampled, such as temperature, pressure, wind, humidity, precipitation, and the like, can make an enormous difference. The length of the sample will shape the results. Using standard thirty-year norms[4] may not show a long-term increase or decrease of what is being measured—such records are simply not long enough. There are also difficulties in taking accurate measurements. Instruments need to be calibrated and properly located, and there are choices to make in regard to sorting the data in terms of mean, highs, and lows. One major problem today is discriminating between natural and anthropogenically-driven changes. These need to be separated, but this if often very difficult, if not impossible. In some cases, measuring only extreme weather events, that is events of eminence, yields clear results and yet reflects only very small variations in the solar constant. Amplifying mechanisms, some of which have already been described,

---

4. Thirty years is a convention, begun in 1901, that is based on a statistical law called the central limit theorem, which states that a sample size of at least thirty will more closely approximate the mean distribution of whatever is being sampled than what would be the case using a lower sample size.

have been suggested and sought for. Given the complexities and uncertainties of the solar-climate relationship, climate modeling, which is completely dependent on the program established by the operator, has its limits. Again, all of this illustrates just how difficult it is to analyze a system using reductionist science.

A side effect of the uncertainties of the Sun-climate connection is that it feeds climate change skepticism and delays movement away from fossil fuels. Some argue that the Sun is obviously the predominant influence on Earth's climate, which it must be, but the devil is in the details. Today, a number of websites exist dedicated to showing that anthropogenic greenhouse gas effects on climate are a deception, and most support this claim with a very small selection of scientific papers that argue for a dominant solar influence on climate. Typical arguments posted on these websites point to correlations between the activity of the Sun and the period of the Medieval Warming, and the Little Ice Age that is correlated with sunspot minimums. Those in the climate community critical of a predominant solar influence on climate argue that the solar constant varies by only about 0.1 percent, not enough to make much of a difference. However, others say that this variation is not realistic, it is more like a 0.5 to 1.0 percent change. Again, if the variability of the solar constant is as small as it appears to be, and this is somehow driving climate in certain ways, then there must be some kind of amplifying mechanism at work, including some previously mentioned. This is an important point and one of the themes of this book: in a self-organizing system, like a cell, an organism, or the climate system, small forces can often have big effects. These small forces may require a chain of events and long periods of time to be effective, or they may kick in rapidly when resonances with other small forces occur.

Most mainstream climate researchers would agree that solar variability is an important factor in climate change and that it must be studied further and incorporated into climate models. But most also agree that it is not enough of a force to explain the warming that is presently occurring (the increase of greenhouse gases is apparently able to account for it). The current thinking is that as greenhouse gas levels rise, solar variability will play less and less of a role in climate change.

This means that anthropogenic greenhouse gas heating has trumped the Sun. Whether or not biotic or other natural mechanisms will be able to limit these human-induced changes is yet another question. The biosphere has apparently produced a species that can interfere with environmental conditions on a vast scale but has so far failed to restrain itself from creating serious problems. We are living through an experiment sustained by a combination of two unsubstantiated beliefs: humans are separate from nature and lightly-regulated corporate capitalism is good for everyone.

COSMIC RAY FLUX

In the last section it was noted that variations in total solar irradiance over the course of the Schwabe 11.1-year cycle are limited to about 0.1 percent of the solar constant and that climatologists have sought amplifying mechanisms to link this very small change with measurable atmospheric data. While the overall results have been mixed and the situation is complicated, some researchers have looked to another feature of Earth's space environment that may link the Sun and climate: cosmic rays. Cosmic rays are high-energy particles, most of which emanate from supernova remnants in the galaxy (galactic cosmic rays) or beyond (extragalactic rays). Their penetration of the Earth system is limited by Earth's magnetic field when, at sunspot maximum, it is being fortified by the solar wind. But when the Sun is calm, more cosmic rays enter the Earth's atmosphere where they may possibly affect tropospheric (low altitude) cloud formation by creating conditions for their production. With increased low altitude cloud cover, stimulated by higher cosmic ray flux, there is an increase in global albedo (reflectivity), which deflects incoming solar radiation. This multi-link process is thought to cool the climate. This series of steps, called an amplifying mechanism for solar variability, was proposed near the end of the twentieth century.

The journey of cosmic rays from supernovae to penetration of Earth's atmosphere, the study of which has been named cosmoclimatology, concludes with the formation of cloud cover (Svensmark and Friis-Christensen 1997; Svensmark 2007). Clouds form when water vapor condenses on very small particles in the atmosphere, these being called cloud condensation nucleii. Cloud altitude is an important factor in regard to heating or cooling the surface. While an increase in high altitude clouds leads to warming (by trapping heat), an increase in low altitude clouds leads to cooling by reflection (higher albedo). During sunspot minimum, when the solar wind is at its lowest levels, cosmic rays penetrate the atmosphere with greater frequency. Cosmic ray collisions with molecules of nitrogen and oxygen in the atmosphere change the isotopic ratios of certain elements (e.g., carbon) that, when incorporated into plants (e.g., trees), can be used as proxies in the detection of climate changes. This is how radiocarbon records from tree rings come about. But if cosmic rays also affect the formation of cloud condensation nucleii, then they could be an amplifying mechanism for the small changes in the solar constant. Therefore, if cosmic ray intensity is inversely proportional to the solar cycle, then the Sun is modulating Earth's climate.

Henrik Svensmark is a Danish physicist who works for the Danish National Space Service on Sun-climate research. In his scientific papers and a 2007 book (co-authored with science writer Nigel Calder) called *The Chilling Stars: A New Theory of Climate Change,* he made a case for cosmic rays having a stronger effect on climate than anthropogenic greenhouse gases. Later, a documentary film about this subject called *The Cloud Mystery* was released. In 2012 he published a paper in which he proposed that cosmic ray variations over long periods of time have affected the evolution of life over the past five hundred million years. Meanwhile the basic idea that cosmic rays create cloud condensation nucleii was still being tested in a small cloud chamber in Copenhagen, directed by Svensmark, and also at the European Organization for Nuclear Research (*Conseil Européen pour la Recherche Nucléaire*—CERN) particle collider on the French-Swiss border. The experiment in Copenhagen found that cosmic rays break up air molecules and release electrons that act as catalysts in the formation of very small clusters of sulfuric acid and water—which may act as cloud condensation nucleii, though critics say they are too small. At CERN, accelerated particles were shot into a chamber in the Cosmics Leaving OUtdoor Droplets

(CLOUD) experiment that also investigated the cosmic ray–cloud connection. The experiment at CERN found that nucleation (the first step in the formation of an assembly of atoms) does occur, mostly involving biogenic compounds, ammonia, and sulfuric acid, but the amount produced is not sufficient to account for the variations of cosmic ray intensity. The international climatology community was not impressed and judged Svensmark's promotion of his ideas to be premature.

The publicity of the cosmic ray–climate connection in print and on screen brought it to the attention of climate skeptics who have embraced it, and, of course, the press has seized on what they imagine to be a Galactic Cosmic Ray versus Greenhouse Gas Warming science fight. All of this has had the effect of making physicist Svensmark a suspicious outsider in the world of climate science. The blog *Real Climate,* a commentary website for genuine climate scientists, argues that Svensmark's work is sloppy and missing critical steps resulting in "by far the most blatant extrapolation-beyond-reasonableness" of a piece of work they have seen. *Skeptical Science,* a blog that gets skeptical about global warming skepticism, says the data doesn't support the hypothesis. But to be clear, Svensmark's critics are not saying there is absolutely nothing to the connection between the solar cycle, galactic cosmic rays, and climate. They are saying the effect is just too weak and inconsistent. The problem was that Svensmark and his team promoted the idea before fixing some major holes in the hypothesis, and meanwhile damaged the campaign for cutting greenhouse gas emissions, an urgent situation in my opinion.

Svensmark's idea that cosmic ray flux over long periods may correspond to ancient climates was picked up by Israeli astrophysicist Nir Shaviv, who proposed that very long climate cycles, driven by variations in galactic cosmic ray flux, occur as the solar system orbits the center of the galaxy (Shaviv 2002, 2003). The solar system, which orbits the galactic center over the course of some 220 to 250 million Earth years (the galactic year), at times passes through the spiral arms of the galaxy where cosmic ray sources are concentrated. Recall that the spiral arms themselves are density waves that can have a destabilizing effect on stars, causing some to explode (called a supernova) and spew cosmic rays. The arms themselves are transient, may last under one billion years, and vary according to distance from the galactic center. There appear to be two dominant arms, and four in the range that our solar system orbits. In Shaviv's hypothesis, cosmic ray flux entering the Earth system is higher when the solar system passes through the spiral arms, causing long periods of cooling due to enhanced low altitude cloud formation. The spiral arms also rotate more slowly around the galactic center than the solar system; the passage of the solar system through all four arms takes about seven hundred million years. Galactic arm passages (the solar system is presently in the Orion armlet) are estimated to occur about every 150 million years, and passage through the arm is thought to have shocking effects. As the solar system enters a galactic arm, acceleration occurs, followed by deacceleration while in the arm. The moving spiral arms themselves leave behind a wake of slowly diffusing cosmic rays that are higher after the solar system crosses the arm, which disrupts it, than before its passage.

The actual timing of spiral arm passages is not known with precision. Shaviv based his dating of spiral arm passages and consequent increases in cosmic ray flux on a combination of density wave theory, observations, and meteorite exposure dating. In this later method, meteorites are dated from the break with their parent body using tracks of cosmic rays on their newly formed surfaces. Shaviv used a sample of eighty iron-nickel meteorites from which he selected fifty and dated them by analyzing isotopic changes due to cosmic ray exposure. Clusterings of dates are interpreted to be an indication of epochs of higher cosmic ray flux. A periodicity of 143 plus or minus 10 million years, close to the period of a galactic arm passage, emerged from this method. The solar system also oscillates above and below the galactic plane in a cycle of about 32 million years, or 64 million years for a complete sine wave-like cycle. (Recall that bolide impacts have been correlated with a roughly 30-million-year extinction cycle.) Shaviv argued that a major flux of cosmic rays is produced in the north crossing of the Galactic Plane, the direction toward which the galaxy is now moving in the local super cluster, and he used a database of fossil diversity, sea level, and large igneous province formations (massive volcanic surface rifts) that purport to show periodicities of approximately 62 million years as correlative evidence. This hypothesis has received strong criticism (Medvedev and Melott 2007; Rahmstorf et al. 2004).

Ice ages in geological history occur on two time scales, in tens of thousands and in hundreds of millions of years, the first generated by orbital variations, the second possibly by galactic arm passages. A reconstruction of global temperatures during the Phanerozoic (the last 541 million years) using ^{18}O paleoclimate data (oxygen isotopes in calcitic shells) suggests a broad correspondence between higher cosmic ray flux and glacial or cool episodes in Earth history.[5] Shaviv used this as support for his argument but also pointed out that climate reconstructions using carbon dioxide do not show a similar correspondence and don't correlate as well with the paleoclimate record, another blow against the consensus opinion regarding greenhouse gases and climate change. Shaviv is a vocal critic of global warming (as is Svensmark), which creates a consensus-reaching problem in the current highly politicized climate crisis. Maybe some of his ideas will hold up in further studies, but his position on global warming only raises suspicion among the majority of climate scientists who are often skeptical of physicists butting into their territory, about which they know a great deal.

One thing that should be apparent by now is that the study of cosmic influences on the Earth system is by necessity multidisciplinary. But this isn't always welcomed by geoscientists and climatologists who have worked long and hard on problems in their fields and are then ignored by a different kind of scientist (usually a physicist) who decides to tackle these same problems on their own terms. This brings to mind an episode in the history of science. In the mid-nineteenth century physicist Lord Kelvin (William Thomson), of thermodynamics and absolute temperature fame, inserted himself into the field of geology and calculated Earth's age. The figure that he eventually settled on was about

---

5. This reconstruction shows a periodicity of cool periods spaced roughly 150 million years apart, one of these being the end of the Ordovician, during which time a mass extinction occurred that is thought to be driven by a massive ice age. Correlations with the other cool periods are not so strong (Shaviv 2003).

twenty million years. His heat-loss model ran against the observations of geologists, who thought Earth was much older, but they only had estimates based on actual observations, not precise calculations. Even Darwin's theory was on the line as a shorter age for Earth meant less time for evolution. But Kelvin, considered to be the greatest physicist of his time, dominated the discussion and kept his critics more or less silenced for decades. Eventually he was proven spectacularly wrong (but he never revised his view) when radioactivity was discovered by chemists and radioactive decay was seen as a source of sustained interior heat. Today, Kelvin is considered by some fundamentalist Christians to be a creation scientist because his calculation is thought (by people who don't understand science) to support Young Earth Creationism. Lesson: the compulsion to be right may produce unintended consequences.

Returning to cosmic rays as a means of regulating climate, the topic continues to be controversial, but it is a fact that more cosmic rays do enter the Earth system when they are not blocked by a strong magnetic field. Normally, Earth's magnetic field strength increases when the Sun is active and decreases when it is not, but there are other ways the field could weaken. A recent investigation of the last geomagnetic reversal (pole shift), which occurred some 780,000 years ago, found that Earth's magnetic field strength decreased substantially at that time and cosmic ray flux increased. Analysis of dust deposits (loess) in China suggests that the winter monsoons during this approximately five-thousand-year period of transition were stronger and that temperatures at the time were lower. What's interesting is that this magnetic field reversal, with accompanying cold climate presumably produced by low altitude cloud cover seeded by an abundance of cosmic rays, is roughly coincident with the shift in glacial cycles from chaotic to entrainment with the one-hundred-thousand-year ellipticity cycle. The present-day weakening of the geomagnetic field, which has been happening for a couple centuries, is now being hypothesized by some geophysicists to correlate with global cooling.[6] And once again

---

6. Magnetic field reversals have historically occurred on average roughly every 250,000 years, so Earth may be a bit overdue for one (Ueno et al. 2019; Courtillot et al. 2007).

the global warming consensus in the climatology community is challenged to show that greenhouse gases will more than compensate for this trend. The media and public can't really follow this complicated situation and, to reduce cognitive stress, seek authoritative statements, but the scientific method is just not designed to do this at the high speed of modern media.

TIDES AND TIDAL FORCING

While Kepler concurred with ancient astronomers and astrologers (e.g., Seleucus, Posidonius, Ptolemy) that ocean tides were Moon-driven, Galileo thought otherwise and argued incorrectly that they were a result of forces caused by Earth's revolution around the Sun, this being one of his arguments for the Copernican hypothesis. Isaac Newton, however, was the first to explain tides as an effect of gravity and was also the first to mathematically describe all gravitational forces in a single formula: the product of the masses of the bodies divided by the square of the distance between them. This means that gravitational attraction drops rapidly as distance increases, and this seems to support primarily Moon-driven, as opposed to Sun-driven, tides. The Sun is about 27 million times more massive than the Moon, but it is roughly 390 times further away, diminishing its pull on the tides to about 45 percent that of the Moon. Given two tide-producing bodies, there are actually two distinct tides that work with or against each other depending on their phase relationship. Tidal fluctuations driven only by the Moon have a periodicity of about 24.8 hours, but those driven by the Sun are 24 hours, the rotation of Earth. The difference is due to the fact that the Moon advances in its orbit around Earth each day and Earth needs to rotate a bit more to catch up.

As Earth rotates, seas are raised as the Moon appears to move above them, this being called a tidal bulge. Meanwhile, on the other side of Earth, another tidal bulge is produced by centrifugal forces due to the orbit of both Earth and Moon around their barycenter, or center of mass, a point located about 1,700 kilometers below the surface of Earth. So as Earth rotates, there are always two tidal bulges: one on the side facing the Moon, which pulls it, the other on the opposite side pulled by centrifugal force generated by circular motion around the center of mass of a two-body system. It is the tidal forces between Earth and the Moon that has locked the Moon into its spin and orbit resonance (tidal locking) where only one side faces Earth. Orbital resonance is common throughout the solar system; Pluto and Neptune are in a two-to-three resonance and many moons that orbit the larger planets are locked into stable orbits with low number ratios.

So far this explanation of tides is fairly straightforward, but many other factors must be taken into account to understand tidal forces and their effects. One of these is lunar declination, which accounts for the elevation angle of the Moon relative to the equator of Earth. Lunar declination changes over an 18.6-year cycle that is marked by the movement of the lunar node through the zodiac. Another is the dampening of the tides when the Sun and Moon are geocentrically 90 degrees from each other, these being the neap tides that result from tidal forces of the Sun and Moon canceling each other out. Higher tides occur when Sun and Moon are in straight-line alignment at full and new Moons (a syzygy). Tides are also dependent on the distance between Earth and Moon, and Earth and Sun, which change as the orbits are slightly elliptical. When the Moon is closest to Earth, at perigee, tides are higher. When Earth is at perihelion, closest to the Sun, solar gravitational force is stronger than at other times. The highest tides occur when Earth is closest to both Moon and Sun and these are in syzygy. This generalized explanation of the tides is called the equilibrium model, but to accurately describe ocean tides, which are shaped by the configurations of shorelines, basins, and other topographic features, a far more complex dynamic theory of tides is necessary. It seems that tidal forces, their genesis and effects, are apparently very complicated and technical discussions on the details continue.

The gravitational pull of the Moon also works on Earth's atmosphere, producing a semidiurnal tide. Not being inhibited by surface features in the same way as the seas, the atmosphere is moved toward the Moon as it passes directly over the surface with a corresponding bulge on the opposite side. The lunar semidiurnal atmospheric tide is

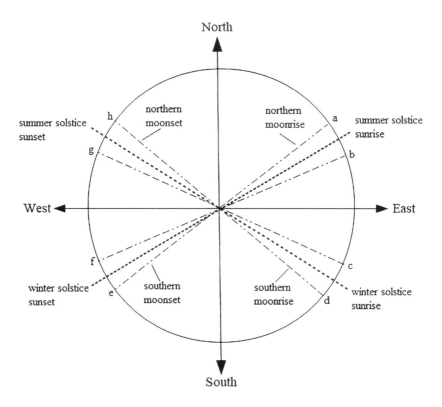

Figure 4. Horizon positions of solar and lunar risings and settings. The cycle of the solar year can be measured in three ways: day/night ratio, elevation of the Sun at noon relative to the horizon, and the rising and setting positions east and west, as shown in the drawing above. All three are effects of the Sun's declination, which varies by 23.5 degrees north and south of the celestial equator over the course of a year. (Notice that 23.5 degrees is the tilt of Earth's axis.) The lunar 18.6-year cycle, driven by the constant shifting of the intersection of the Moon's orbital plane relative to the Earth's orbital plane (the lunar nodes), is likewise measured in these three ways. These variations modulate the track of the Moon through the sky and consequently its elevation and time above the horizon—which has gravitational implications for differing terrestrial latitudes. A number of ancient archaeological sites have alignments that track the lunar nodal cycle by marking the extremes of lunar risings and settings, as this information is also of use in predicting eclipses. In the drawing, points a, d, e, and h mark the extreme rising and setting positions of the Moon at major standstill, while points b, c, f, and g mark the rising and setting positions at the minor standstill. Both biological and other environmental cycles have been found that match the 18.6-year period of the lunar nodal cycle.

equivalent to the lunar ocean tides, except that the Moon is moving gases, a more refined fluid. These tides have been difficult to measure. There is some displacement of the atmosphere measurable at the surface with a barometer, and correlations have been made with weather patterns, but unlike the oceans, there is no precise top boundary layer, though some sense of this can be found in barometric pressure readings at altitude. The amplitude of the lunar diurnal atmospheric tide is small, measured in microbars, and varies by geographical latitude. These differences are accounted for by the gravitational pull of the Moon being stronger at low latitudes where the plane of the Moon's orbit is closer to the local zenith (Chapman and Lindzen 1970). In addition to the semidiurnal tide, the atmospheric tides also have larger periods driven by the apogee orbital period of 8.8 years and the oscillation of the orbital plane of the Moon and the orbital plane of the Earth's orbit (ecliptic), which is 18.6 years. This is the nodal, or declination (distance north or south of the celestial equator), cycle of the Moon, which has many correlations with terrestrial phenomena and can be seen in terms of rising and setting positions that many ancient societies have measured (see diagram on page 77).

The second daily atmospheric tides, the solar daily atmospheric oscillations, are driven by the Sun. These are diurnal thermal tides caused by the heating and expansion of the atmosphere as it passes under the Sun's rays, and they are about twenty times stronger than lunar atmospheric tides. During the day, the Sun heats the atmosphere and it expands, causing a very measurable displacement of gases that occurs at a twenty-four-hour periodicity—a thermal tide. The tidal bulge remains stationary as Earth rotates under it (counterclockwise as seen looking down at the north pole) producing a westward propagating wave. Because the atmosphere becomes less dense with height, solar heating tidal effects are greater in the stratosphere and above. The effects of this tide are evidenced in wind patterns and, as with the lunar tides, have a higher amplitude at lower latitudes.

The high and low tides of the seas are obvious to anyone living near a shoreline, but there are also tides on land. These are called body tides, effects of the pull of Moon and Sun that may raise the crust by a meter or so. As already noted, ocean tides are complicated by surface features (continents, coastline irregularities, bays, islands, etc.) and tidal effects are not always immediate. Body tides are broader in effect and follow the motions of these two astronomical bodies more closely, and patterns of surface displacement also move with the changing declination of the Moon. Volcanic events are known to be triggered by body tides, and vulcanologists use this information when monitoring such events. The gravitational effects of the Moon and Sun can also trigger earthquakes. Again, its not as simple as a full Moon causing a quake. Tidal effects on the seas and land will vary according to the elevation of the Moon relative to Earth's equator as shown by its declination, the distance to the Moon, the distance to the Sun, the proximity of the Moon and Sun, and many more subtle orbital and physical factors. In fact, there are at least three hundred components of tidal force that need to be calculated in order to assess the possible influence of the Moon and Sun on the Earth at any given time, which explains one of the reasons why the tidal connection to earthquakes has not been reduced to a formula. Just to get a computer model to reasonably simulate these tidal effects requires at least thirty of those factors. Another reason is, of course, the issue of actually understanding the geology of any given earthquake-prone region. This is a situation where astronomers, physicists, and geologists need to collaborate.

Jim Berkland (1930–2016) was a controversial geologist from California who predicted earthquakes using Earth tides. Before retirement, he was the geologist for Santa Clara County, and he had also worked for the U.S. Geological Survey (USGS). What he did was devise a predictive system based on his observation that earthquakes are most likely to occur at specific points in the synodic cycle of the Sun and Moon, points that he called seismic windows. These points are the days around new and full Moons, the straight-line alignment of astronomical bodies (syzygy), which are amplified when the Moon is at perigee (closest to the Earth). Unlike most geologists who work in a lab or in the field and cautiously publish scientific papers, Berkland made a name for himself in 1989 by predicting in a newspaper article that a major earthquake would strike the San Francisco area between October 13 and 21. The Loma Prieta earthquake, 6.9 in magnitude and near

San Francisco, occurred on October 17, 1989, just as the third game of the World Series was about to begin. In this case there was a Full Moon a few days before the event that occurred close to lunar perigee, and this indicated a much stronger than usual Earth tide that Berkland thought could cause a fault to slip. Berkland's methodology also went beyond astronomy and included noting the number of lost pet ads found in local newspapers. Animals can apparently sense an earthquake coming; they become disoriented and exhibit strange behaviors, including getting lost days before it happens. This component of his method was not Berkland's original idea; it was recognized by city officials in Haicheng, China, too, and saved thousands of lives on February 4, 1975, when a massive quake struck that area.

Berkland's earthquake predictions were not well received by the town government he worked for, nor by other geologists or the USGS, and he was temporarily suspended from his job for offering information that some thought could cause a public panic. But this didn't stop him, and he continued to work with his earthquake prediction model and keep records of earthquakes. He had a book written about him, and also set up a website where he could publish predictions and engage in discussions with like-minded earthquake enthusiasts (Orey 2006). Berkland didn't write any scientific papers, and at least two critical evaluations of his methods found them not to produce statistically significant results, so he remains an outlier in geocosmic studies. He wasn't as far out as he was made to be, however.

Ocean tides can apparently trigger earthquakes. The sloshing of the oceans as they respond to the tidal pulls of the Moon and Sun changes the weight of the water column over the coastal floor. Large tides that occur at a new or full Moon, and especially when one of these occurs at the same time that the Moon is at perigee, are particularly strong and are now thought to increase the frequency of earthquakes. A relatively recent scientific report found a high correlation between tidal peaks near coastal subduction zones and shallow-fault earthquakes (University of California Berkeley 2004). Subduction zones, where an oceanic tectonic plate is being pushed under a continent, occur along much of the Pacific Rim. The western United States and the west coast of South

America are places where many faults are found, both shallow and deep, and earthquakes are common. Just a few additional meters of water depth pulled by the Moon and Sun are apparently enough to trigger a fault that is ready to break and consequently generate an earthquake.

The Moon at perigee has been shown to be an earthquake trigger, but so has the Moon's declination. In a paper published in 1968, Gurgen Tamrazyan, a Russian geologist, demonstrated that the frequency of larger earthquakes varies with lunar perigee, subtle variations in the rate of the Earth's rotation (spin rate), and also the declination of the Moon at its culmination. Apparently, more than 75 percent of all earthquake energy released occurs when the Moon is at its most northerly declination. This may be due to the fact that the Northern Hemisphere simply has more land mass and is therefore more affected by lunar gravity when it is overhead. Tamrazyan studied earthquakes for many years and suggested that certain regions have their own cycles. He actually predicted earthquakes, including the December 7, 1988, Armenian quake, and proposed a sixty-two-year cycle of earthquakes for that region (Tamrazyan 1967, 1968, 1988). All of this is considered Earth physics, and the mathematical complexities are beyond the understanding of most people, but he was basically on the same track as was Berkland, he just added more astronomical factors and more numbers.

There is other evidence that tidal forces may trigger earthquakes. The Earth is spinning rapidly "under" the Moon and Sun, and the gravity from those bodies tugs on the Earth, causing a tidal bulge in the direction of the Sun and Moon. The tidal bulge, an east-to-west wave moving around the planet on a daily basis, extends well below the surface and can produce an earthquake when it triggers a fault that is about to slide, which was Berkland's thinking. One study has found that this "Earth tide" caused by the gravitation of the Sun and Moon definitely stimulates fifteen-mile-deep tremors along the San Andreas Fault (University of California Berkeley 2009). Apparently, the water at this depth is highly pressurized and lubricates the rock to such an extent that it takes very little force to generate movement.

Because both astronomical and geological variables must be accounted for, prediction is very difficult and it is not considered prudent among scientists to make predictions unless they are extremely confident—and that's how it should be. Those that are not so cautious may damage the reputations of others in the same field by confusing the media and public with unauthorized and potentially inconsistent predictions and notions that lack consensus in the scientific community. Science is a highly specialized social activity and rogues are not tolerated, even if they turn out to be right later on. Today the control of scientific information is often attacked by the anti-expert propaganda machine of right-wing media.

Back in 1959, a very unusual scientific paper appeared in *Nature,* one of the top scientific journals in the world. It was written by a Russian named R. Tomaschek who made a case for a connection between the planet Uranus and terrestrial earthquakes (Tomaschek 1959). He took a total of 134 earthquakes with a magnitude greater than 7.75 and calculated where Uranus was located in its diurnal cycle at the time of the quake. (Due to the rotation of Earth, the Sun, Moon, and every planet in its diurnal [daily] cycle rises, crosses the meridian, sets, and crosses the lower meridian.) For the period 1904 to 1906, in a total of twenty-three cases, he found that Uranus was within an hour of crossing the upper or lower meridian at the epicenter of the quake. After this two-year period, Tomaschek reported that the correlation became weaker but was still significant according to his statistics. He suggested that a possible reason for the higher significance of the 1904–06 period was that Uranus was then moving in opposition to Pluto and Neptune, and therefore these additional bodies would, in all cases, be at the opposite meridian. Tomaschek didn't think that gravity could be a key factor behind this correlation because there were no similar findings for the Sun or Moon, both of which exert force on the Earth several orders of magnitude greater than the outer planets do. He did note the fact that Uranus's axis is unusual, at 90 degrees to the plane of the solar system, and that this may enhance its magnetic field in some strange way. His paper was criticized for its statistics, but Tomaschek defended his work.

Another case of a proposed link between planets and earthquakes came in 1974, when astrophysicist John Gribbin and astronomer Stephen Plagemann published a book called *The Jupiter Effect: The Planets as Triggers of Devastating Earthquakes* (Gribbin and Plagemann

1974). In it they pretended not to be doing astrology, but did so anyway, and predicted that a very loose conjunction of planets in 1982 would cause dire events. They even suggested that the San Andreas Fault would give way and the big one would finally hit California. For starters, the span of planets in 1982 was far from a conjunction. The outer planets were spaced between 60 and 70 degrees apart during that year, and at the date an earthquake was predicted, March 10, 1982, the entire solar system, from a geocentric view, fell within a range of 180 degrees. (Just to note, in astrology the standard limits for a conjunction [0 degrees] are much tighter, usually given as about 15 degrees along the ecliptic.) Gribbin and Plagemann proposed a mechanism that ran like this: The gravitational influences of the planets all coming from one general direction would disturb the Sun and consequently fortify the solar wind. The solar wind, a fine stream of charged particles, would smash into the Earth's atmosphere, disturbing the weather and affecting the Earth's rotation rate, and this would trigger an earthquake that was waiting to happen. So, as we know, the big California quake didn't happen. Gribbin and Plagemann then put out another book, titled *The Jupiter Effect Reconsidered,* in which they sort of took credit for predicting the Mount Saint Helens eruption in 1980. Years later, Gribbin, an otherwise excellent science writer, admitted that this was all one big mistake. Oddly, he didn't attribute a spike in the geomagnetic index and the volcanic eruption of El Chichón, Mexico, in April of 1982 as possible effects of their grand alignment. It was a very significant eruption, just a month after their prediction, and the resulting dust blocked sunlight and changed the weather all around the world.

Not long after the Gribbin and Plagemann debacle, a prediction of a potentially massive earthquake in the Midwestern United States was made by a maverick zoologist and business consultant named Iben Browning, who identified himself as a climatologist (Winkless and Browning 1980). The fact that this prediction was made for the Midwest and not California is significant. Back in the winter of 1811–12, three major earthquakes occurred with an epicenter near New Madrid, Missouri. The third one was estimated to have been a whopping 8.8 or so on the Richter Scale and is reported to have rattled church bells in Boston. Few people were living near the epicenter at the time, but those who were reported land rolling in waves and rivers being diverted. More recent research has shown that shocks this big have occurred in this region in the past, though not too frequently. Why they occur here is not well known, but it is thought that an ancient fault lies deep underground. Browning predicted that a large quake may occur on December 3, 1990, and, of course, the fear-mongering media publicized it and the scientific establishment discredited it.

Browning thought that it was tidal effects from a lunation (full or new Moon) when both the Sun and Moon were closest to Earth that would trigger seismically sensitive areas, including the Midwest, and he noted this would happen in 1990. Browning also thought that the approximately 179-year recurrence cycle of Jupiter and Saturn noted earlier, where Jupiter makes 15 cycles and Saturn 6, could be a factor as well. Add 179 to 1811, and you get 1990, the year of his prediction. A lot of Midwesterners did prepare for a quake and held readiness drills that were mocked by skeptical scientists. Obviously, there was no quake in the Midwest, but that year the number of earthquakes greater than 6.0 globally was higher than average. Of note is the death toll that single year, which was much higher than the years in the decade before or after, though a single earthquake in 1990 accounts for most of that total.[7] If there is something to planetary tides triggering earthquakes, a lot more needs to be known. In this case we have just another example of a half-baked prediction, using just a few variables, promoted by a zoologist who got media attention.

Although the above predictions by an astrophysicist, an astronomer, and a self-appointed climatologist were spectacular failures, their general assumptions about a mechanism that produces earthquakes is not complete pseudoscience. It appears to be the case that earthquakes may be triggered by a multitude of forces, including gravitational tugs from the Moon and Sun, the weight of water along a coast, minute changes in the Earth's spin rate caused by solar activity, any number of tidal

---

7. A single earthquake in Iran that year on the day of the summer solstice accounted for about 80 percent of the total.

pulls and air mass displacements, and possibly even the angularity of the planet Uranus. When a system is at the edge of stability, as in a fault ripe for sliding, the last straw, so to speak, may make all the difference. Exactly what this last straw happens to be is the problem, and reducing earthquakes to a simple formula is just not possible. Geologists, who actually study the subject, know that earthquakes are very complex, and the proof of this is that precise prediction is elusive.

If the Moon and Sun cause tides on Earth, do the planets cause tides on the Sun? As already noted, English astronomer Percy Seymour proposed a theory of magneto-tidal resonance in which the gravitational pull of the planets on the Sun may modulate the solar cycle by generating sunspots and flares (Seymour, Willmott, and Turner 1992). Seymour's theory, which incorporates the ideas of Paul D. Jose also mentioned earlier, begins with standard dynamo theory—that the longitudinal magnetic field lines on the Sun begin to distort and form magnetic canals parallel to the solar equator due to the differential rotation of the Sun (faster at the equator, slower at the poles). Seymour then argued that a planet orbiting the Sun will pull the plasma in these equatorial canals, creating a tidal bulge. On the Sun's surface, this tide of gas will not have to deal with coastlines that would dampen its strength, and it will therefore be strongly responsive to other gravitational pulls, especially ones synchronized with it. Seymour argues that two or more planets moving around the Sun together in syzygy or some other harmonic resonance should be able to produce a tidal force much greater than the simple sum of their gravitational influence.

The explanation for simultaneous tides on opposite sides of Earth, as previously noted, is that the Moon and Earth revolve together around their common center of mass, or barycenter. The two astronomical bodies are held together by gravitational attraction, but are simultaneously kept apart by an equal and opposite centrifugal force produced by their individual revolutions around the center-of-mass, a motion that may also have geological consequences. Similarly, the orbit of the Sun around the barycenter of the solar system produces differential forces that cause very tight loops, called jerks by those who study this phenomenon. These are thought to disturb the Sun at depth resulting in abrupt changes in solar radiation and the solar wind. Likewise on Earth, barycenter movements and forces may affect the spin rate, which is known to have consequences geologically and meteorologically. Again, we see here examples of small forces becoming amplified and, through a chain of events, producing strong effects.

Our effort to understand planetary influences on solar cycles has a long history. Galileo was the first to suggest such a thing, and much research on this subject was conducted during the nineteenth and early twentieth centuries. Most of this work is ignored by modern scientists who are usually more focused on the major problems of their time, but there is one body of alternative research that is worth reviewing. By the middle of the twentieth century, the radio industry recognized that strong solar events could disrupt shortwave radio communications taking place on the side of the Earth facing the Sun, the daylight half. Shortwave signals are very sensitive to solar activity, and one radio company, RCA, realized it could manage its communication network more efficiently if disturbing solar events could be predicted. Around 1950, one of RCA's radio engineers, John [H.] Nelson, perfected a method for forecasting such disturbances by considering the influence of the planets on the Sun. His method, which he called Astro-Physics, was to locate the planets on a 360-degree graph with the Sun at the center, basically a heliocentric astrological chart. Next, he located times when the planetary positions formed certain kinds of alignments with each other. When they did line up in specific ways, based on observed correlations with past solar events, he would predict a solar storm. His forecasts were claimed to be consistently near 90 percent accurate, and his job at RCA was secure for many years.

Nelson's findings in regard to heliocentric planetary alignments and disturbances in short wave radio signals are complex but include the following (Dean and Mather 1977, 307–312): Radio wave disruption occurred when at least three or four angular separations (astrologers call these phase angles "aspects") of a certain class formed within a twenty-four-hour period. Among those angular separations, conjunctions (0 degrees between two or more planets) were observed to have a weaker effect (in disturbing radio signals) than squares (90 degrees) or oppositions (180 degrees).

Linkages of trines (120 degrees) and squares occurring simultaneously were observed to be strong, but if these aspects were not linked with other planets the effects were weak. Major radio wave disturbances also involved planets separated by multiples of 30 and 45 degrees. Distance above or below the plane of the ecliptic (celestial latitude) did not seem to make much of a difference in the effects of these angular separations. Planets near their node (intersection of the planet's orbital plane with that of Earth's) had increased strength, while planets near aphelion (farthest from Sun) had decreased strength. All planets, including Pluto, have an effect, but the slower planets tend to establish a temporary field, which the faster planets then activate (Dean and Mather 1977, 307–312).

Although Nelson's ideas were published in reputable journals, he never presented his work as formal scientific research; it was more like a report from an engineer. His method was basically a kind of heliocentric astrology, and his skill set was developed entirely on his own. Nelson's heliocentric planetary alignment maps were so similar to astrological horoscopes, and his interpretive skills comparable to those of practicing astrologers, that for many years he was a welcome speaker at astrology conferences. Because solar system configurations change constantly and never repeat themselves, reduction to a simple formula was impossible and interpretation was crucial. Today Nelson's work is mostly forgotten. His methodology has only been tested a few times, and these have produced conflicting results. One particularly savage astrology debunker claims to have shown Nelson's work was meaningless, though RCA, which one would think was concerned with concrete results, did employ him for many years. At least one test of his method, published online, has shown some convincing results, but these were accompanied by snarling dismissive comments from those who apparently smelled astrology lurking just below the surface (Dean 1983; Dalton 2004). We will revisit Nelson in a later chapter.

The search for cycles and periodicities in nature has long been a part of science. Knowledge of cyclic behaviors offers predictability that, along with pattern perception, is part of the scientific method. Cycles are a series of events regularly repeated, that is intervals during which a set of processes or events occurs, such as the formation and decay of sunspots or glaciation and deglaciation. The time taken to complete a cycle is its period, its recurrence at specific intervals. The sunspot cycle has a periodicity of 11.1 years on average, and over the past eight hundred thousand years glacial periods have averaged about one hundred thousand years, each glacial cycle having phases of icing and de-icing. What is of interest to some researchers, and is also a theme of this and the preceding chapter, are cases where cyclic phenomena in different organisms or natural processes share the same periodicity. When these are identified, the logical places to look for a commonality, or perhaps a driver of the period, directly or indirectly, are geophysical and astronomical movements.

The commonly cited economic cycle of boom and bust is an interesting case. In his 1879 book *Progress and Poverty,* political economist Henry George outlined an economic cycle that he thought was driven by changing differences between the value of land (which can't be produced) and general economic growth as shown by commodities (which change in value). When land costs too much, labor will respond in ways that amount to less production and a slowing economy, which in turn lowers the cost of land and stimulates real estate activity. In the twentieth century, other economists (e.g., Homer Hoyt) found that this cycle was roughly eighteen years in length, though it can be distorted by events like the Great Depression and also government intervention in the markets. Today what is called the eighteen-year real estate market (or property) cycle is divided into four main stages, and one can find numerous articles online that discuss the details—but not much is said about why it averages around eighteen years. On the other hand, there are a number of modern financial astrologers who link this cycle with the 18.6-year cycle of lunar declination, which has been referred to previously.[8] Since scien-

---

8. The cycle of lunar declination can be tracked with the Moon's nodes, the intersection points of two planes, that of the Moon's orbit around Earth and Earth's orbit around the Sun, which are tilted relative to each other by about 5 degrees. The Moon's nodes retrograde through the zodiac. When the node enters a zone roughly 60 degrees before and after the vernal equinox (the zodiac signs Taurus, Aries, and Pisces), the real estate cycle drops off from its high a few years early when it was near the summer solstice point. This cycle was also used by stock trader W. D. Gann and astrologer Louise McWhirter. As previously noted, the lunar nodal cycle tracks lunar declination variations, which do affect lunar gravitational forces on Earth.

tific studies have correlated this lunar cycle with rainfall variability and temperatures of air and sea, it is possible that the real estate cycle may be an indirect result of weather that affects agricultural output.

Edward R. Dewey (1895–1978) made a career out of the search for cycles. Working as an economist in the Hoover administration shortly after the Great Depression, Dewey was charged with determining the causes of this catastrophic event. His approach was to study the actual behavior of the economy rather than the causes, something that most economists were not doing at the time. Dewey began his work by collecting information on hundreds of cycles in nature and business, looking for those that shared the same period. His work crossed disciplines: geology, biology, astronomy, economics, and the social sciences were all sources of information. Although Dewey did not attend it, an international conference on biological cycles was held in 1931, in Canada. Its organizing committee later worked with Dewey, and together their association led to the formation of the Foundation for the Study of Cycles (FSC) in 1941, with Dewey at its head. The FSC began publishing a magazine in 1950, and at about the same time Dewey and a co-author published a book titled *Cycles: The Science of Prediction*. Another co-authored book, *Cycles: The Mysterious Forces That Trigger Events,* appeared in 1970 (Dewey and Dakin 1949; Dewey and Mandino 1971). After its founding, with membership then consisting of mostly scientists and economists, the FSC began to attract traders who were interested in cyclic patterns in the stock market. This movement away from the concerns of science changed the organization, and not long after the death of Dewey, its chairman, Martin Armstrong, was arrested for commodities trading fraud, an event that caused the FSC to slip into something like an organizational coma. A few years later a spinoff group was founded, the Cycles Research Institute, which continues the work.

One useful side effect of the decline of the FSC was that its archives became available online. Included are many articles and scientific studies of various natural phenomena including a wide range of sunspot cycles, lake levels, grasshopper outbreaks, tree ring patterns, and the well-known claim for a 9.6-year cycle in Canadian lynx abundance.

Economic cycles include those in cotton and steel production, cheese consumption, building construction, an 18.3-year real estate cycle and, of course, stock and commodity prices. There are also reports on sociological cycles such as church membership, immigration, marriages, and warfare. What's remarkable about all this is how wide-ranging certain cycles appear to be. For example, the 9.6-year cycle is not only found in lynx abundance, its also found in salmon abundance, outbreaks of war, heart disease, and ozone levels in London and Paris, among other things. If these cycles are real, then what is behind them? Most of the time those involved in the FSC pointed to sunspots as the culprit. But the fact that several books on business and financial astrology, a topic that connects the economy with solar, lunar, and planetary cycles, were published about the same time as the FSC got going, and that many of those practicing this branch of astrology were members of the organization, is suggestive of an overlap. While the FSC was founded by a number of reputable scientists and economists, it eventually became a source of information and a haven for market traders and people interested in fringe subjects, including astrology, but not the kind of astrology that concerns individuals.

Dewey himself suggested that the movements of the planets were a possible cause for at least some of the cycles he was documenting, and for this he was accused of doing astrology. His response was to say in public that he knew that astrology was completely discredited, that its premises are absurd, and that actually investigating it would ruin a scientific reputation. (Dewey and Mandino 1971, 159). But according to people I've met who knew Dewey, this was just cover-up. Dewey may not have been an astrologer, but astrology was considered a viable body of knowledge in the hard task of unraveling the many cycles collected by the FSC. What it illustrates more than anything else is the fear and loathing toward that subject, which had to be avoided if one was to retain any sense of respectability at the time. This hasn't changed.

3

A History of Natural Astrology

*Astrology is a pseudoscience that claims to discern informa-
tion about human affairs and terrestrial events by studying the movements and relative positions of celestial objects.
Astrology is no more than a test of chance and it is not a
reliable way to predict personality. According to American
astronomer James J. Heckman, the reason why people rely
on horoscopes is explained by a psychological phenomenon known as "self-selection bias" which is the tendency
of humans to look for interpretations or confirmations for
what they already hope to be true.*

Wikipedia, accessed September 2021

*. . . [B]elief in astrology is one thing, and its scientific dem-
onstration, the proof that it is more than just a belief but
also a reality of Nature, is another.*

Michel Gauquelin, *Neo-Astrology*

We now turn to the history of a tradition that has long observed
and studied correlations between the sky and Earth. While this
tradition came to be concerned mostly with correlations between the

sky and human behavior and destiny, it began with the recording of correlations between astronomical motions and the phenomena of the natural world, including weather. This general orientation, what has been called universal, natural, or mundane astrology, was cultivated over the course of three millennia, and this chapter will focus on its origins, history, general form, and methodologies. One of its areas of inquiry and practice, astrometeorology, will be the focus the next two chapters.

ORIGINS OF WESTERN ASTROLOGY

The origins of Western astrology can be traced to ancient Mesopotamia.[1] It was in this region of the world that humans first established large permanent settlements, developed agricultural techniques, and became organized into towns and cities. Social and cultural evolution accelerated by population growth, which was made possible by agriculture, led to organized religions, laws, and other means of controlling large groups of people. The history of this region comes into focus by the fourth millennium BCE when the Sumerians, peoples of uncertain origin who spoke an unclassified language and who had migrated to southern Mesopotamia (now Iraq), took over a number of older settlements and established major centers at Ur, Uruk, and Babylon. Over the next millennium, a complex religion/mythology was formalized, and cuneiform writing, a kind of notation on clay tablets, became widely used to record kinship linkages, ownership, commercial transactions, mythology, and

---

1. Astrology also originated, or at least developed in its own characteristic way, in India, China, and Mesoamerica. Indian (Joytish), also known as Hindu, or more recently Vedic astrology in English-speaking countries, clearly owes some of its qualities to Western astrology, these attributed to Alexander's intrusion into the region bringing with him Greek scholars. There is, however, an indigenous astrological tradition, mentioned in the Vedas, that is based on the lunar cycle. Burmese astrology is informed by Hindu astrology, and Tibetan astrology is a mixture of both the Hindu and Chinese systems. The astrological system of China is almost completely different in that it tracks planetary cycles, generalizes them into round numbers, and applies these on several temporal scales. While there are some similarities between Chinese astrology and the astrological tradition that developed in ancient Mesoamerica, particularly in the use of blocks of time as significators of specific processes, they differ in regard to key cycle lengths.

astronomical data, the latter sometimes with interpretations. Other accomplishments include the use (and possibly invention) of the wheel, intensive agriculture based on massive irrigation projects, the sexagesimal counting system used today for time and circular measurement, the first calendars, and the naming of constellations. The Sumerians were the major cultural founders of the Mesopotamian region, but they were conquered about 2400 BCE by Sargon, King of Akkad (now northern Iraq), a Semite, who built the region's first empire by joining Sumer in the south and Akkad in the north. For the next two millennia the power of city states rose and fell and empire followed empire. Of note were the empires of Babylon, the city itself being a major intellectual and cultural center, and Assyria (in place from 1800 to 800 BCE, during which time systematic observations and record keeping of natural phenomena were made, and many regard this as one of the roots of science). Change and decline of this long age of empires began with Persian dominance over the region beginning in the sixth century BCE, followed by Alexander's conquests two hundred years later in the fourth century.

Of special interest to historians of science are the contents of the Royal Library of Nineveh. The learned king Ashurbanipal, King of Assyria from 668 to 627 BCE, searched for, collected, and organized historic tablets (of clay, wood, and wax) for his library, which became a time-capsule for Mesopotamian intellectual achievements. The library was discovered in 1853 and, in addition to intricate relief sculptures lining the walls, it was found to contain thousands of tablets that were then shipped to the British Museum where they have been repaired, organized, and translated. The contents of these tablets are mostly omen literature, astrology, and entrail divination, but also included are the Epic of Gilgamesh, ritual texts, astronomical data, medical formulas, dream books, and more. Among the roughly thirty thousand tablets found in Ashurbanipal's library, and also at other archaeological sites in Mesopotamia, are two kinds of astronomical collections. One consists of star catalogs of astronomical events and star positions. Another is a set of tablets called the *Enuma Anu Enlil,* or book of Anu (god of heaven) and Enlil (god of Earth). This is the primary astrological text, essentially a compilation of interpretations of astronomical phenomena from ancient Mesopotamia that was organized around 1000 BCE, though it contains information dated to as early as the seventeenth century BCE. The *Enuma Anu Enlil* is composed of sixty-eight to seventy tablets (the total depending on how they are analyzed) and about seven thousand omens, depending on which of the existing copies is referred to.[2]

It is apparent from the dating and contents of the *Enuma Anu Enlil* that correlations between sky and Earth were observed and recorded over a long period of time, probably in several traditions that were integrated around the time the compilation was created. The astrological components contained in it were concerned with the affairs of the king, the state, the value of commodities, and the weather. The people who produced the astrological interpretations were organized in teams composed of both experts and trainees who systematically recorded astronomical data in great detail and correlated them with observations of weather and current events. These technical experts, which are probably best called astronomer/astrologers, maintained an unbroken watch of the sky and used sighting devices and water-clocks to measure the positions and the timing of astronomical events such as risings, settings, conjunctions, and oppositions of the Sun, Moon, planets, and stars. Omens, basically astronomical events with interpretations based on past correlations with terrestrial events, were then reported to the king. The astronomer/astrologers, being principal advisors to the royal courts, were king-supported, and this professional institution lasted well into the second century BCE (Baigent 1994; Rochberg 2007).

The tablets of the *Enuma Anu Enlil* are organized in the following manner. The first twenty-two are concerned with the Moon, its appearance, and its eclipses, and the next eighteen are based on the Sun. Meteorological omens, interpretations based on phenomena such as fog, clouds, storms, and thunder, comprise the next nine tablets. The individual planets, constellations, and stars form the basis of the

---

2. Readers can see early translations of tablets, including many that are meteorological (Thompson 1900) and a history of the collection and organization of the tablets that make up the *Enuma Anu Enlil* (Baigent 1994, 59; Gehlken 2012).

remaining tablets. The oldest part of the *Enuma Anu Enlil,* a complete astrological text in itself, are the Venus Tablets of the Babylonian King Ammisaduqa, a listing of Venus phenomena and respective interpretations that date from the mid-seventeenth century BCE. These tablets list Venus's first and last visibility above the horizon, exactly the same phenomena that the Maya recorded in their codices over two millennia later and then used as a fundamental rhythm for their astronomical, astrological, and calendrical systems.[3] The full synodic cycle of Venus, the cycle of relationship between Venus and the Sun, takes 584 days and consists of two primary phases that we observe as morning and evening stars. The astrologers of these Mesopotamian Venus tablets, like the Maya astrologers much later, saw these phases as having differing influences on terrestrial phenomena such as rainfall, food supply, outcome of wars, and the king's affairs. The existence of these Venus tablets indicates that by roughly 1900 to 1600 BCE astrology had become a systematic, specialized, and descriptive body of knowledge in Mesopotamia.

In the *Enuma Anu Enlil* each of the various combinations of planetary appearances (alignments, risings and settings, number of days into a month that a sighting occurred, etc.) had specific omens attached to them that described social, economic, and meteorological conditions. Lunar eclipses (the exact Sun-Earth-Moon syzygy) were associated with winds, lightning, thunder, and earthquakes. The Sun's eclipses (Sun-Moon-Earth syzygy) were generally considered sinister; they were thought of as predictors of invasions and the destruction of both people and crops. Mercury was a planet that was correlated with commerce, and also rain, but this depended on exactly where in the sky it might be found. Venus brought either prosperity or the destruction of crops by winds and floods, again depending on where in the sky it might be found. Jupiter was considered a good planet that brought successful harvests and rains, but Mars brought plague and, when in conjunction with Saturn, famine. This approach to data

---

3. The Maya Dresden Codex contains elaborate tables that allow computation of Venus's 584-day synodic cycle and offer specific astrological delineations for each of the five eastern/morning star appearances of the planet. Other codices and inscriptions also refer to this planet and its synodic cycle (Aveni 1980, 184–95).

collecting amounted to something like a cookbook: planetary configurations and sky locations could be looked up to see what had happened during a previous occurrence. Here are some examples of these omens that give specific information on the location of planets relative to stars and constellations, their synodic cycle with the Sun, and the calendar months:

> When Mercury culminates in Marcheswan, the crops of the land will prosper. When Scorpio is dim in the center, there will be obedience in the land. When (Mercury stands) in the flaming light of Scorpio (Ishara) its breast is bright, its tail is dark, its horns are brilliant, rains and floods will be dry in the land; locusts will come and devour the land; devastation of oxen and men; the weapon is raised and the land of the foe is captured.[4]

> When Jupiter grows bright, the king of Akkad will go to preeminence. When Jupiter grows bright, there will be floods and rains. When Jupiter culminates, the gods will give peace, troubles will be cleared up, and complications will be unraveled.[5]

> Venus is now disappearing at sunset. When Venus grows dim and disappears in Ab, there will be a slaughter of Elam. When Venus appears in Ab from the first to the thirtieth day, there will be rains, the crops of the land will be prosperous.[6]

It is apparent from the *Enuma Anu Enlil* that the omens for the planets could be dependent on what month their phenomena occurred in and how they were positioned relative to the constellations and each other in terms of distance along the zodiac. Over time the individual

---

4. Marcheswan is a month; Ishara is a goddess identified with Istar (Thompson 1900, lxxii–lxxiii).

5. The brightness of Jupiter is related to its varying distance from Earth (Thompson 1900, lxxii–lxxiii).

6. Elam was another ancient culture. Ab was a summer month in the Babylonian lunisolar calendar (Thompson 1900, lxxii–lxxiii).

qualities of the planets, Sun, and Moon were gradually established: the Sun as king, the Moon feminine, Venus as regulating crops, Saturn with law and order, Mars with heat and strife, Jupiter a savior, and Mercury a scribe. Exactly how and why these descriptions, which gradually evolved in the region, came to be is not clear. It has been suggested that the color of the planet, its brightness (which is an indicator of distance from Earth), the rate of its motion relative to the stars, and the nature of the myths and gods associated with it figured in establishing its astrological qualities. It may also be that certain gods were linked to a planet and then their original mythological characteristics evolved over time to be more like how the planet was being interpreted. Marduk, the patron god of Babylon, began as a minor and local cult god associated with thunderstorms and the slaying of the ocean goddess Tiamat, but later became prominent and associated with Jupiter. We can't know what these ancient people were thinking, and our knowledge of the stories, myths, and gods with multiple names in those times is limited; but it is interesting that the planets each came to be seen as having definite and stable characteristics. Further, what each planet signified wasn't changed by decree or religious reform. Astrologers have always stated that the astrological qualities of the planets are the result of a sustained multi-millennia empirical approach of noting correlations between astronomical and terrestrial phenomena, and not the projection of a theory onto the phenomena. It is indeed the case that each of the individual planets, and the Sun and Moon, have retained a more or less specific set of descriptors with little change over at least three millennia (Gauquelin 1982).

The Mesopotamian tradition of celestial divination based on recorded correspondences evolved over the course of at least two millennia in that region, ending sometime during the first century BCE. During most of that time it was the general conditions of the kingdom and its king, including weather and agriculture, that was of concern and only in its last phase were the planetary positions of individuals noted and interpreted. Mesopotamian astrology was a practical tradition. It lacked an orbital theory and it didn't require complex mathematics, but from it, and the observational astronomy that it was built on, we have inherited the 360-degree circle and sexagesimal notation that are still used for time and for degrees of arc in navigation, cartography, geometry, and trigonometry. Other surviving conventions include the twelve-sign zodiac, dated to the end of the fifth century BCE; and the first known "horoscopes," which were lists of planetary positions for a person's birth; and certain lunar phenomena, dated to 410 BCE. The notion of conjunctions and oppositions of planets (which includes the new and full Moons and eclipses) are also found in this tradition. The above phenomena, all of which became core components of Hellenistic astrology and consequently the Western astrology tradition, were thought to correlate with natural and meteorological events, and also developments in human affairs, in ways consistent enough to establish specific "rules" that were passed down through the centuries.

GREEK AND ROMAN ASTROLOGY

In early Greece, a separate, though far less sophisticated, type of astrological tradition had developed. It is best illustrated by Hesiod's *Works and Days,* an almanac-like text in the form of a poem that was written in the seventh century BCE, roughly the same time that Homer wrote about Troy and Odysseus. *Works and Days* considers the cycle of the year and that of the month in the context of agricultural life (Hesiod 1982). The first, and larger, portion describes the cycle of the year and how a resourceful and honest person might live in attunement to natural rhythms; astronomical references in the text inform the reader as to when specific agricultural activities should be commenced. Having outlined the annual cycle, which is based on the movement of the Sun (photoperiod) and its relationships to the prominent constellations as seasonal markers, Hesiod then addressed the synodic cycle of the Moon. This 29.5-day cycle was counted beginning with the first appearance of the crescent after new Moon (when it is first visible after new Moon) and was divided into waxing and waning halves, and also into thirds. Various points in the count were deemed favorable or unfavorable for one activity or another, the positive qualities of the first, fourth, and seventh days of the Moon being noted (the first day being the day on which the crescent appears after the New Moon, about 1 to 1.5 days after the New Moon). Hesiod then shifted into another way of counting the cycle and stated that the sixth day of the mid-month is bad for plants. This day, six days into the second third of the cycle, is also the sixteenth day of the lunar cycle, the day after the full Moon. Hesiod's "good and bad" days of the lunar month appear to be a kind of indigenous Greek astrology that offered information for farmers. In this sense it could be considered the world's first farmer's almanac.

A cultural diffusion event of immense proportions resulted from the conquests of Alexander the Great (356–323 BCE). One effect was that Mesopotamian astrology entered the Greek world, where it rather quickly became a more rigorous discipline and took on a general form that has survived to the present day (Tester 1987; Barton 1994; Neugebauer and Van Hoesen 1959; Oll 2010). In 280 BCE on the eastern Mediterranean island of Cos, some four decades after the deaths of Alexander and Aristotle, an important school facilitating the transmission of Mesopotamian intellectual culture was established by the Babylonian priest, historian, and astronomer/astrologer Berossus. He was apparently a highly regarded person, enough so that the city of Athens erected a statue of him. The actual mixing of Mesopotamian astrology with Greek geometry, mathematics, and four-element theory began to take place about a century later in Alexandria, Egypt, which had become the intellectual center of the Hellenistic world. During the first two centuries BCE, crucial astrological components, including the zodiac, planetary symbolism, astronomical calculations to predict future planetary positions and movements, and, especially, a time-slice sky-mapping technique, were incorporated into a system designed for evaluating births, illnesses, events, and the weather, among other things. By the first century BCE, astrology, as a unified body of knowledge and a methodology, had become a part of Hellenistic intellectual culture, and for the next several centuries it exerted an influence on Roman society.

The actual sequence of events behind this synthesis, now called Hellenistic astrology, is obscure, but it has been suggested that one or two individuals were responsible for the combination of methodologies that are still at the core of Western astrology. An astrological manual frequently referenced by ancient astrologers may have held clues to this origin problem, but only a few small fragments of it have survived. This mysterious document is usually dated from the first or second century BCE and attributed to Nechepso and Petosiris, the former referring to an Egyptian king of the sixth century BCE and the latter possibly a high priest of the fourth century BCE. While not much about either of these characters is known, Egypt did emerge during the two centuries preceding the common era as the center of astrological studies, and for centuries afterward this tended to obscure the earlier Mesopotamian origins of astrology. During this same period other mixtures of exotic practices, philosophies, divinatory techniques, and star knowledge flourished, much of which formed the Hermetic tradition that later came to have a powerful influence on Renaissance thought, to be discussed in a later chapter.

In the next section we look at the roughly eight-hundred-year period during which the practices and methodologies of the Greek and Roman astrologers flourished throughout the Mediterranean region, and simultaneously challenged certain philosophical and theological notions. Historians generally divide this history into two sections: the Hellenistic period, which is dated from the death of Alexander in 323 BCE to the Battle of Actium in 31 BCE, and the Roman Empire, which began in 27 BCE and extended to 476 CE, when the last emperor was deposed. Astrology took form during the Hellenistic period and became a social force during the Roman Empire.

PTOLEMAIC DISTINCTIONS

From the Roman Empire to the Renaissance, a period of roughly fifteen hundred years, the scientific writings of a single author, about whom little is known except his works, defined both astronomy and astrology. Claudius Ptolemy (c. 150 CE), often considered the greatest scientist of the ancient world, was the author of the *Syntaxis Mathematica,* or *Almagest,* which was a detailed, mathematically sophisticated work on the structure and movements of the solar system. In it, the circular motions of the Sun, Moon, and planets around an Earth-centered universe were explained and demonstrated. The geocentric cosmos that Ptolemy had mathematically modeled could predict with reasonable accuracy where among the background stars and constellations the Sun, Moon, and planets could be found in the future, so it was of great practical value. To calculate the orbits of the planets, and all the epicycles that he had to add to make the model work, Ptolemy developed an early form of trigonometry; he is credited, along with Hipparchus, with being one of its founders. While Ptolemy's geocentric model was extremely complicated, it did work and it preserved the obvious, common-sense motions of sky phenomena to an Earthbound observer—it "saved the phenomena." Improvements were therefore not pressing, and so the Earth remained at the center of the cosmos in the minds of nearly all natural philosophers for centuries. Even Nicolaus Copernicus (1473–1543), whose heliocentric model produced only slightly better results, hesitated to topple this elaborate astro-mathematical edifice.

In addition to astronomy, Ptolemy authored books on optics, geography, music theory, and a major work on astrology. The latter work is generally known as *Tetrabiblos* or *Quadripartitum* (four books on astrology), though the title *Mathematical Treatise in Four Books* is found in some manuscripts.[7] In it, Ptolemy introduced the subject material of astrology, organized it into sections, and discussed it theoretically. He treated astrology as a demonstrable system, with consistent rules and methods, but he gave no specific examples or any indication that he actually practiced it. His masterful description came to define the boundaries of astrology, raised it to the status of a science, and placed its contents in an order that has been followed, for the most part, ever since. Ptolemy explained astrology as a systematic description of natural processes that requires prerequisite knowledge of astronomy, mathematics, and natural philosophy. Astronomy,

---

7. Several English translations exist. The nineteenth-century Ashmand translation, often used by astrologers, relied on Latin translations of a paraphrase attributed to Proclus. The twentieth-century Robbins translation (found in the Loeb Classical Library) was based on Ptolemy's Greek text but suffers from errors and a condescending and judgmental attitude toward the subject material. A more recent translation of books 1, 3, and 4 by R. Schmidt is an alternative to these previous translations.

which he expounded in great length in the *Almagest,* he defined as the apprehension of "the [con]figurations of the motions of the sun, moon and stars relative to each other and the earth"; astrology he defined as the investigation of "the changes in that which is encompassed by these figures, as produced by the individual physical characteristics of these figures themselves."[8]

According to Ptolemy, astrology is a less exact, and less self-sufficient, science than astronomy, which deals with, for the most part, perfect spheres in a non-material realm. Astrology is characterized by the unpredictability of the material qualities found in individual things, and it presents problems because certain parts of it are so difficult for some to understand that they come to regard the subject itself as incomprehensible. In regard to the astrology of people, Ptolemy was not an astral fatalist and stated that genetics, environment, and child-rearing, as well as planetary influences, each contribute to a particular course of life. Having stated these problems and assumptions, Ptolemy then covered the subject matter with an intention to examine both the possibility and usefulness of astrology.

At the start of the second book of the *Tetrabiblios,* Ptolemy made an important distinction in regard to the subject of astrology—its division into two fundamental categories: universal, or general, and genethlialogical. The former is concerned with natural phenomena such as regional and collective factors, climate, weather, agriculture (planting and harvests), plagues, and the like; the latter is concerned with the affairs of individual humans. Universal astrology would include the most obvious effects from the two dominant astronomical bodies: the Sun (heat, the seasons) and the Moon (tides). This is how Ptolemy defined universal or general astrology:

> The foreknowledge to be acquired by means of Astrology is to be regarded in two great and principal divisions. The first, which may be properly called General, or Universal, concerns entire

---

8. In Ptolemy's definition of astrology the planets are seen as surrounding (encompassing) Earth (the sublunar realm), which is, in Aristotle's cosmology, the part of the cosmos where changes occur (Ptolemy 1994, book I-1, 2).

nations, countries, or cities—war, pestilence, famine, earthquakes, inundations—the revolution of the seasons; their greater or less variation in cold and heat; the severity or mildness of the weather; the occasional abundance or scarcity of provisions; and other like occurrences (Ptolemy 1822, Tetrabiblios Book II-1, p. 71).

The only area of inquiry within the domain of universal astrology, as defined by Ptolemy, that concerns humans has to do with the experiences faced by groups, cities, and regions. This component regards mass human behavior and the experiences of human collectives as similar categorically to natural phenomena, including changes in weather and earthquakes. Both human collectives and the Earth are then subject to the same influences from the Sun, Moon, and planets and the same astrological methodologies are applied to both.[9]

After noting that there are many starting points for inquiry in regard to universal astrology, Ptolemy turns to genethlialogical astrology, which pertains to individuals, the topic taking up the entire second half of *Tetrabiblos*. He makes it clear that universal astrology encompassed genethlialogical astrology; human-focused astrology ultimately yielded to the larger and more general influences of groups, weather, and other natural conditions. He also addressed the value of the time of conception, and the difficulty of knowing it, but he regarded the birth moment to be most descriptive of the individual:

> But we do investigate both one and many beginnings for the events of individual men. The single one is the beginning of the composite [body/soul] itself. . . in the case of those who do not know [the moment of conception], which happens for the most part, it is fitting to follow the beginning for the moment [of birth]. . . . (Ptolemy 1996, book III-1, 4. pp. 2, 4)

---

9. In many respects the domain of natural astrology has affinities with the field of geography, that subject taken to include geological and climatic processes as well as human social processes. In the astrological communities of today, these topics are classified under the heading "mundane astrology."

The astrology of individuals (called nativities by Ptolemy), which concerns the temperament and life history of individual humans, had links to other categories not discussed by Ptolemy, including interrogations, which answer questions that are the immediate concerns of humans, and elections, the times during which humans consciously choose to take action. Related to both of these are inceptions, the analysis of the astrological conditions operative at the time of a given event. Elections and inceptions, and sometimes interrogations as well, were classed as katarchic astrology, from the Greek word *kartache,* which means "beginning or commencement." These techniques, which have some qualities associated with divination, have raised questions about exactly what astrology is and how it works, and these questions will be considered in later chapters of this book. Although these were not covered by Ptolemy in his *Tetrabiblios,* elections were a service offered by other practicing astrologers during Hellenistic and Roman times, and probably interrogations were practiced as well, though references are few and debatable.

The distinction between universal and genethlialogical astrology was accepted for the most part for the next fifteen hundred years, though by the time of the Renaissance these two branches were often referred to as natural (also called world or mundane) astrology and judicial astrology.[10] This division of the subject is generally accepted by historians not only because it was traditionally defined that way but also because it best explains the problems the subject faced that led to its decline. Other divisions have been proposed. Robert Hand has argued for a five-fold division based on the actual practices and methodologies of astrologers. His distinction works well for his thesis, which is focused on the Medieval period, and is consistent with historical sources, but doesn't emphasize divisions in terms of astrological subject matter.[11] In

10. That may be true, but the distinction between astrology and astronomy was not always clear. Plato called the study of the planets "astronomia," while Aristotle called it "astrologia." In the Middle Ages astronomia was often used as a term to encompass knowledge of the stars, both measurement and interpretation.
11. Hand divides astrology into five categories: introductive material (the general schema outlined), revolutions (cycles), nativities, elections/inceptions, and interrogations (Hand 2014, 26).

my view, the historic distinction between an astrology that analyzes the individual and their choices, and that which considers natural phenomena other than individual humans, is more useful. It is not a perfect division, however, as inceptions, and also medical astrology, appear to cross these boundaries in certain ways.

The astrology that concerns humans and their choices—judicial astrology, which includes nativities, interrogations, and elections—is extremely anthropocentric and has always caught the attention of religious authorities because of free-will issues and the implications of a reduced role for God in personal life. Astrology has always been a problem for religion—the interpretation of a person's fate could run against theological doctrines, and interrogations and elections were frequently associated with divination. Related to elections are inceptions, an approach to studying events that utilizes astrological charts in the analysis of both natural disasters, like the start of a war, or personal events such as the time of an introduction to another person. They are similar to charts for elections, but no human choice is involved in the timing, which places this practice somewhere between natural and judicial astrology. In sharper contrast to the affairs of people is the astrology of weather, seasonality, agriculture, biology (such as plagues), and geological phenomena. These were generally not targeted by religion because they weren't seen to affect human choice and they provided practical knowledge. Then there is the astrology of collectives of people (cities, countries, etc.) and mass behaviors. I would argue that the astrological analysis of collectives such as cities and states, as they are affected by the natural conditions of climate, seasonality, weather, and biology (mass behaviors), is essentially a kind of objective natural astrology. For the most part, the grouping of topics making up natural astrology was taken to be a kind of natural science even into the seventeenth century.

Methodology can't be the sole basis of distinctions between natural and judicial astrology because the horoscopic (time-slice) method, that is the conventional astrological chart calculated for a specific moment that, like a seed, is seen to contain potential, has been applied to both divisions. However, the astrological chart is not the only way astrology has mapped its subject material. What are called revolutions in traditional astrological literature may involve simply listing sequences of angular separations (called aspects) between planets over time, these marking phases of synodic cycles. The divisions of the lunar synodic cycle found in Hesiod or repetitions of the Jupiter-Saturn cycle found in Medieval Arabic astrology are examples. In some cases, charts are calculated for significant moments in a cycle such as a conjunction or the crossing of the Sun over the vernal equinox. Another example of this alternate methodology would be use of the aspectarian, commonly found in Renaissance almanacs, a listing that places the angular separations (phase angles) of the planets, often the parts of a synodic cycle (i.e., lunar cycle from the new to full Moon and back), in chronological order usually on a daily basis.

There is also the matter of medical astrology, called iatromathematics, which is concerned with the astrological signatures of illness and with the choice of appropriate remedies, including therapies and the use of healing herbs and minerals. From antiquity through the Renaissance, it was common for astrologers to practice a kind of holistic medicine that used astrology and Greek four-element theory for diagnosis and treatment.[12] The key assumption behind medical astrology is that the parts of the body are described by planetary and zodiacal symbols, this concept being consistent with the notion of humans being a microcosm of the larger cosmos. The functions of the individual planets, each representing parts of a whole, then correspond to the appropriate parts of the body that express those functions. For example, the Sun has dominion over the heart (life force), while the Moon rules ingestion and digestion. As the planets have their own links to the zodiac signs, the latter are also used in diagnosis of a person's predispositions to health problems and the determination of proper treatment. Diagrams of the "zodiac man" showing body parts and signs linked have been around for millennia.

Ancient medicine incorporated certain astrological methodologies along with some more general lunar and solar factors. The Greek

---

12. This holistic astro-medical tradition is the Western equivalent of ayurvedic and traditional Chinese medicine (Tobyn 1997).

physician Hippocrates (c. 460–c. 370 BC) viewed the body as a whole and treated it accordingly using a scheme of four humours and temperaments, closely related to the Aristotelian four elements theory: fire, earth, air, and water (see diagram below). His approach to healing included some features of natural astrology, one being the identification of critical days, timed by the phase angle of the Moon relative to its location at the start of a health problem, these being the 90- and 180-degree angles that correlate with 7 and 14 days. Another is an emphasis on the changing environmental effects brought on by seasonal changes caused by the Sun (i.e., photoperiod), something that is now being investigated by chronobiologists. Astrology and medicine remained connected for centuries. Galen (129–200 CE), the physician and medical researcher of Roman times, wrote a treatise on critical days timed by the lunar cycle and also incorporated parts of natural astrology into his work (Cooper 2011). Some practitioners of medicine used astrology extensively, particularly during the Renaissance, and astrological charts called decumbitures were calculated for the time a person took to the sick bed. Medical astrology, which was thought to only work on the material body and not on the soul, protected it from the judgments of the clergy. This, along with medical astrology's concern with the astrological classification of plants and minerals as medicines, are arguments for placing it under the general category of natural astrology.

ANCIENT SCHEMATA OF CORRESPONDENCES

Element	Qualities	Season	Humour	Temperament
Fire	Warm & Dry	Summer	Yellow bile	Choleric
Earth	Cold & Dry	Autumn	Black bile	Melancholic
Air	Warm & Moist	Spring	Blood	Sanguine
Water	Cold & Moist	Winter	Phlegm	Phlegmatic

Classifying types of astrology is very useful, in fact essential, in understanding its history. But a point I want to emphasize is that the subject matter of both branches of astrology appears to be limited to self-organizing systems. It is not the height of a tree or the temperature of a body of water that is being described and measured by astrology. What Ptolemy considers astrology's subject material are really processes that are in a constant state of change. Self-organizing systems are ephemeral, relatively speaking, but can be quite resilient over time and will make adjustments so as to maintain homeostasis—and most don't lend themselves to analysis by reduction because the whole is more than the sum of the parts. Referring to Ptolemy's definitions, the Earth system (Gaia), including the climate and weather systems (all of which affect agriculture), the behavior of human collectives (which would include politics, history, economics, warfare, and plagues), the physical body, the personality and mind of individuals, and the emergent collective mind of a community or population can all be classified as self-organizing systems (Scofield 2019).[13] From this perspective judicial astrology is then concerned with (1) the non-physical emergent properties of individual humans, specifically the mind and personality (from which the likely life trajectory may be deduced) as shown in the nativity or seed moment of air-breathing existence apart from the mother, and (2) the fluctuations of the network of social communications that an individual may be linked to, as measured by the analysis of moments of inquiry and choice. Medical astrology is concerned with the system of the physical body, its parts and issues related to its vital homeostasis. Natural astrology would comprise the more objectivist applications of astrological methodologies to mass behaviors and events such as plagues or wars, weather, climate, agriculture, and other environmental processes. Inceptions might be seen as windows through which, depending on the situation, the astrologer may assess issues pertinent to both judicial and natural astrology, a higher order perspective perhaps. All of these systems are interrelated, the larger encapsulating, and influencing, the lesser. I will return to this theme in later chapters.

In the first book of his *Tetrabiblos,* Ptolemy briefly addressed the problem of exactly how the Sun, Moon, and planets, and also stars, produce effects on the Earth and its processes and inhabitants.

---

13. See Chapter 14.

He explained astrological phenomena in terms of Aristotle's model of the cosmos, itself based on ideas from Plato's pupil Eudoxus. The mechanism involved a transmission of motion and energy through a series of layered (like an onion) celestial spheres moving downward toward Earth, positioned at the center of the universe.[14] Ptolemy added that the planets harbored within these spheres have their own life force, can move themselves, and they move with respect to each other like a flock of birds, each pacing themselves without touching the others. The planets also move in perfectly circular orbits by their own volition within a space-filling ether that is a medium through which motion-energy is transmitted. The motions of the Sun, Moon, and planets then transfer their energy through the ether down to the sublunar region that is surrounded by the primary elements fire and air, which, in turn, transfer this energy to the more Earthly elements water and earth (Ptolemy 1994, 50–57). Ptolemy's explanation for the efficacy of astrology did not contradict Aristotle's model, and because of the enduring prestige both had, the search for a more testable model did not become a pressing matter for well over a millennium.

THE STOICS

For about five hundred years, from the third century BCE to the late second century CE, Stoicism may have been the most widely-embraced philosophical tradition in the ancient Hellenistic and Roman worlds. Zeno of Citium (c. 336–263 BCE) founded the lineage in 300 BCE in Athens, the name Stoic deriving from the public colonnade, a painted porch (the Stoa Poikile), that was Zeno's preferred teaching location. Over the following centuries Stoicism evolved into a complete intellectual system and lifestyle that included not only ethics, for which it is best known, but also logic and physics. Historically, the Stoic lineage-tradition began roughly two centuries after the foundations of Greek philosophy had been established by the pre-Socratics, a century after

---

14. A more extensive account is given in another text by Ptolemy, *On the Hypotheses of the Planets,* a translation of which is contained as Appendix I in *Tetrabiblos* (Ptolemy 1994, 50–57).

Socrates, and shortly after the conquests of Alexander, which stimulated the movement of ideas between Greece and the Near East. This was probably a distinguishing factor in Stoicism, as most Stoic teachers were not from Greece proper; they were from lands closer to Mesopotamia. Zeno was also a contemporary of Berossus, the transmitter of Babylonian culture to the Greek world, information that included astrology.

Historians have pointed out that the Stoics brought little that was completely new into Greek philosophy, which is not surprising since Zeno lived after the founders of organized rational thought and inquiry into nature: Socrates, Plato, and Aristotle, and their predecessors. But it is also recognized that the power of Stoic philosophy was in its synthesizing and extensions of pre-existing doctrines, including those of Platonism and Aristotelianism. Stoicism appealed to the intelligent of all classes, and its doctrines evolved and stayed current with intellectual developments in the Greek and Roman worlds for hundreds of years. The authors of the major surviving texts on astrology from the Roman world, including Marcus Manilius of the first century CE, Ptolemy and his contemporary Vettius Valens in the second century, and Firmicus Maternus in the fourth, all had Stoic leanings. This might be expected as at least some Stoic teachers were accepting of divination and astrology because the kind of fatalism implied in both were consistent with Stoic doctrines of causal determinism. Exactly when and how astrology and Stoicism influenced each other, or if they simply developed parallel but complimentary notions in regard to fate, is not completely known. Not all Stoic teachers were advocates of astrology, however, and some were critics.

Only fragments of early Stoic writings have survived, and most of these were preserved by critics, some hostile. One thing known for certain is that the Stoics believed the cosmos, that is the orderly, physical universe, to be a living, intelligent being in itself, but not quite the same as Plato's anima mundi. The source of Platonism's world soul, or animating principle, was an external god, the architect of the cosmos. For the Stoics, the world soul was all there was; it was God in essence, a pantheistic monism. Stoic natural philosophy ignored Plato's distinction between the world, the creator, and transcendent forms and instead produced a materialistic vitalism that some historians of philosophy have labeled a "cosmobiology" (Hahm 1977, 136). Zeno taught that the cosmos is like a biological organism and its animating and cohesive principle was pneuma, the force of life, breath, and soul. Later Stoic teachers added that the cosmos, seen as a dynamic continuum, grows continuously, gradually incorporating non-living matter into itself. Zeno and most Stoic teachers held that the universe does not completely die. Periodic destructions occur that are followed by a renewal of growth, and these were said to be structured by long-term astronomical cycles marked by rare conjunctions of all the planets. Building on earlier ideas of cyclic time, Zeno explained the universe as a pulsing, cycling organism that is eternally created and destroyed, a view supporting a cosmological doctrine of both catastrophism and eternal recurrence (called ekpyrosis), one that was later adopted by Friedrich Nietzsche.

A major figure in the Stoic lineage was Posidonius of Rhodes (c. 135–51 BCE). He was a synthesizer of ideas, an international figure, a polymath, and considered by many to be the greatest intellectual of his time. His influence on Roman intellectuals, including Seneca and Cicero, is attested to by frequent references to him found in their writings. The great Roman physician Galen called him the most scientific of the Stoics. The few surviving fragments of his writings, which were apparently voluminous and influential, reveal him to have been a powerful writer with a wide-ranging and scientific (in the sense of laws pervading a physical cosmos) view of the world. Posidonius, who Augustine said was "much given to astrology," apparently took the subject seriously and gave it some credibility (Augustine 2009; Ulansey 1991, 70–73).

One Stoic doctrine that came to be associated with Posidonius was that of *cosmic sympatheia* (meaning sympathy), which describes the interactions and relationships between the different parts of the cosmos. In Stoic physics, the universe itself was seen to be a coherent and unified whole, analogous to the body of an animal or plant. It was composed of parts in sympathy with each other; that is, the parts have affinity for each other, they communicate, and they interact. An affliction to one part, even if that part is distant, may then be felt by the whole. This implied a transmission of information on a cosmic scale (a bit like entanglement in quantum physics) that was used to explain how divination and astrology worked, one implication being that the same laws operate everywhere in the universe. In search of evidence of sympathy, Posidonius traveled to Gades (Cadiz) in Spain, and for a time carefully observed the tides of the Atlantic Ocean, these having far more extreme highs and lows than those of the Mediterranean. He regarded his field observations, that the sea responds to the movement of the Moon, as proof of an interconnected cosmos (Sandbach 1975, 131).

Posidonius was as much an astronomer as a philosopher. In regard to the Sun, he thought it to be much larger than the Earth and composed of pure fire, the animating principle of Stoic physics. He made calculations for the circumference of Earth, though his method was flawed, and also the size and distance of the Sun, which was more accurate than those of Aristarchus, the original proposer of the heliocentric hypothesis. He thought the Sun was the source of vital energies and the cause of many Earthly phenomena, including the generation of plants, animals, and crystals. He was also interested in volcanoes as evidence of a subterranean fire that also stimulated life. Posidonius was a significant ancient scientist, possibly more modern than others given his experimental attitude toward nature, and it is unfortunate his works only exist as fragments or as references by other writers. (There are some today who think he may even have been connected in some way to the Antikythera mechanism).[15] The combination of the "mechanism" of cosmic sympathy he promoted, along with traditional Stoic doctrines on causal determinism, served as explanations for astrology, not much of which was provided by individual astrologers as the subject was mostly an analytical craft based on astronomical data.

The idea of fate, or causal determinism, was important in Stoicism and for many in the Stoic lineage; fate (often capitalized as it was considered to be like a god) had connections to astrology. Chrysippus, an early Stoic, argued that a person's individual nature (which could be described by astrology) supported causal determinism because it limited

---

15. The Antikythera mechanism is a mechanical planetary and calendrical computer found in an ancient Greek shipwreck near the island of Antikythera (Marchant 2009, 274). Cicero reported that Posidonius had built a mechanical representation of the heavens (Cicero 1933).

choices. While a person could be conscious of their actions, and should take responsibility for them, they were still bound by the inclinations of the person they were. The earliest mostly complete text on astrology, Manilius's poetic *Astronomica,* praised Stoic ideas and actually celebrated fate. The general idea was that the planets, with their unalterable motions, spelled out one's fate, and the wise person recognized and worked with it, thus improving themselves—it was in this conscious choice to participate in the unfolding of one's true nature that free will was to be found. Ignoring one's fate was only asking to be dragged around by it. This topic will be considered again in a later chapter.

THE STRUCTURES OF ASTROLOGICAL CHARTS

By the time of Ptolemy in the second century CE, astrology had become an integral component of Roman life. It was taken seriously by some philosophers and became deeply embedded in the general understanding of the times, and it was practiced by both professionals and charlatans. Emperors, including Tiberius and Hadrian, studied and used it. As noted, the poet Manilius expressed the Stoic philosophical perspective in his long poem on astrology, and Vettius Valens, a contemporary of Ptolemy, wrote a major text describing his methods that also took Stoic positions. Both of these works survive. In recent decades other writings on astrology from the time of the Greeks and Romans have been translated into English, and it is widely accepted by historians that astrology occupied a prominent and significant position in the Hellenistic and Roman worlds. Even words from these times have survived in our language, for example *consider,* "to use the stars to make a decision"; *disaster,* "unlucky star"; *desire,* "to wish for what the stars bring"; *influence,* "emanations from the stars"; and *aspects,* "to look at the planetary angles, which describe the qualities of a given time or situation." Descriptors of personality based on astrological planetary typology are, in order of distance from the Sun: mercurial, venal, lunacy, martial, jovial, and saturnine.

The form that astrology took during the Hellenistic period and the Roman Empire involved five categories of data: planets, aspects,

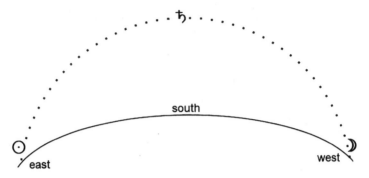

Figure 5. Snapshot (time-slice) of the diurnal arcs of Sun (shown here rising), Saturn (halfway through its diurnal arc), and Moon (shown here setting).

symmetries, houses, and the zodiac. The positions of the planets were organized by a reference point, calculated from observations and spherical geometry, that is called the horoscope (from the Greek *horoskopos* for "time observer"). Historically and technically, horoscope only refers to the degree of the zodiac (the Sun's observed path through the stars) rising in the east, which is today known as the Ascendant. From here on I will only use the term *horoscope* when appropriate and will otherwise refer to the actual map of the sky as an *astrological chart*. (Other traditional terms for the astrological chart include figure, genesis, and radix.) The challenges of producing an accurate map of the sky, which moves and is never the same, were considerable. In fact, the needs of astrology were a major driver in the development of spherical trigonometry, a process lasting well over a millennium, to which many astrologers made important contributions (Scofield 2018, 39–74).

The astrological chart is a snapshot, or time-slice, of the sky calculated for a specific place and point in time, as seen facing south, with east on the left and west on the right. With this framework, basically a 360-degree grid, the positions for each of the five visible planets and the Sun and Moon in their diurnal cycle were placed (due to Earth's

---

16. Francoise Gauquelin gives a scientific study of keywords for the planets from ancient to modern times and their correlation to their diurnal cycle positioning in the book *The Psychology of the Planets* (Gauquelin 1982).

rotation each planet appears to rise, culminate, set, and pass under the earth, called lower culmination). If the astrological chart was calculated for dawn, as shown in the diagram in figure 5, it would depict the Sun as ascending in the east directly on the horoscope (Ascendant), and if the Moon happened to be full at this time (opposite the Sun), it would be depicted as descending, or setting, in the west. Also shown is Saturn (which would be invisible after sunrise and before sunset) at culmination, the highest point in its diurnal cycle.

Planet is the Greek word for "wanderer" and was applied to the five visible planets and also the Sun and Moon. These bodies were organized in several ways, one according to their rate of motion against the stars, as perceived geocentrically, which established what was thought to be their distance from Earth. The fast-moving Moon was closest to Earth, while Saturn was furthest, the order being Moon, Mercury, Venus, Sun, Mars, Jupiter, and Saturn. Each of these moving astronomical bodies symbolized basic themes and concepts that have remained more or less consistent since the origins of astrology in Mesopotamia to the present day, with only minor differences attributable to language and changing cultural values. This planetary taxonomy extended well beyond individuals and included weather conditions and parts of nature such as metals and plants that were used in alchemy and medical herbalism. It was this qualitative but structured distribution of planetary correspondence and influence in nature that also rationalized Stoic cosmic sympathy. In the table on page 116 I have listed the principles of the planets in order of their daily motion as seen from Earth (known as the Chaldean order) and in a form that synthesizes some of their fundamental concepts from ancient to modern times.

Certain angular separations between planets, the Sun, and the Moon measured along the ecliptic (the Sun's apparent path, actually Earth's orbital plane) are called aspects. These angles signify important phase relationships between planets, allowing for a blending of qualities that yields additional descriptive power. Ptolemy recognized the conjunction (0 degrees between points) and four aspects: the opposition (180 degrees), the trine (120 degrees), the square (90 degrees), and the sextile (60 degrees). These are based on division of the 360 degrees by an integer (two, three, four, and six).[17] The nature of a given aspect was then related to a number and this was a source, or confirmation, of Pythagorean notions of the properties of numbers. Aspects also serve as a means of structuring a cycle by measuring phase angle. The conjunction of two bodies was normally regarded as the beginning of a cycle, and the aspects divided it into successive sections. The opposition, for example, marks the midpoint of a cycle, considered a significant point of phase shift in astrology.

THE FUNDAMENTAL PRINCIPLES OF THE PLANETS

Planet	Function	Gender	Behavior	Actions	Weather	Metal
Moon	Instinct	Feminine	Reacting	Protecting	Change	Silver
Mercury	Cognition	Neutral	Changing	Messaging	Winds	Mercury
Venus	Attraction	Feminine	Bonding	Socializing	Moisture	Copper
Sun	Vitality	Masculine	Leading	Directing	Drying	Gold
Mars	Competition	Masculine	Asserting	Initiating	Heat	Iron
Jupiter	Extension	Masculine	Believing	Expanding	Thunder	Tin
Saturn	Contraction	Masculine	Skepticism	Judging	Cold	Lead

In addition to aspects, at least two kinds of planetary symmetry were considered important. Each planet has what is called its antiscion, or shadow point, one that is equidistant from the solstices (seen as an axis), but on the other side of it. Two planets equidistant on each side of the solstices were considered to be connected to each other, with positive or negative consequences for individuals or groups depending on the combination of planets. There is an astronomical basis for this doctrine. As the planets travel in more or less the same plane as the Earth's orbit around the Sun, planets equidistant from the solstices would have

---

17. Other angular separations have been added since Ptolemy's time. Note that the zodiac, which has twelve 30-degree signs, accommodates the Ptolemaic aspects, which are all multiples of 30 degrees. Aspects based on 360 divided by 5, 8, 9, and so on are not consistent in this way.

a similar declination, that is they would be about the same distance north or south from the celestial equator, rise and set at the same points east and west, and have similar diurnal and nocturnal arcs (explained below). A second type of symmetry is found in what are called the lots, or parts; these are zodiacal positions derived from the positions of the Sun, Moon, Ascendant, and other planets. In this technique of Hellenistic astrology an axis is established with two points symmetrical to it (e.g., Sun and Moon). Next, a third point (e.g., Ascendant) is used to derive, relative to the axis, a fourth point completing a symmetry. This found point, designated in degrees of the zodiac, is the lot, or part, and it was considered a viable point that could be activated by planetary motions and used to base interpretations of a person's life events. The most commonly calculated lot was called the Lot (or Part) of Fortune, which completes a symmetry of the Ascendant, Sun, and Moon.

ASPECTS AS PHASE ANGLES

Aspect	Divisor	Angle
Conjunction	1	0
Sextile	6	60
Square	4	90
Trine	3	120
Opposition	2	180

Houses (domiciles) are the methodology that quantifies diurnal motions, an effect of Earth's daily rotation. As Earth rotates, each planet moves through its diurnal cycle, the most obvious being the Sun as its diurnal motion defines the day. The Moon's diurnal cycle is what drives the tides. Every day the Sun, the Moon, and the planets rise, culminate, set, and pass through what's called lower culmination. When a planet is above the horizon, between its rising and setting point, it is said to be traveling through its diurnal arc; when below the horizon it moves through its nocturnal arc. These two arcs are bisected by the meridian that runs from north to south through the zenith and nadir (see diagram) and marks the culmination, or lower culmination, of a body. A semi-arc is the distance from horizon to meridian, or from meridian to horizon, there being two diurnal semi-arcs and two nocturnal semi-arcs. Additional zones were found by dividing the arcs by six, or the semi-arcs by three, for a total of twelve domiciles or houses that could be occupied by planets. Various methods of house division have been proposed and used throughout the history of astrology, but there has never been a persistent consensus on which ones work best, probably because, in most cases, the differences aren't really that great. The use of houses to plot diurnal cycle positions was basically a way of classifying a planet's approximate distance from the horizon or meridian. Another way of looking at these divisions is in terms of a temporary zodiac, a spatial grid against which a planet could be located. It is in regard to the diurnal cycles of the planets that the best statistical evidence for astrology has been found, a comprehensive series of statistical studies that will be discussed in Chapter 11.

In tracking the diurnal cycles of the planets, Sun, and Moon, special attention was placed on their rising in the east, the point at which they cross from beneath the horizon to above it. As previously noted, this rising point was called the horoscope. In modern times it is generally referred to as the Ascendant, and the zodiacal sign positioned there is called the rising sign. The use of the term horoscope in popular astrology to refer to a short predictive paragraph is therefore wrong and misleading. The rising point, being the boundary between the diurnal and nocturnal arcs, also functions as the cusp of the first or beginning house in most systems of diurnal cycle division. The calculation of the horoscope/Ascendant/rising degree was one of the great astronomical and mathematical challenges of ancient mathematics. In order to find this point, the use of multiple coordinate systems is required (ecliptic, equatorial, and horizon), and the problem is only solved precisely with spherical trigonometry. While it is true that celestial mechanics are also necessary for cartography, navigation, and the compiling of star catalogs, it is also true that the needs of astrology were a force behind advances in trigonometry. Hipparachus may have been the first to calculate tables of ratios for angles by way of drawing chords between angles on a circle, with Ptolemy improving on

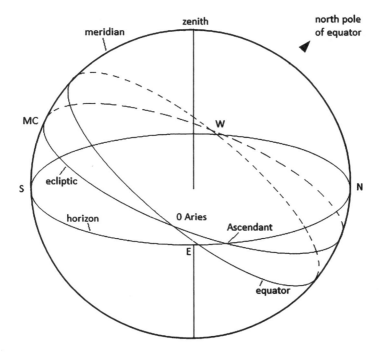

Figure 6. The celestial sphere. Earth rotates daily around its axis (north pole of equator), which is the cause of the diurnal cycles of astronomical phenomena. In a separate motion, the planets, Sun, and Moon travel (orbit) along the ecliptic (celestial longitude), and as Earth rotates each will rise at the Ascendant (horoscope) and culminate at the Midheaven (Medium Coeli or MC), this distance being the diurnal semi-arc. By dividing this, and the other three semi-arcs, into thirds, twelve individual compartments, called domiciles or houses, are created. Other methods used in traditional astrology divide the diurnal and nocturnal arcs into sixths. In these ways the stages of the Sun's, Moon's, and each planet's diurnal cycle and distance from the horizon are quantified.

the method, but the evolution of trigonometry continued into the seventeenth century, and along the way the list of contributors included a very high percentage of the best mathematician-astrologers.

In addition to locating a body in its diurnal cycle, there is also a need to position it relative to the background stars or constellations, which are more permanent spatial markers. This observational challenge evolved over time in Mesopotamian astronomy and astrology, and it resulted in the zodiac, perhaps the best-known element in the astrological tradition. In ancient Mesopotamian astronomy the Moon and planets were observed to move along the plane of the ecliptic (the Sun's apparent path), and they were located relative to both individual stars and groupings of stars along this path, called constellations. The naming of constellations began very early with the Sumerians, but it wasn't until around the seventh century BCE that names and positions of the constellations became more or less standardized. By the fifth century BCE the set of constellations along the path of the Sun, the zodiac constellations, were established. The name zodiac comes from the Greek *Zōidion,* which means "circle of animals" or "circle of life." Of the twelve zodiac signs, seven are named for animals (Aries/ram, Taurus/bull, Cancer/crab, Leo/lion, Scorpio/scorpion, Capricorn/goat, Pisces/fishes), three are humans (Gemini/twins, Virgo/virgin, Aquarius/water bearer), one is a centaur (Sagittarius), and one is symbolized by scales (Libra).

A zodiac anchored on constellations is called a sidereal zodiac. In the fifth century BCE the zodiac constellations were located in such a way that the star Pollux in Gemini was set to be at 90 degrees of celestial longitude from the vernal point (the vernal equinox or first day of spring). This placed the vernal point at about 10 degrees of the constellation Aries in that period (Britton and Walker 1996, 48–50).[18] There was a problem, however, due to the precession of the equinoxes, which caused this point in Aries to move earlier in the constellation at the rate of about a degree every 72 years, or 1.4 degrees per century. By the second century BCE, when the vernal point was at about 5 degrees Aries, the Greek astronomer Hipparchus (credited with discovering precession) fixed the vernal point to 0 degrees of the constellation Aries, thus linking the zodiac to the seasons, not the stars. This created two zodiacs, one based on stars, the other on spatial divisions between the equinoxes and solstices, with each sharing the same names. Hipparchus's innovation didn't catch on right away, and minor adjustments to the starting point of the zodiac were later made by Posidonius, his pupil

---

18. Note that the vernal and autumnal equinoxes are the only points on the ecliptic that are also on the celestial equator, this being so because they are on the intersection of the plane of the ecliptic and the plane of the celestial equator.

Geminus, and then Ptolemy in the second century CE. At this time the equinox would have been at roughly one degree of the constellation Aries, and a century later they would have been coincident. A feature of this new, fixed-to-the-seasons zodiac was that adjustments to the stars, because of precession, were no longer needed.[19]

The zodiac anchored at the equinoxes and solstices is called the tropical zodiac. One natural feature of this zodiac is that it allows for an easy measurement of photoperiod, the annual wave-like changing ratio of light to dark. The signs themselves may then be seen as the photoperiod wave broken into twelve discreet light-level packages. The tropical zodiac signs near the equinoxes will always have a ratio of day to night near twelve to twelve, at the solstices the ratio (at 40 degrees north latitude) will be near fifteen to nine (summer) and nine to fifteen (winter). The change from a zodiac of constellations (sidereal zodiac) to one of signs, while retaining the names of the constellations for the twelve equal sections of the ecliptic measured from the equinox, is found in Ptolemy's writings. He was clearly a strong advocate for the tropical zodiac because it makes sense, is supported by observations, and must be used if descriptions of the signs are to be a part of accurate astrological interpretation.

> It is indeed reasonable to start the twelfth-parts and the boundaries [zodiac] from the tropical and equipartite points. . . . This is both because the writers in a certain fashion make this clear, and especially because we see from the previous demonstrations that the natures and powers and [planetary] affiliations of the twelfth-parts and boundaries derive their cause from the tropical and equipartite origins and not from any other starting points. For, if other starting points are assumed, we will either be forced no longer to use the natures of the zodiac [signs] in prognostication, or else, if we use them, we will be forced to make mistakes. . . . (Ptolemy 1994, book I, 22, 45)

---

19. The astrologer Cyril Fagan attempted to make some sense out of the zodiac problem, with mixed results (Fagan 1971, 10; Fagan 1951; Rogers 1998).

This new framework for the zodiac, now called the tropical zodiac, was more or less coincident with the constellations for a few centuries. It was not immediately adopted by all astrologers in the Greek and Roman world, as evidenced by a number of surviving ancient horoscopes that are based on slightly different versions of the sidereal zodiac of constellations. There does seem to have been a general adoption of the tropical zodiac by around the middle of the fourth century CE, however, when precession had separated the two by several degrees. Confusion over the two zodiacs continues, especially for critics of astrology today, many of them astronomers, who seem to have no fears about demonstrating their ignorance of the issue in public. More on this later.

This summary of the primary components of ancient astrological charts suggests how a set of precise measurements and qualitative descriptors can be used in a practice. Some of these descriptors, such as the use of the four elements that form the internal structure of the zodiac, are the same as those used by the natural philosophers at the time who were speculating on the nature of reality. Others, such as the aspects, are mathematically precise in the tradition of Pythagoras and Plato. It is through the use of these descriptors that information is produced from the data. Consider that each of the planets holds a specific set of meanings. Those who have bothered to look closely will discover these to be far from superficial; they are structures astrologers use in a highly disciplined pattern-processing thought mode that penetrates quite deeply into the perception and experience of natural phenomena. The principles of the planets, which could be seen as universal forms, begin at a fundamental level and serve to group related themes. For example, because Saturn symbolizes at the most basic level concepts like structure, hardness, and constraint, things of the world such as stone building blocks and social artifacts like government would then fall under its domain. The planets are the primary forces or causes in astrology, but they are guided in their expression by the pure geometry of the aspects (phase angles) and the zodiac (photoperiod quantified), this being an early example of mathematics in natural law. The planetary symbols, and those of the aspects, zodiac, and others used in astrology, function like words in a language, each a source of information that contributes to the construction of meaning. The glyphs used to represent these focal points of meaning are therefore packed with information making the interpretive part of astrology a true symbolic language. This use of symbols should not be considered primitive or purposely concealing; it comes under the heading of semiotics, the study of signs and the production of meaning. Semiotics has a long history in philosophy and linguistics but only emerged as an interdisciplinary academic field of study in the late nineteenth and early twentieth centuries.

Since astrology uses a symbolic language to produce information, the resulting knowledge is hard to quantify or reduce to a formula. This is not science as we know it today, but it was a methodology, based on real data, that people used to locate the patterns of information needed for making a specialized knowledge two thousand years ago; it has lasted in this form, with some improvements, to the present day. One should be able to see why astrology was generally highly regarded before the Scientific Revolution—it was an organized set of procedures that utilized precise measurements, was based on mathematics, and it conformed to some of the leading cosmologies of the time. Even the methodologies of astrology are not at all unreasonable, these being based on measurement of the phases of astronomical cycles (aspects), diurnal motions (houses), and photoperiod (zodiac and symmetry), all of which are relevant phenomena in the scientific study of biological rhythms and tides. The point I want to make here is that astrology has qualities that straddle both precision measurement and creative synthesis. The best comparison might be to a healer or physician who uses scientifically-produced knowledge, tools, and remedies but relies on years of practice in making a diagnosis. No two cases are exactly the same, but the creative and synthetic processing of experience in formulating a treatment can often deliver the best results for the patient (as opposed to rigidly following one standard procedure that only kills 40 percent of the time).

To be sure, there were objections to astrology in Greek and Roman times, but these were directed toward judicial astrology. Carneades (214/3–129/8 BCE), born in Cyrene, North Africa, was a skeptic who became the head of the Academy (Plato's lineage) in Athens. Since he held that no knowledge whatsoever was certain, astrology was only one of many subjects that did not fare well in his view. His arguments included some classic ones: that making precise planetary observations at a birth was impossible, that twins can have different destinies, and persons that die at the same time will likely have different birthdays. Sextus Empiricus (c. 160–c. 210 CE), a physician and philosopher, was also skeptical of knowledge in general but argued for suspension of judgement rather than outright dismissal. He wrote a series of books, commonly called *Against the Professors,* one of which attacked astrologers. Like other ancient and Medieval writers, he called them mathematicians because they had to do math in order to construct an astrological chart. In it he argued against Stoic sympathy, said astrology couldn't be true unless there is such a thing as destiny, and he questioned why the time of conception wasn't used for a horoscope. Saint Augustine of Hippo (354–430) had some experiences with astrology in his early years but rejected it after converting to Christianity. He believed the determinism implicit in natal astrology ran counter to free will and personal responsibility, that twins born at the same time were different, and that astrological predictions were a kind of divination that was aided by demons. All of the above criticisms of judicial astrology, to be discussed in more detail later, were addressed intelligently by the most serious astrologers from Roman times through the Renaissance, but their responses were mostly ignored by skeptics and religious believers, so the same arguments kept coming up again and again. Still, from ancient times through the Renaissance, natural astrology was taken to be completely reasonable and considered a part of science, a sort of natural philosophy.

To summarize, astrology, by itself and uncontaminated by religious or magical doctrines, was an empirical science in ancient Mesopotamia where centuries of observations were synthesized into a systematic body of knowledge used to interpret events and to make predictions. Astrology was understood to be a kind of divination, a way of knowing the will of the gods. A wave of cultural diffusion, particularly in Egypt, followed Alexander's conquests and, during the Hellenistic period, astrology came to be improved by Greek geometry and four-element theory. It then developed into a formal working system and was rationalized by Ptolemaic astronomy and Stoic cosmology, allowing it to become a subject of great interest. Astrology eventually became common knowledge in the Roman world, and its practice included both the natural and judicial branches of the subject, although writings on the latter are more abundant than the former. Surviving works by astrological purists, such as Manilius, Dorotheus of Sidon, Ptolemy, Valens, and Firmicus, illustrate the empirical, practical, and logical (i.e., not religious or mystical) Hellenistic foundations of the subject. In the next section I consider the historical techniques employed in the practice of natural astrology, specifically astrological weather forecasting, which came to be called astrometeorology.

4

ASTROMETEOROLOGY

Whenever Saturn is joined to the Sun the heat is remitted and the cold increased, which alone may be a sufficient testimony of the truth of astrology.

GEROLAMO CARDANO, IN *ANIMA ASTROLOGIAE*, 1676

The techniques of weather forecasting by astrology were more or less the same as those required for the analysis of birth charts, the charts of cites, elections and questions. This meant that astrometeorology was not a specialized subject in the hands of experts, but that it was practiced by those knowledgeable of astrology in general.

Four meteorological traditions, each with ancient roots, can be discerned from historical writings. Applied weather lore, built mostly on the Classical writings of Theophrastus, Pliny, Virgil, Seneca, Lucretius, and others, is concerned with obvious, visual phenomena correlated with weather. Weather signs such as clouds, halos, rainbows, and other sky phenomena, and also bird, frog, and insect behaviors that occur with certain types of weather, are examples. This knowledge was of use to farmers and sailors. Another related tradition tracked annual weather or seasonal patterns using regular and cyclic astronomical phenomena on a calendrical board sited in a public place. An example of this kind of meteorological information linked to the stars would be the flooding of the Nile, which occurs when the star Sirius rises heliacally (before the Sun), a correlation that locates the date of that event each year.[1] These timed correlations, called *parapegmata* in ancient Greece, were in the form of a device with moveable parts and accompanying textual data. The term *parapegma* refers to the moving of calendar pegs on a block of wood or stone against a listing of regular weather patterns, necessary because the Greek calendar, like others of the time, was not a solar calendar and therefore not synchronized with the seasons. Here it is regular astronomical and meteorological phenomena that help keep a social and cultural calendar in place; it is basically a compilation of markers for the cycle of the year and its seasonal changes.

Meteorology as an independent subject is different. Aristotle included meteorology as one of the major topics of natural philosophy, others being physics, astronomy, zoology/botany, and the transformations of the four elements in growth and decay (generation and corruption). He defined meteorology as the study of things that happen naturally, but with less regularity, in the ever-changing sublunar region that extends from Earth to the sphere of the Moon. The phenomena in this region include comets, meteors, and those of air and water (winds, earthquakes, thunderbolts). Although much of his book *Meteorologica* is theoretical and correlational, Aristotle does make a few statements that bring astronomy into his exposition. He noted that the Sun's annual movement through the zodiac regulates heat and the rise and fall of moisture, including rain, which he called a wet exhalation (Aristotle 1952, I-9, 71). In other words, the Sun drives the cycle of the year, which displays seasonal variations of moisture. He also wrote that earthquakes are caused by winds (a dry exhalation) that get trapped in hollows in the Earth—but some occur at lunar eclipses as they cause winds to run into the Earth (Aristotle 1952, II-8, 215). Other than

---

1. A heliacal rising, that is when a star is visible for only a few minutes before being obliterated by the light of the rising Sun, is an excellent timer for the annual cycle. As the Sun moves ahead in the zodiac at the rate of about 1 degree per day, due to the Earth's changing orbital position, there will come a point at which a known star will cease to be visible as the Sun approaches it. The last day on which the star is visible just before dawn is a reliable calendar marker as it will happen on the same day each year.

these statements, there is nothing in his writings that is concerned with making forecasts and nothing astrological, which would be expected as he lived before the influx of Babylonian culture and knowledge that occurred after Alexander's conquests. The roots of modern meteorology, in the sense that observations are made, mechanisms proposed, and theories developed, however, lie in Aristotle.

The description and forecasting of weather using astrological charts and planetary alignments is something completely different from the above meteorological traditions. As the scope of astrology was so vast, and as practitioners moved easily from nativities, medical judgments, weather forecasting, and other applications, a separate name for this branch probably did not seem necessary, and it was considered to be just one part of universal (i.e., natural, or mundane) astrology. However, the name astrometeorology was in use by the Renaissance and considered to be a sub-discipline of astrology in general. There are differences of opinion today as to what the term *astrometeorology* precisely refers. Gerrit Bos and Charles Burnett suggest the topic should simply be called "weather forecasting." They argue the name astrometeorology is unnecessary because most weather forecasting of this type, while it was done by astrologers who also worked in the other branches of the subject, was excluded in the condemnation of astrology, an argument that apparently defines astrology as only judicial astrology. Daryn Lehoux confused the matter by using the term as a designation for the kind of lunar cycle information contained in Hesiod's *Works and Days* and also the parapegmata, the established Greek practice of linking seasonal phenomena to the regular risings and settings of stars on public displays. On the other hand, Stuart Jenks regards astrometeorology as a useful term that describes the attempt to predict weather astrologically. Anne Lawrence-Mathers, in her book *Medieval Meteorology,* uses the term, but only as a new science (yet simultaneously a part of traditional mundane or natural astrology) that became established in Europe as part of the Renaissance of the twelfth century and then dominated weather knowledge until the end of the seventeenth century (Bos and Burnett 2000; Lehoux 2007; Jenks 1983; Lawrence-Mathers 2020, 74). Considering that fact that Ptolemy clearly set foundations for this specialty, and also that the most important work on the subject in the seventeenth century was John Goad's *Astro-Meteorologica,* it is this name, without the hyphen, that I will use for the subject; a reasonable choice that is similar to Jenks's usage but extended timewise in both directions from Lawrence-Mathers's historically constrained labeling.

With these definitions in place, we could say that the origins of astrometeorology are to be found in the weather omens of Mesopotamia and that during Hellenistic times these, and possibly other contributions, were formalized into an organized methodology. While certain parts of this tradition of planet-related meteorological phenomena influenced compilers of weather lore, such as Pliny (first century CE) in Rome, or Isidore of Seville (seventh century) and the Venerable Bede (eighth century) in the early Middle Ages, its core principles were rigorous and required all the mathematics normally employed in astrology to generate data for interpretation. In the second century CE, Ptolemy, in *Tetrabiblos,* Book II, left an early description of techniques used in astrometeorology that served as a foundation for later refinements (Ptolemy 1940, II-12, 207–213). He lists four approaches to the subject, the first three involve the calculation of astrological charts, the fourth being similar to the other non-technical meteorological traditions described above.

Ptolemy's first method utilizes charts calculated for the time of the new or full Moon that most closely precedes the equinoxes and solstices. This requires a knowledge of precisely where the zodiac boundaries are located relative to the equinoxes and when the syzygy of Sun and Moon will occur, followed by the calculation of a map of the sky for that moment and that location. From this "time slice," or astrological chart, an analysis of the "condition" of the Sun and Moon is made based on zodiacal position (signs), position in terms of diurnal cycle and relation to the horizon (houses), brightness (orbital distance), and angular distances (aspects/phase) from the other planets. From this data the weather for the next quarter of the year would then be determined and a forecast made. This may sound strange, but consider that the equinoxes, the points of balance between the extremes of winter and summer, are a time of rapid changes in solar radiation for the hemispheres. Sailors know that winds shift at this time of year and meteorologists know that air mass patterns change, which makes for unstable weather. The solstices present a different situation, one in which solar radiation, at maximum or minimum, peaks and then reverses the trend. By focusing on these times of year, when the climate system is in transition, and then zeroing in on the nearest new or full Moon, both of which raise the highest atmospheric tides, Ptolemy may have been on to something. This method, though still used by a few, remains to be tested.

Ptolemy's second technique involves the calculation of astrological charts on a monthly basis for either the new or full Moon (both are called lunations). Which of these two should be used is determined by which one falls nearest the beginning of the sign that starts the season, or quadrant, as Ptolemy calls it, these again being the equinoxes and solstices. The planets in these lunation charts are also to be considered, specifically in regard to their sign position, which, in addition to the latitude of the Moon, was said to describe the wind patterns for the next month. The third technique is another extension of the basic principle of astrological charts calculated for new or full Moons, but in this case done for the quarter Moons. Presumably, this third method added some shorter-term, weekly information. The use of the lunar cycle in these three techniques is an early example of what is known in traditional astrology as revolutions. Besides the Moon, the positions (in the zodiac and their diurnal cycle) of the Sun and planets were considered important in the charts calculated for these first three methods. Planets located in certain positions expressed weather conditions: Saturn for cold, Mars for heat, Jupiter for storms, Mercury for winds, and so on. This mathematical and interpretive work was complex, and practitioners had to be expert astrologers.

In his section on weather, Ptolemy talks of the hour-to-hour tensioning and relaxing of the weather, and how this is related to the Moon and tides. The conditions of the air are said to change when the Sun and Moon occupy the angles, that is the rising/setting and culminating/lower culminating positions relative to the observer, these being positions in the diurnal cycles of these bodies that are relevant to tidal schedules. The same principle was applied to the planets. An extension of these more general ideas makes up Ptolemy's fourth methodology in which phenomena such as the Sun's appearance at sunrise gives indications for the weather of the day, and at sunset for the weather at night. He also gives attention to the Moon's appearance (clarity, halos, etc.), plus or minus three days before the quarters, as an important sign of coming weather. Ptolemy brings in the stars as indicators of weather by their color and magnitude, and also comets as indicators of winds and droughts. This fourth section clearly has far more in common with the weather lore meteorology of Theophrastus, Pliny, and others than it does with astrometeorology proper.

Weather predicting from planetary positions and alignments themselves, not using astrological charts, which was basically the traditional Mesopotamian approach, continued as an element in the tradition of Hellenistic astrology. Much of it doesn't seem quite as "scientific" as what Ptolemy suggested, however. An example is found in an introduction to astrology written by Paulus Alexandrinus, a late Roman Empire astrologer, which contains a section on predicting winds (for Alexandria) based on the position of the Moon in signs of the same element, and also by its application (movement toward the formation of an exact aspect) to other planets. For example, consider the Moon moving through one of the fire signs (Aries, Leo, or Sagittarius) at a time when, let's say, Jupiter was located in Leo. The Moon would then be moving toward forming either a trine (if it were in Aries or Sagittarius) or conjunction (if Leo) to Jupiter. This condition was thought to produce easterly winds, their strength depending on other details (Alexandrinus 1993, 39).

ASTROLOGY IN ARABIC CIVILIZATION AND THE MIDDLE AGES

With the fall of Rome and the decline of Mediterranean civilization just a few centuries after Ptolemy, the practice of astrology, which requires regular precision in astronomical observations, declined as well. It was revived in the Middle East beginning in the seventh century with the rise of Islam. Arab learning, centered in Baghdad (founded in 762 CE under the direction of an astrologer), was extensive and included astrology as a subject of central importance (Bobrick 2005,

74–75). The House of Wisdom in Baghdad, established by the Caliph Al-Mansur and aided by the learned Jew Jacob Tarik, concerned itself with the translation and assimilation of Greek, Persian, and Indian philosophical and scientific writings. One of the great names in this rich intellectual tradition was Al-Kindi (c. 796–873), a mathematician, astronomer, astrologer, musician, and physician from Baghdad known as the Philosopher of the Arabs. Al-Kindi translated many Greek and Roman works on a variety of subjects and also wrote extensively himself. Not many of his writings have survived, but the first work by him to be printed in Latin was a treatise on astrometeorology. It was combined with one on the same topic written by the influential astrologer Abu Ma'shar (c. 787–886, Latinized as Albumasar; Thorndike 1923, vol. I, 641–652). For religious reasons, Arab astrology was mostly concerned with those parts of astrology that did not deal with individuals, though interrogations, elections, and medical astrology were studied and practiced. Astrometeorology was safe astrological territory, and astrology was also applied to history, with a focus on Jupiter-Saturn conjunctions.

An astrometeorological work by Al-Kindi titled *De Mutatione Temporum* ("On the Change of the Weather") was apparently a compilation of other writings, including two letters ascribed to him.[2] Its contents are almost exclusively practical techniques and methodologies for predicting weather, especially rains, which makes good sense given the dry conditions of the eastern and southern Mediterranean region. Al-Kindi stated that the weather forecasting techniques he outlined were relative to the normal climate and weather of that particular region, and at points in the text he suggested modifications for different latitudes. Overall, his writing in this text is completely astrological, highly technical in methodology, with little theory and few examples. According to Al-Kindi the probability for rain in the Middle East becomes greater near a new or full Moon when the planets are retrograde in a specific quadrant of the year, usually winter. Also, the general motion of the planets must be moving toward the Sun and Moon, which should be ahead

---

2. See the book *Scientific Weather Forecasting in the Middle Ages: The Writings of Al-Kindi* for translations of Al-Kindi's writings (Bos and Burnett 2000).

in the zodiac, meaning that, in the diurnal cycle, the Sun or Moon, or both, will rise after the planets do. Further details amplify or decrease the possibility of rain. Al-Kindi thought the Moon, Venus, and Mercury were responsible for moisture and rains, more so when they are located in specific 13-degree sections of the ecliptic. These sections, taken from Indian astrology and called lunar mansions, evidence the eclectic nature of Arabic astrology. Other techniques, including zodiacal sign positioning, aspects between planets, aspects to the quarters of the Moon (which are always 90 degrees to the Sun, so an aspect to one is simultaneously an aspect to the other), and entrances into the autumn equinoctial sign Libra, all contribute to his system of weather forecasting.

Al-Kindi wrote that the Sun, Moon, and planets generate heat and light due to their friction with the air. The closer they are to Earth, the hotter they become, and this leads to dryness and certain types of winds. The heat produced by planets varies according to the distance they are from the zenith at specific times (i.e., at the precise time of a new or full Moon) and also due to proximity to Earth, which is greatest when they are retrograde. Further, heat is modulated according to the rate of motion and position in the zodiac of the individual planets. Planet locations in the quadrants of the year (seasons, which are framed by the equinoxes and solstices) and retrogradation are also factors to consider in predicting rains (Bos and Burnett 2000, 164–166). Al-Kindi's rules of astrometeorology are complex as they are based on the ever-changing dynamics of the solar system, where planetary configurations never recur in exactly the same way, but they are internally logical and consistent.

Theory has never been a well-represented topic in historical astrological writings. One exception is Ptolemy's theory of astrological effects, which were framed mostly in an Aristotelian context using the four elements. In this explanation described earlier, causality moved down through the planetary spheres to the sublunary realm of the central Earth. In a work titled *On the Stellar Rays,* Al-Kindi offered something different in regard to astrological theory, a single mechanism involving rays (from the word *radius*), or light beams, that are propagated along straight lines like vectors from the planets to Earth. In his physics, the cosmos was seen more as a living continuum, rather than a series of bounded spheres, that allows the rays, essentially the astrological forces or energies that drive change, to directly penetrate into the sublunary realm. He stated that every planet or star has its own nature, which is projected by rays to specific objects under its influence (as in cosmic sympathy), but combinations with other rays from other stars or planets (facilitated by the aspects) are also possible. Further, rays from the center of stars or planets vary in strength according to the obliquity of their angle to the horizon (placement in their diurnal cycle), but can be fortified by the rays of other planets or stars (phase angles). This is clearly in line with classical astrological chart analysis in which planetary strength, a way that signals (influences) from the planets are ranked, is related to latitude, declination, angularity, and the like (Al-Kindi 1993, 7–10). Al-Kindi's astrological mechanism was also a supportive metaphysics behind natural magic and alchemy, other subjects of great interest to Arab scholars.

During the ninth century (Al-Kindi's time), astrological studies were far more limited in early Medieval Europe. Writers on astrology were not well acquainted with the subject, and church law placed it among the diabolical arts. The last great compiler of classical knowledge, Isidore of Seville (in the seventh century), provided a limited source of information on astrology in general, describing the subject as being partly natural and partly superstition. The former, natural astrology, was acceptable and included astrological medicine, but he completely rejected judicial astrology and his judgments on the subject were influential for centuries (Thorndike 1923, vol. I, 632–33). In the eighth century, the Venerable Bede wrote on a group of related subjects: astronomy, computations for calendar dates, weather lore, and the idea that planets have individual effects on the weather; all of it was considered explanatory information on nature (God's creation) as a system that could be understood by natural philosophy. As late as the early twelfth century, astrology was still only an academic discussion, as few real texts on the subject were known. But this changed when, in the middle of the twelfth century, the bulk of Aristotle's writings entered Europe via the Arab world, and along with them came the important works on astrology, most of which included astrometeorology. Ptolemy,

Al-Kindi, Abu Ma'shar, Alchabitius, and Messahala were the important writers on astrology that entered Europe at this time; Abu Ma'shar's *Introductorius Maius* was the first major astrological work to enter Europe, and its synthesis of Greek astrological (Ptolemy) and astronomical science (Aristotle) had a profound influence on Medieval thought (Wedel 1920, 27; Lemay 1987, 65–69).

The theory of the great conjunctions, an important astrological topic that falls into the general category of natural astrology and the methodology called revolutions, also came to the West via the Arab world. It was the Stoics who had elaborated on the idea of the universe cycling over and over again (called *ekpyrosis*), this rhythm being established by great conjunctions of planets; but it was Al-Kindi and Abu-Ma'shar, in particular, who developed the idea of historical markers set by planets. Mostly using the cycles of Jupiter and Saturn, which meet in conjunction every twenty years, these astrologers found presumed correlations with historical events, and in doing so gave meaning to the idea of a historical period. Crises in history such as the beginnings and endings of kingdoms, the rise and fall of leaders, and the occurrence of natural disasters were thought to be synchronized to the drumbeat of slow-moving and relentless planetary cycles. The planets were seen as signals of change in the sublunary world—indicators of causes, actually. This historical model also circumscribed religion in that it located the origins of prophets and the rise of believers in a cyclic framework, something that was perceived by religious authorities as sacrilegious. The idea of recurrences marked by planetary cycles that repeat themselves (called revolutions) was later an influential concept in regard to the idea of the Renaissance itself, which means a return or revival of former greatness (Pingree 1968; Garin 1976, 1–28).

During the thirteenth century natural astrology was integrated into the natural philosophy of the time; it became more respectable, while judicial astrology raised problems for Christian theology. In their writings, the Dominicans, Albertus Magnus (1193–1280), his pupil Thomas Aquinas (1225–1274), and the Franciscan Roger Bacon (1214–1294), discussed astrology in some detail. Albertus addressed it theoretically in a work dedicated to the subject, *Speculum Astronomiae,* and found ways to reconcile it with Christianity. He saw the planets not as causes, but as signs and also instruments of God's will that could influence physical and even psychological conditions. In general, both Albertus and Aquinas concluded that Aristotle's cosmology legitimized the rule of the stars over nature and corporeal bodies, which meant astrometeorology and astrological medicine were safe to study and practice. On the other hand, judicial astrology presented problems for Christian beliefs. Aquinas wrote that the stars and planets have no sway over the human will and intellect because these are not corporeal. But the stars do affect the physical body and most people are governed by their passions—hence they can be affected in that way by the stars and planets (Wedel 1920; Thorndike 1923, vol. II, 517–615). Given this view, precise predictions would have to be impossible—or, as Augustine had said, they would have to be mediated by demons. Roger Bacon, who, as an Aristotelian, valued observation, saw astrology as more of an environmental factor in the context of natural philosophy and thought it should be studied carefully. Influenced by the Arabic writings on planetary cycles, he wrote that astrology had a kind of organizing and taxonomic role in the history of religions, and that the mechanism behind cycles of religious change were groupings of the twenty-year conjunctions of Jupiter and Saturn (Thorndike 1923, vol. II, 672–673). Bacon justified all this with the notion that God had created correlations between planets and people as a way to increase the sense of wonder and love for him.

By the fourteenth and fifteenth centuries (the late Middle Ages), European intellectuals were reading and writing manuscripts on astrometeorology authored by both Arab and Western Latin writers. One of the centers of this early renaissance was Merton College, Oxford, where astrometeorological manuscripts were studied, copied, annotated, and commented on by knowledgeable writers including university professors, court astrologers, princes, monks, and friars. Some manuscripts were introductory treatises that demonstrated how the determination of weather patterns could be accomplished using a weighted astrological scoring system. In this methodology, the relative strengths of planets in a weather chart, like those described by Ptolemy, were numerically quantified, added up, and then compared with each other in order to reach a conclusion. More comprehensive and technical works written for professionals offered even more complex and subtle methodologies (Jenks 1983). William Merle of Oxford (d. 1347) studied astrometeorology by collecting his own weather data and comparing this with planetary motions. He even published a report of a seven-year astrometeorological weather study, certainly one of the first attempts at what would later be considered a (Francis) Baconian scientific research program (Thorndike 1923, vol. III, Ch. 8).

The astrometeorological treatises of the Middle Ages were not concerned with weather phenomena of economic significance for Northern Europe. Rainfall was a topic, as it was in Arab astrometeorology, but so were earthquakes and the aurora borealis. These treatises were intellectual exercises for students of what was seen as the science of astrology, not merely a source of useful information for farmers or merchants. During the course of the fourteenth and fifteenth centuries, the level and complexity of serious astrometeorological writing rose in comparison to the beginner treatises. The subject, having become extremely technical, though not separate from the other branches of astrology, was practiced by the best astrologers and academic authorities who apparently were consulted by clergy and aristocracy in regard to future weather. Records of the success or failure of specific forecasts have, unfortunately, not survived, though the method was criticized by some as being unreliable. Simultaneously, interest in other parts of astrology was growing, and many astrologers wrote general texts on the subject and offered prognostications based on conjunctions of planets, the technique of revolutions found in Arab texts. For example, a conjunction of Jupiter, Saturn, and Mars, coincident with a lunar eclipse, in 1345 was interpreted as a sign of bad things to come including cold weather, violent winds, crop failures, long diseases, and wars. Because this conjunction occurred in the air sign Aquarius, some astrologers predicted corrupted air (or miasma), which was thought to have caused the Black Death that struck Europe beginning in 1346. At a time when other explanations for the pandemic were entangled with religion, a conjunction of planets, which all could see, was easy to accept.

As previously discussed, meteorology had four divergent lines:

weather lore, parapegmata, astrometeorology, and Aristotle. The latter tradition, considered a branch of natural philosophy, involved descriptions of all natural processes that occurred in the region of air including clouds, winds, lightning, meteors, comets, rainbows, and so on. By the mid-sixteenth century, this physical science of meteorology had become more distinct from astrological weather forecasting and became the study of the causes and the description of the effects of phenomena in the sphere enclosed by the Moon, known as sublunary occurrences. Astrometeorology, a subcategory of natural or mundane astrology had, since the late Middle Ages, become integrated into the intellectual culture of Europe and was classed as a separate kind of scientific knowledge that was not challenged by the church. This respectable classification persisted into the sixteenth century. Evidence of this is that, between 1545 and 1555, the Swiss naturalist and bibliographer Conrad Gesner published the four-volume *Bibliotheca universalis,* a work that organized all existing knowledge into twenty-one books. Book VIII was on astronomy, Book IX was on astrometeorology, and Book XIV was Aristotelian meteorology in the context of natural philosophy (Havu 2005; Henninger 1960). Weather lore, which included non-technical information that could be used to make forecasts, lived on among sailors, farmers, and commoners.

ALMANACS AND ASTROLOGY IN THE RENAISSANCE: 1450–1650

Interest in all branches of astrology spread throughout Europe during the early Renaissance. From the elite to the vulgar, astrology and its notions became deeply embedded into the cultural fabric, even appearing as a theme in art and literature. Astrological references and themes are found in fresco paintings and in many of Shakespeare's plays. The term *influence,* which had entered the general vocabulary sometime in the late fourteenth century, was an astrological word that referred to emanations from the planets and stars. Individual astrologers were supported by highly placed patrons, and intellectual discussions on the subject were carried on in the universities. Astrology as a component of a generally accepted worldview pervaded scientific thought, including physiology, medicine, botany, metallurgy, psychology, weather, and agriculture. But this didn't last. In the late fifteenth century the humanist Giovanni Pico della Mirandola (1463–1497) launched a massive attack on the subject that put astrologers in a defensive situation for the next 150 years. This event will be considered in a later chapter.

Natural astrology, as well as judicial astrology, became even more familiar to the public due to the rise of print technology in the second half of the fifteenth century. Prior, it had been customary for individual astrologers to offer manuscripts containing tables of planetary positions as well as prognostications for the year ahead to their patrons, who may have included royalty, the wealthy, a town council, or the university where they were employed. The printing press changed this situation by enlarging the potential audience for such information. The first printed almanacs containing tables of planetary positions appeared in the mid-fifteenth century, and the numbers printed grew annually during the 1470s. Prognostications, a separate kind of publication that offered predictions, including weather forecasts, appeared in about 1470, and publications containing both tables and predictions soon became a common format. During the late fifteenth century almanacs also appeared in Germany, France, Italy, Hungary, the Netherlands, and Poland. Not only were almanacs and prognostications among the very first works printed (Gutenberg began printing almanacs in 1448), they were among the best-selling. A tradition of Flemish almanacs and prognostications, established by the master astrologer Johannes Laet, became very successful, and these were published annually from 1469 to the mid-sixteenth century. The Laet almanacs and prognostications were shipped to England, where English printers arranged for translations, which also stimulated the growth of a separate English almanac tradition.[3]

Continental almanacs, and later English almanacs, maintained a similar form. They were dedicated to the year at hand and were thus

---

3. The *English Almanacs 1500–1800* provides a history of the merger of the almanac and prognostication, which is seen as reaching its fully developed standard form in England by the later part of the sixteenth century (Capp 1979, 25).

disposable, requiring a new copy to be purchased annually. The form typically included the following: astronomical tables, political predictions for the coming year, disease and weather forecasts, medical notes often involving astrology, times for agricultural activities, a listing of religious holidays and their dates during the year, and other miscellaneous information. Individual writers of almanacs competed with each other, and the quality of the product varied. Astrological predictions were sometimes outlandish, but most almanac writers at least sought to publish the best astronomical and calendrical data available. Almanacs were a kind of mass media and, due to their widespread circulation and the scientific interests of their writers, they spread new and progressive ideas to every level of society. Some almanac writers popularized Copernicanism, argued that Earth was a sphere, and stated that astronomical bodies were spaced at great distances. The astrologers who wrote the almanacs were often advocates for the new science and even hostile to ancient geocentric astronomical notions. A detailed look at the contents of a mid-sixteenth-century almanac from England is instructive in regard to the content and tone of these publications.

The *Prognostication Everlasting of Right Good Effect* was an early English almanac written and published by Leonard Digges (c. 1515–c. 1559), a mathematician, cartographer, surveyor, and the inventor of the theodolite, a precise surveying instrument. He was also the father of the astronomer Thomas Digges, an early Copernican who was credited with correcting and augmenting the annual almanac's astronomical sections. The *Prognostication Everlasting* was first published in 1553 but was reprinted many times, its everlasting quality referring to the fact that the almanac contained the standard methodology for predicting the weather. With this almanac came the year's planetary positions along with a "do-it-yourself" manual. I have examined two versions of this almanac, one from 1555 and the other from 1605, both being very similar.

In the introduction, where readers were addressed as "dear Christian," Digges immediately launched into a defense of astrology by dropping well-known names such as Ptolemy, Bonatus, Gerolamo Cardano, and Melanchthon, all of whom were associated with astrology. This near rant, supposedly sanitizing the publication by associating

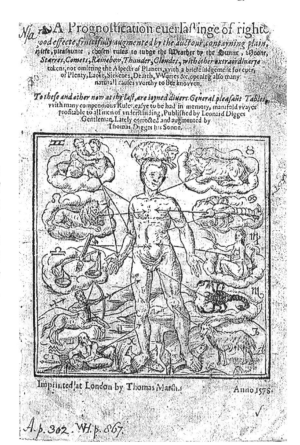

Figure 7. *Prognostication Everlasting of Right Good Effect* for 1596 by Leonard Digges with "zodiac man" showing correlations between the zodiac and parts of the human body on the first page.

it with intellectual celebrity and pious Christianity, evidences the defensive position taken by astrologers toward their subject in the second half of the sixteenth century. The introduction is followed by a geometrical diagram/template for a sighting device, one that Digges suggests be made out of sheet metal measuring a square foot, to be used for locating planets and stars. Following this is a diagram showing the relative sizes of the Sun and planets, the inner planets being small but Jupiter and Saturn being about two-thirds the size of the Sun. Next is a long section showing how to judge the weather by each of the five planets, Sun, and Moon, including some weather lore. First is how to judge weather by the color of the body, for example a red Sun in the morning implies wind and rain. Omens of future weather from comets, clouds, and rainbows comes next and is then followed by weather judged from planetary aspects, that is the Ptolemaic aspects of 0-, 60-, 90-, 120-, and 180-degree spacing between planets in the zodiac. Saturn, the planet that moves slowest, is listed first and there is a delineation for each of its aspects with the other bodies. For the aspects (conjunction = 0 degrees, quadrature = 90 degrees, opposition = 180 degrees) between Saturn, Jupiter, and Mars with the Sun, Digges writes the following:

> The conjunction, quadrature and opposition of Saturne with the Sunne, chiefly in cold signs, shows dark weather, hail, rayne, thunder and colde days.

> The conjunction, quadrature and opposition of Jupiter with the Sunne, great and moist vehement winds.

> The conjunction, quadrature and opposition of Mars with the Sunne in fiery signes, drought; in watry, thunder and rayne (Digges 1605, 10).

While all this appears straightforward—the user simply finds an aspect in the ephemeris section of the almanac and then looks up the forecast—it is actually far more complicated. To begin with, there are often multiple combinations of aspects forming at the same time, some of them involving the slower-moving planets that were said to have effects lasting days or weeks, a situation that makes the existence of a single aspect separated by a few days from others an infrequent occurrence. Further, the Moon's aspects with the planets form and dissolve over a matter of hours due to its high rate of motion, about 13 degrees a day. Digges offered some clues as to integrating this "minute hand" factor in making weather predictions by paying attention to the movement of the Moon along the zodiac, that is the sequence of aspects to other planets:

> The conjunction, quadrature and opposition of the Moon with Saturne, in moist signs, bryngeth a cloudy daye, colde ayre, according to the nature of the signe: If they go from Saturne, to the Sunne, by conjunction or otherwyse, harder weather ensueth.

The conjunction, quadrature and opposition of the Moon with the Sunne in moist signes, rayny weather: the more if the Moon go from the Sunne to Saturne (Digges 1605, 11).

What these branching, recipe-like descriptions require is a substantial knowledge of astrology. First, the zodiacal sign must be incorporated into the procedure; like filters, each has its own specific modification for each planet, the Sun, and Moon. Then, the sequence of astronomical phase events must be considered. The Moon passing from a conjunction with the Sun to a conjunction with Saturn is apparently different, weatherwise, from the Moon conjuncting Saturn first, then the Sun. This assumes a lot on the part of the reader and is an indicator of the high level of public knowledge of astrology at the time.

After a few pages of aspect delineations, Digges then described each planet's effects in the signs of the zodiac. Following this is a method of making a prognostication for the year ahead based on the day of the week on which New Year's Day falls, a very simple technique used at least since the Middle Ages. The procedure is described and short delineations for each day of the week are given (which are, of course, named for the Sun, Moon, and planets in the Romance languages). Further discussion of meteors, thunder, earthquakes, rainbows, and others leads to some very specific delineations of trends coincident with lunar aspects. A table showing what sign the Moon is in on any day of the year also states whether or not that day is good for purging, bloodletting, or bathing. A woodcut showing a man with his body parts linked to the twelve signs of the zodiac (called the zodiac man) is also part of this quasi-medical astrology section. Calendrical tables follow that locate events such as Lent over a range of years. More tables show the length of the day and night and the time of sunrise and sunset throughout the year for several localities. There is then a "peculiar kalendar," as it's written, that treats each month at a time astronomically, followed by a "general kalendar" that lists fairs and events. Next are tables of the Sun's altitude and a diagram of a quadrant that may be employed by the user for this measurement. Finally, in a section called "Brief Collections," are a series of pieces of information on topics like how to track moveable feasts and how to know how long the Moon will shine on a given day. The almanac concludes with a section by Thomas Digges in which he made a few corrections to his father's work and then promoted the Copernican model of the solar system, complete with a diagram.

Digges's almanac is not a simple read aimed at the masses. It is far more complicated than present-day farmer's almanacs. Even if it were re-written in modern English, very few people today would actually understand much of it, let alone do their own calculations and weather forecasting. The fact that this widely-selling common publication, full of astrology, was promoting Copernicanism not long after Copernicus's work itself was published also raises questions about who was educating the masses, at least in astronomy.

William Lilly was the most successful astrologer of seventeenth-century England. He was consulted by people both poor and royal, he published an annual almanac for nearly forty years, and he wrote books on astrology, including an autobiography. Lilly wrote perhaps the definitive astrological textbook of his time, titled *Christian Astrology*. This again says something about the pressure astrologers faced in regard to religion. Although most astrologers were religious to some degree, the clergy was almost unanimously set against the judicial branch of the subject. However, Lilly's almanacs were extremely popular, selling in the tens of thousands, and in addition to offering the usual astronomical, weather, and calendar data, he regularly published political predictions that were often quite accurate. The title of his 1646 almanac (published over a year earlier) reads:

> Anglicus, or ephemeris for 1646. Delivering mathematically the success of this years actions, between the king and parliament of England. With astrological aphorisms, expedient for physicians and others, useful for students in this science. To which is added the nativity of prince Rupert (Lilly 1644).

In a total of eighty-eight pages is found an apology for astrology, an ephemeris and aspectarian (a listing of the aspects made by the planets), and a listing month by month in which Lilly makes both political and meteorological predictions. Also included is a table of houses used in calculating astrological charts, and a political section called "a general judgment of the affairs of England." The almanac concludes with fifty astrological aphorisms relevant to the practice of medicine, another fifty relevant to the practice of interrogations (horary astrology), and a delineation of the horoscope of Prince Rupert. For a periodical selling roughly thirty thousand copies per year, this annual almanac was far from light reading for the consumer; it was more of a serious reference work.

JOHN GOAD'S *ASTRO-METEOROLOGICA*

While the practicing astrologers who published almanacs were versed in astrometeorology, few deviated from the standard methodology that was based on accepted correlations between planetary aspects and types of weather. Johannes Kepler (1575–1630) was one exception as he kept weather records and gave some thought to his observations. Although he never published a book on astrometeorology, his almanacs and writings discuss the subject in some detail. The one serious study of the subject of the time was John Goad's (1616–1687) *Astro-Meteorologica*. In 1686, the year before Newton's landmark *Philosophiae Naturalis Principia Mathematica* ("The Mathematical Principles of Natural Philosophy"), a major work on astrology with a hefty title appeared in print: *Astro-meteorologica, or, Aphorisms and discourses of the bodies cœlestial, their natures and influences discovered from the variety of the alterations of the air . . . and other secrets of nature/collected from the observation at leisure times, of above thirty years.* In it, Goad compared weather records with the angular separations of the planets, thus establishing himself as one of the very few astrologers who attempted to apply the newly emerging experimental method to his subject. The *Astro-Meteorologica* is a comprehensive work of over five hundred pages, many of them samples of his weather log, and it presents some difficulties for the reader not fully versed in the language and concepts of astrology (Goad 1686). John Goad was the only significant follower of Kepler's astrometeorological studies and, although he disagreed with him on many points, Goad understood the value of testing astrology. He embraced the empirical method that Francis Bacon advocated and Kepler practiced, compiled one of the earliest detailed weather diaries, and published his results in both English and Latin. *Astro-Meteorologica* was a report on what was the most ambitious scientific study of astrology undertaken during the entire seventeenth century.

Born in London, Goad was the headmaster of the Merchant Taylors School, an extant private boys' school with ties to St John's College, Oxford, which he had attended himself. In 1680, during the Popish Plot when anti-Catholic sentiments were at a peak, he was charged with Catholic leanings that were believed to be concealed in the content of some material he had written for the use of his students. He was dismissed from his post, and Goad's opponent in this affair, Dr. John Owen, was successful in placing his nephew in Goad's former position. Goad next opened a private school, and in the year before his death, he declared himself a Roman Catholic. Among Goad's other writings are some sermons, a comment on the catechism of the Church of England, a comment on monarchy as a form of government, and a method of teaching Latin to students. He had also written a treatise on plagues, but it was destroyed while in press during the Great Fire of London (1666). His major work was the *Astro-Meteorologica,* which earned him a great reputation, one leading to even discussing his findings with Charles II. A version of this work in Latin without the weather records, titled *Astrometeorlogica Sana,* was published in 1690.[4]

Goad was a contemporary of numerous contributors to the Scientific Revolution, and he was familiar with the scientific currents of his time. In his book he mentioned the Royal Society's call for navigators to make notes of ocean currents and asserted that astrology could help solve such problems.[5] Elias Ashmole (1617–1692), a founding member

4. Biographical information on Goad is found in *The Dictionary of National Biography* (Smith 1921–22, 8, 18–19; Thorndike 1923–1941, vol. 8, 347–349; Thomas 1971, 327–329). The full title of the Latin edition is *Astrometeorologica sana sive Principia Physico-Mathematica, quibus Mutationum Aeris, Morborum Epidemicorum, Cometarum, Terre-Motuum, Aliorumque Insigniorum Naturae Effectuum Ratio reddi possit.*
5. Goad noted that conjunctions of Mars and the Sun seem to affect ocean currents (Goad 1686, 213; Curry 1989, 69; Capp 1979, 185).

of the Royal Society and for whom Oxford's Ashmolean Museum is named, regularly corresponded and collaborated with Goad. Joseph Williamson, the second president of the Royal Society, was impressed with Goad's research (Bowden 1975, 186). Like his contemporary Joshua Childrey (1623–1670), another astrology reformer, Goad applied a Baconian method to astrometeorology. He collected data for some thirty years that he then used to check traditional doctrines in regard to the planetary aspects. This is perhaps the most significant feature of Goad's astrology and one he shared with Kepler, the de-emphasis of the zodiacal sign positions of the planets and luminaries and his emphasis on their angular separation (aspects), which basically turns astrology into a phase angle analysis of vectors (rays) from planets converging on the Earth. Goad's methodology in testing astrology was logical and consistent and led him to believe that, for the most part, it conformed to traditional doctrines and was therefore a valid body of knowledge. In her Yale Ph.D. thesis on the Scientific Revolution and astrology, Mary Ellen Bowden commented that Goad's work was "the most conscientiously executed research of any work inspired by Bacon in the 17th century" (Bowden 1975, 187). For this reason, I will go into his thought and work in some detail.

The *Astro-Meteorologica* reflects the overlapping of ancient thought and modern method that was occurring when Goad was working on his project. Aristotle and Copernicus were still being debated during the mid-seventeenth century, and weather instrumentation was very new. The first thermometers had been invented but were one of a kind and unreliable. It wasn't until Daniel Gabriel Fahrenheit, who in 1714 made the first mercury thermometer and in 1724 proposed his temperature scale, that a decent attempt to quantify daily temperatures could be made. As for the barometer, Galileo's assistant Evangelista Torricelli (1608–1647) is generally credited with inventing it in 1643, though it wasn't until the 1870s that these became available as a weather instrument. But instrumentation, which provided data points in weather studies, was where amateur scientists were headed, and by the mid-seventeenth century most writing on meteorology was about instrumentation and the application of the new physics to the atmosphere and related phenomena (Symons 1893, 338–351). Goad did most of his work just before the cusp of this change. His weather log was mostly subjective; he used the Aristotelian four elements to explain how the planets affected the sublunary world; and he frequently referred to Ptolemy, Pliny, and other ancient authorities. Had he been born a few decades later and had access to reliable and standardized instruments with which he could have made a quantitative weather log, employed better mathematical analysis of his data, and also kept in contact with others in the field, his work may have become better known.

In addition to his own qualitative weather diaries, Goad used other non-instrumental data, including records made by Elizabethan astrologer John Dee (1527–1608/9) and those from Kepler's nine-year diary of the weather in Linz and Ulm (1621–1629), to confirm some of his findings. He did observe some thermometer readings and barometric fluctuations, which he compiled around the time that these inventions were in developmental stages, but these do not appear in his published weather diary. He apparently owned some instruments, but he only recorded the temperature when it was out of the ordinary, and he also regretted that he did not use a horizontal plate with a compass for measuring the wind direction exactly. Goad's methodology on which he based his results was then completely descriptive and, in spite of his meticulous consistency in making weather records, in some cases hourly, all of it was subject to his own judgment. In contrast, his younger contemporary and contributor to a wide range of scientific subjects, Robert Hooke (1653–1703), suggested a method for making a record of the weather that employed three existing primitive measuring devices: the thermometer, barometer, and hygroscope (a guage that measures humidity). Hooke, while critical of astrology, had astrologers as friends, visited Goad, and also advocated recording the longitude of the Sun and Moon and their aspects alongside the weather data for each day. It has been suggested that, due to Hooke's associations with astrologers such as Goad and Childrey, who were both compiling scientific weather records a decade before him, he borrowed some of their ideas on data collection (Sprat 1667, 173–179;

Thomas 1971, 351–352; Capp 1979, 189–190; Curry 1989, 68).[6]

In the middle of the seventeenth century, critics of astrology were increasing in number and the reform of the subject became more urgent, a fact reflected in the emotionally charged defenses and apologies for the subject that introduce typical works on astrology published at this time (Lilly 1644; Wilsford 1642). Goad believed that his findings, resulting from an analysis of many observations, represented a correction of previous defects, or at least a summation of planetary effects consistent with the climate of England, as opposed to Ptolemy's Egypt (Goad 1686, 29). He remarked that his critics, who he says want nothing less than exact effects, do not study nature closely enough and consequently will be disappointed and likely to reject his work. He implied that his non-instrumental, qualitative work was not unlike that of the naturalist studying organisms in nature, though the influence of the planets is a far more slippery subject to observe, sketch, and dissect. Further, he said the principles of astrometeorology cannot be reduced to absolute laws that will work consistently, but are rather patterns that can be perceived when one examines the data over long periods of time.[7] He pointed out that his conclusions about the various planetary combinations are not meant to be a total explanation of natural phenomena, but they reveal a natural predisposition in nature. Goad believed that astrology could be upgraded with some attention and funding, but affluent and concerned persons willing to fund research in astrology were apparently scarce in Goad's time; without resources not much gets done, then or now. Why this was so will be discussed later in more detail.

Goad used Aristotelian explanations for meteorological phenomena. An alternative would have been Greek atomism, which was adopted by Descartes, Gassendi, and other seventeenth-century scientists, but

6. In her book *Earth Cycles: The Scientific Evidence for Astrology,* Valerie Vaughan points out that Hooke's methodology did not specify taking weather records at regular intervals. Goad began his daily weather diary in 1652, a year before Hooke was born (Vaughan 2002, 60–61).

7. This is the case with the weather system. Reducing it to parts and getting results acceptable to a linear, mechanical science is a problem the science of meteorology still deals with.

that materialistic philosophy was considered atheistic, not something a serious Catholic would be interested in. Goad argued that the Sun, Moon, and planets emit varying amounts of heat that is transmitted by light, but moisture is an effect that is pulled from the Earth, evidenced by dew. For Goad, as for Ptolemy, the principal cause of the weather was the Sun. He argued that the Moon and planets carried heat via reflected light and that this light altered the air by warming, more by Mars but very little in the case of distant Saturn. He viewed the Earth's atmosphere as a "terrestrial spirit, regulated according to its vicissitudes, from the modification of the light celestial" (Goad 1686, 40). This explanation of planetary effects on the Earth from light differed from Kepler's notion that the angular separations of the planets produced a non-physical harmonic resonance that could be "heard" by the living Earth, which it then expressed through weather and other phenomena (Kepler 1987, 14–17).

In the *Astro-Meteorologica,* weather records are correlated with the traditional Ptolemaic aspects or angular separations between the planets based on division of 360 by integers 1, 6, 4, 3, and 2 (resulting in aspects of 0, 60, 90, 120, and 180 degrees). Goad argued that the aspects were not founded on harmonic proportions, which is what Kepler thought, but followed physical and optical principles. He discussed the aspects added by Kepler (30, 36, 72, 150, and others) but regarded them as either being too weak or superfluous, called them pseudo-aspects, and mentioned that other astrologers were "sick of them" as they increased the already abundant daily aspects produced by the rapidly moving Moon. Ultimately, he rejected all but two of Kepler's additions, 30 and 150 degrees (based on the twelfth harmonic or 360 divided by 12), which he claimed his observations confirmed (Goad 1686, 39–40, 61). In his published data, however, Goad listed only the traditional Ptolemaic aspects, calculated to the date of exactness, which he called the aspect's "acme." These *partile* (meaning "exact" in terms of astrology) aspects were the points in time against which the weather was observed. Further, his methodology involved tracking the weather before and after an aspect formed (Goad called this the aspect's "access" and "recess"), the stretch room, called the orb of influence in astrology.

This orb he regarded as variable, depending on the daily motions of the planets aspecting each other, and also other coordinates such as declination and latitude.

> Confining therefore the conjunction, and with that the rest of the configuration to the same sign and degree, and allowing the Acme of the aspect to take place the precise Astronomical Time, with proportional allowance of vigor or abatement, according to the scruples of access and recess; yet it is true that the physical influence of an aspect, exerts itself before and after, i.e. as long as the Heavenly Movables keep within the terms of the definition (Goad 1686, 41).

The *Astro-Meteorologica* systematically considers the angular separations of the various planetary combinations in the order of their frequency of occurrence, beginning with the Sun and Moon and ending with Jupiter and Saturn. In Book I, after a lengthy introduction covering his basic principles, Goad examined the various Sun-Moon aspects beginning with a seven-year period of eighty-seven conjunctions (new Moons) that he correlated with the weather. He next considered the opposition between the Sun and Moon (full Moon), the quarters (90 degrees), the trines (120 degrees), and the sextiles (60 degrees). Goad sought to examine the frequencies of various kinds of weather patterns, which he standardized with short descriptions, that occurred during the range of time he thought each aspect was effective. If there was a correlation between aspect and weather pattern more than half the time, Goad maintained that the influence of the aspect was demonstrated. This was as close to statistics, which wasn't a developed branch of mathematics at the time, as Goad could get. It wasn't until the second half of the eighteenth century that modern probability and statistical theory began to take form.

A closer look at Goad's Sun-Moon data will better serve to illustrate his methodology. He treated each aspect as a separate study and included the relevant portions of his weather diary alongside his discussion. The eighty-seven Sun-Moon conjunctions (new Moon) during a seven-year period were listed in twelve monthly tables. Using a range of three days (from midnight to midnight) for the aspect, the weather conditions in London during the designated periods were recorded, often to the hour, over a total of 251 days. The resulting data are qualitative; he used a descriptive vocabulary of the variety of meteorological phenomena to specify the type of weather at that time (rain, mist, frosty nights, etc.). The direction and nature of the winds, notes on the temperature, and the exact time of any weather change, or as Goad understood it, the effect of the aspect, were also noted. The results were then tallied in the following list of eighty-seven Sun-Moon Conjunctions (Goad 1686, 54).

The discussion and analysis of these results in the *Astro-Meteorologica* is quite thorough. First, Goad defended his Baconian methodology in a number of ways. He pointed to the care he took in recording the hour of meteorological events, but also the value of extending the observation period for the conjunction to three days to capture weather trends that were indicated by the aspect but did not appear when it was partile. He also took care to record weather events relative to the season, that is, what would be considered cold in summer would be warmth in winter. Goad's findings contradict, in a few places, traditional notions of planetary influence, and he was not above criticizing Kepler for discounting the influence of the new Moon, even referring to Kepler's own weather diary from 1621–1629 to prove that his findings were confirmed in Kepler's own records (Goad 1686, 62). Everywhere Goad shows both a commitment to the data he produced and delight in his discoveries, most of which support traditional astrological notions, with some minor exceptions.

The full Moon investigation was presented next, and Goad's records show that some form of moisture was recorded in his weather log on seventy-five of the eighty-seven aspect events during the same seven-year period. After a lengthy analysis of the findings and a comparison to the new Moon record, Goad moves on to the squares, the trines, and finally, the sextiles. At the conclusion of his study of the five Sun-Moon aspects, Goad summarized his findings in a table. Each type of weather condition is listed in a vertical column on the left and the number of days that each kind of weather had occurred at the time of each aspect is noted horizontally at the top (Goad 1686, 115). The weather record suggested

GOAD'S DATA FOR 87 SUN-MOON CONJUNCTIONS

Cold Frost Nights	63	Mist	47	Wind Change	29
Clouds Pregnant	72	Northeast	30	Wind Tempestuous	37
Fog or Grosser Mist	2	Northwest	31	North Wind	40
Fila	2	Rain Moderate	109	East	45
Frosty Days	34	Violent	28	West	44
Hail	4	Serene, Fair	31	South	18
Halo	0	Trajections	10	Southeast	16
Hot Days	28	Thunders	3	Southwest	58
Nights	8	Warm	31	Northeast	36
Lightning Nocturnal	2	Wind	101	Northwest	12

Table 1. Goad's qualitative descriptors and number of instances for eighty-seven Sun-Moon conjunctions 1671–1677 in London. Conditions occurring more than forty-four times he would consider significant.

that there were more wet days at the full Moon than at the new Moon and that the second half of the Sun-Moon cycle—or the later trine, square, and sextile—was generally warmer than the first half. Goad suggested that this was possibly due to the fact that the Moon rises before the Sun in these positions and adds warmth to the air.[8] He observed that it rained consistently on the full Moons falling in April and August, and that the actual meteorological events occurred much closer to the time of the exact opposition (within four hours) than was the case with the conjunction. A further observation concerning rain was in response to the high figures for the first sextile (60 degrees after the new Moon) and the second trine (60 degrees after the full Moon). This result of more rain about five days after the new and full Moon surprised him; he had expected that the traditionally more powerful aspects, the conjunction

---

8. A study published in *Science* actually found a slight global warming occurring around the full Moon (Balling and Cerveny 1995).

GOAD'S WEATHER DATA FOR EACH ASPECT IN THE
SUN-MOON SYNODIC CYCLE

	Con	Opp	1.Squ	2.Squ	1.Tri	2.Tri	1.Sxt	2.Sxt
Frosty Day	16	26	34	27	26	16	28	19
Frosty Night	37	27	31	26	29	27	30	26
Hot Day	28	11	13	24	25	16	20	36
Trajections	19	4	12	20	5	6	17	21
Lightnings	0	0	1	2	1	0	5	5
Thunder & Lightning	2	4	4	4	5	7	3	6
Stormy Winds	37	69	34	43	44	31	33	35
Varying Winds	3	5	3	3	2	5	1	1
Changing Winds	29	55	71	53	43	43	32	41
Rain	109	103	143	132	111	162	149	144
Violent Rain	28	47	47	42	48	52	60	27
Snow	5	14	16	12	12	15	13	10
Hail	9	8	3	6	4	4	7	6
Iris	1	1	0	0	1	0	0	0
Halo	0	3	4	6	5	6	3	6
Grosser Fog	38	23	31	29	17	28	21	38
Winds East	45	53	56	35	42	44	50	44
Winds West	44	49	50	47	31	49	31	43
Winds North	40	33	36	41	28	27	41	44
Winds South	18	38	23	20	39	21	21	31
Winds Northeast	30	29	42	37	34	34	35	42
Winds Northwest	31	26	24	40	21	27	20	15
Winds Southeast	16	15	7	17	20	26	18	14
Winds Southwest	55	80	73	103	90	69	91	51

and the opposition, would account for the most moisture. We'll come back to this particular finding, for which there is modern scientific confirmation, later.

In Book II of the *Astro-Meteorologica,* Goad analyzed the aspects of the inferior planets, Mercury and Venus, to the Sun and the other planets. Each planetary pair has its own chapter that begins with an overview of the traditional interpretations along with frequent references to the opinions of Kepler and others. A discussion of Goad's findings follows along with his weather records, which unlike the lunar tables, cover a wider range of years. His Sun-Mercury study (1668–1680) records earthquakes, thunder, and lightning, in addition to correlations with the usual weather events. His observations suggested that Mercury was a rainy planet, though the ancients considered it windy, and he wrote that since he is confined to London, he cannot speak confidently for its influence in other climates. To further examine this possible contradiction, Goad introduced some nautical observations made on a voyage to the East Indies, the ship's captain having been a friend of his. This additional data did not prove to Goad's satisfaction that Mercury was a windy planet, and he was inclined to think the bulk of evidence supported his correlation of the Sun-Mercury conjunction with rain.

While on the subject of Mercury's influence, Goad clearly articulated his most basic position regarding planetary influence on the weather. He noted that, while a few planetary pairs in aspect tended to show very high correlations with a certain type of weather, in most cases the weather was conditioned by several planetary patterns. In examining each aspect pair separately, Goad's only recourse was reductionism, to try to narrow the range of days around the exact aspect in hopes of catching a glimpse of its purest meteorological manifestation, but that was not always successful. In some cases, particularly with the conjunction of Sun and Saturn, Goad found the effects to be strongest when the planets were separated by several degrees. He clearly realized that this situation, the impossibility of finding a consistent effect for each single aspect, was not likely to convert many investigators of nature. What Goad was essentially saying is that aspects were measuring a system, in this case the weather system, that is not reducible to discrete parts. In his own words:

We have said, we make no one aspect an adequate cause of the effect; only Eminent and Considerable; which much be assisted with its neighbors: We have other aspects which put in for their share of the business; we shall see them in the following chapters, and surfeit on them. There is scarce conjunction or opposition, yea, sometimes trine or square, but steps in to help at a dead lift (Goad 1686, 147).

With the Sun-Venus conjunctions (records from 1655 to 1681), those with Venus in retrograde motion (near inferior conjunction) are separated from those that occurred when they were in direct motion. In this section, Goad referred to Kepler's diary for further data and noted that when the aspect occurred in a "state of destitution," that is not involved aspect-wise with other planets, cool, clear air was produced. If the aspect was "assisted" by others, that is if other aspects were forming at the same time either to the Sun-Venus conjunction or separately, the trend was toward warmth, clouds, and rain. He also suggested that, since Venus was closer to the Earth when retrograde, its influence might be more potent when in that position. His records showed a strong connection between conjunctions of Venus and Mercury and rains and storms. Again, a foreign diary is produced to support what he insisted was an obvious positive correlation.

In addition to the Mercury-Venus conjunctions, Goad reported in Book II the following findings. Mars was found to be associated with varied effects, though it tended toward storms and tempests when with Venus, and toward dryness when with Mercury. He found that Saturn produced cold when in aspect to the Sun (more on this later) but did the same when in exact aspect to Venus. Jupiter was found to be cold in some positions and also appeared to be associated with drought. As to the recently discovered satellites of Jupiter, Goad speculated that they do not influence our weather but perhaps they affect Jupiter itself. In Book III, Saturn and Mars were observed to show a correlation with storms and thunder in both his diary and those of foreign records. He noted that when Mars and Saturn were separated by exactly 30 degrees—Kepler's semisextile—there was usually a violent effect. Goad pointed out that with these slower-moving planets, three or four days for effect meant nothing; they defined trends over longer periods of time, even months. Jupiter and Mars in aspect brought "monstrous frosts" and, when in opposition and with assistance from other planets, thunder and lightning. The longest cycle of the visible planets is the Jupiter-Saturn cycle of twenty years to which Goad devoted many words. One observation he made with some confidence was that drought commonly occurred when these planets were in conjunction and that, nearly half the time, they "produced" comets. Maybe there is something to this observation. Some modern studies have reported a roughly twenty-year drought cycle, and it is considered possible that the aligned and combined gravity of the two biggest planets may sometimes disrupt comets in the Kuiper Belt or Oort Cloud, deflecting them into the inner solar system.

Goad was of the opinion that astrometeorology was ultimately interpretive because only rarely was the pure effect of an aspect observable; more often the blending of several was the rule. Further, the aspects often did not have to be mathematically exact at the time of the observed meteorological effect; in certain cases the window of effect could cover several days. Overall, he was making a case for an interpretive astrometeorology that accepted less than exact angular separations between planets, but still within a few degrees of their exact focus, what he called "platic distance." This was a tenet of traditional astrology, which for centuries had doctrines of planetary orbs, meaning effective distances. In sharp contrast to this was Kepler, who stressed that only partile aspects could produce effects, including the case of Jupiter-Saturn conjunctions (Goad 1686, 446). In Goad's view, Kepler had proposed numerous additional aspects to account for changes in the weather when no traditional aspects were forming in order to save his theory of astrology.

In summary, the evidence Goad gathered showed few consistent clear-cut effects from any single planetary combination because most of the time several planetary combinations were simultaneously in effect. His conclusions left him with an art of estimates and not a science, which is what astrology, as a practice, has always been. But this is not to say this kind of interpreting of patterns found in data is entirely unscientific. Modern meteorologists, seismologists, economists, medical diagnosticians, etc., when making predictions, find themselves working with complex systems and noisy data that must be filtered before interpretation becomes possible. And most of these fields (which have replaced the role of astrology in society for the last few centuries) have yet to develop consistently accurate long-term prediction techniques, and they often fail in short-term predictions. All of this illustrates that self-organizing systems like the weather are very hard to constrain to the point where accurate and specific predictions are easily made. It is more in the discovery of patterns and relationships in the data that progress is made.

Astro-Meteorologica is really a remarkable piece of astrological writing both in content and in historical place. Goad's careful correlation between planetary alignments and observed meteorological events, a Baconian way of dealing with the material he was investigating and the scale of his study, was unique in seventeenth-century astrology. Goad had no instruments, no real units to work with, and therefore no way of mathematizing any correlations he found. He thus lacked the kind of proof that was becoming fashionable in scientific circles during his time, and his lengthy weather records only showed somewhat higher frequencies of a particular kind of weather when pairs of planets were in some aspect to each other. Evaluating his observations in terms of probability and statistics, which appears to be the next logical step from a modern vantage point, was impossible as such approaches were not to develop until much later. In the final analysis he was testing a subject with a set of qualitative rules that had evolved over the centuries to measure the weather system. Goad's attempted Baconian reform of astrometeorology came to a dead end, and by the latter part of the seventeenth century the subject was in steep decline. However, the long tradition of keeping weather records in astrometeorology (e.g., Merle, Dee, Kepler, and Goad) was adopted, and vastly improved with instrumentation, by the founders of modern meteorology.

Kepler was a key figure in the scientific revolution, and his work in astronomy influenced Isaac Newton, the greatest reductionist scientist of the seventeenth century. Kepler's astrology was not a major influence on astrologers in general, but it had an effect on Goad. Although Goad disagreed with Kepler on many things, he was the one who took Kepler's reformed astrology, with its emphasis on planetary aspects and de-emphasis on the zodiac, as far as it could go, given the limitations of the time. Although he was respected in his time by both astrologers and experimental scientists, Goad's influence was minimal and his reputation quickly faded. Only much later was he recognized for his pioneering work by the British astrologer A. J. Pearce (1840–1923) who, in his own writings on astrometeorology, quoted him frequently and referred to him respectfully throughout as "Dr. Goad." It wasn't until the early twentieth century that Kepler's ideas on the primacy of planetary angular distances once again began to influence astrological theory and methodology in the teachings of the innovative German astrologer Alfred Witte.[9]

LUNAR INFLUENCES ON WEATHER

Astrometeorology has not been taken seriously by the scientific community since the late seventeenth century. The idea of planets affecting the Earth's weather is today regarded as preposterous, and any association with the idea, or even with those who hold such ideas, could ruin an academic reputation. However, a number of scientific studies have confirmed a lunar influence on weather. A study of rainfall and lunar phase pointed out earlier and described below produced results consistent with one of John Goad's findings. Recall that his seventeenth-century weather diary recorded more moisture at the first sextile (60 degrees after the new Moon) and the second trine (60 degrees after the full Moon). This result, basically indicating more rain at roughly 4.6 days after the new and full Moons, surprised him; he had expected that the traditionally more powerful aspects, the conjunction and the opposition,

---

9. Alfred Witte (1878–1941), founder of the radical Hamburg School of Astrology, utilized some of Kepler's astrological reforms and techniques in creating a highly original method of doing astrology. This school influenced the practice of astrology throughout Europe and the United States (under the name Uranian Astrology) during the twentieth century and continues to have some influence.

would account for the most moisture. This observation of significant rainfall roughly five days after new and full Moons was confirmed (though Goad was not mentioned) in a 1960s study published in the prestigious journal *Science*. Donald A. Bradley and Max A. Woodbury examined precipitation data over the continental United States for a fifty-year period (1900–1949) and found that maximum precipitation events over broad areas were related to the Sun-Moon cycle. The study showed above average precipitation just after the new and full Moons at points very close to Goad's first sextile and second trine (Bradley and Woodbury 1962).

Donald A. Bradley (1925–1974) was an engineer, and also a technical astrologer who published articles in astrology magazines and journals under the pseudonym Garth Allen. In the astrological field he was a strong advocate of precession correction, that is adjusting for the 1 degree shift every seventy-two years between the tropical and sidereal zodiacs. He also published a statistical study of birthdates and professions and is known among financial astrologers for the Bradley Model of stock market forecasting. In the 1962 study mentioned earlier, Bradley and Woodbury analyzed precipitation data in the context of the lunar synodic cycle (Bradley and Woodbury 1962). The study utilized U.S. Weather Service records of the dates and places of maximum twenty-four-hour precipitation. When these dates of maximum rainfall were plotted against the lunar synodic cycle, the result was a curve with peaks of rainfall occurring near the middle of the first and third weeks of the synodic month, that is three to five days after new and full Moons. The second and fourth weeks were found to be deficient in heavy precipitation. This study was duplicated in New Zealand with similar results (Adderley and Bowen 1962). Not only does the work by Bradley and Woodbury examine the same phenomena as does Goad in a section of the *Astro-Meteorologica,* but their findings three hundred years later were similar to his. It should be made clear, however, that Goad's data measures total days on which rain occurred while the modern study measures dates of maximum precipitation, a case of extreme sampling.

In a second study published in 1964 Brier and Bradley reported a precipitation cycle half the length of the lunar synodic cycle in ninety

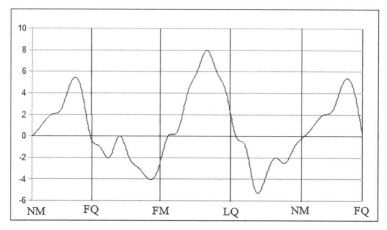

Figure 8. Lunar phase (synodic cycle) and maximum rainfall 1900–1949. Adapted from Bradley and Woodbury (1962). Maximum rainfall events were found to peak about four to five days after new and full Moons.

years of data for the United States (Brier and Bradley 1964). They also found a correlation between the synodic cycle and the lunar nodal and declination cycle of 18.6 years. In 1986 a study tested these results, which were confirmed, and also found that precipitation peaks and lunar phase apparently shift somewhat across the continent and vary by season (Hanson, Maul, and McLeish 1987). How the lunar cycles could be related to rainfall is unknown, but a possible explanation may involve ice nucleii abundance in the upper atmosphere on which water (as ice) condenses. A year after Bradley and Woodbury's first study was published, the journal *Nature* reported a finding that ice nucleus concentrations coming from the upper atmosphere varied according to the lunar cycle, with peaks just before first and third quarter, the same parts of the lunar cycle where Goad and Bradley found more rainfall (Bigg 1963). It was hypothesized that the Moon modulates the incoming meteor rate, and consequently the abundance of debris from burning meteoroids (meteoric dust). The dust serves as ice nucleii in the upper atmosphere, which then falls into the lower atmosphere becoming rainfall.

A similar pattern for rainfall was found in India, and several studies that analyzed heavy rainfall events found a peak several days after the full Moon (Roy 2006). One study, using data from 129 stations covering a large part of India, looked at rainfall events in July and August from 1910 to 2000, these months being the start of the summer monsoon season. The findings were again much like Goad's, that maximums in frequency and amount of precipitation occurred a few days after new and full Moons. Among other observations noted was that the west coast of India had higher precipitation after a new Moon (which begins the ascending phase of the lunar cycle), though precipitation peak in northern India occurred after the full Moon (the descending phase). Explanations proposed for this effect again include lunar gravity affecting meteoric dust in the atmosphere on which water condenses, or possibly a lunar modulation of circulation in the troposphere.

The formation of hurricanes and tropical storms is apparently also influenced by the lunar cycle. A comprehensive study published in 1972 using the formation dates for 1,013 hurricanes and typhoons over a 78-year period reported a correlation with lunar phase (Carpenter, Holle, and Fernandez-Partagas 1972). This was not a completely new finding as earlier studies, beginning with one in 1928, had found a 29.5-day cycle of cyclone development. The study found that storms tended to form near syzygy (new and full Moons), with more near the new than the full Moon, and with fewer forming at the quarters. In analyzing only storms that formed at perigee (minimal lunar distance), it was found that more of them originated when perigee was coincident with syzygy. These effects were attributed to a combination of tidal forces and other pre-existing atmospheric conditions.

Thunderstorm frequency has been found to correlate with lunar phase and lunar declination as reported in a 1970 paper (Lethbridge 1970). The author of the paper notes that quite a few investigators had looked into this and related phenomena in the late nineteenth and early twentieth centuries and had set the topic, formerly regarded as folklore (which, presumably, includes astrometeorology), on a factual basis. In the study, central and northeastern United States thunderstorm data for three sets of years (1930–1933, 1942–1965, and 1953–1963) was smoothed with a running mean and prepared in such as way as to cancel out any effects of solar activity. Peaks of activity were found when the

Moon was at maximum northern declination and two days after a full Moon. A possible mechanism for this effect was proposed: the passage of the Moon when it is full through Earth's magnetic tail may cause a disruption of energetic particles that may initiate a chain of events leading to thunderstorm formation.

Other studies indicating lunar correlations with storms, diurnal pressure changes, and cloud cover, among other meteorological phenomena, have been, and continue to be, published in reputable science journals. A study reported in 1995 by Balling and Cerveny found a correlation between lunar phase and lower tropospheric global temperature anomalies. Using daily satellite data they showed a temperature modulation between new and full Moons with the later phase being slightly warmer (Balling and Cerveny 1995). This study was criticized and it was suggested that the very small temperature difference was due to the Earth and Moon barycenter placing the Earth closer to the Sun when the Moon is full. The authors responded by stating it may not be that simple because there are strong geographic variations to account for.

No specific mechanism is currently accepted in regard to these correlations between the Moon and weather. Several hypotheses have been proposed, one being the meteoric dust hypothesis previously mentioned. Lunar distortions of the Earth's magnetic tail have been proposed as a mechanism to explain correlations between lunar phase and thunderstorm activity, which appears to be more intense at full Moon. Lunar tidal effects on Earth's atmosphere and spin rate have been suggested as factors affecting pressure fields and basic circulation patterns, especially subtropical high-pressure belts. Another mechanism may involve increased infrared emission from the Moon when it is full. Modulation of Earth's weather by the Moon and Sun have therefore been studied to some extent and, while elusive, are assumed to have a physical basis that will eventually be worked out in more detail. The influence of planets on the weather is a much longer stretch, and yet knowledge and application of their real or imagined effects was central to the doctrines of astrometeorology in the late Middle Ages and the Renaissance, and also the methodologies of Digges, Kepler, and other almanac writers, and of course, John Goad.

ASTROMETEOROLOGY REVIEW

From Renaissance times through the nineteenth century, annual farmer's almanacs published in England and New England typically contained weather forecasts based on astrometeorology.[10] For the most part, the methodology used in making almanac weather predictions was originally based on geocentric planetary alignments, or the aspects that Kepler and Goad used, though in more recent years other approaches have been used with mixed results.[11] What aspects (the angular separation of planets) actually measure is phase angle, the temporal and spatial distance into any given synodic (two body) cycle. The first quarter of the Moon is the square aspect between Sun and Moon (when they are 90 degrees apart) which occurs about seven days after the new Moon. Likewise, a first quarter square between Mars and Saturn marks the 90-degree angle between the two bodies in the Mars-Saturn synodic cycle. The next paragraphs continue this descriptive summary in order to lay a foundation for the following chapter.

In astrometeorology aspects are described in various ways. First are the different "families" of aspects, which are those based on division of the 360 degrees of the circle by two, three, four, five, six, and so on. Second, there is what is called the "orb of influence," which attempts to measure the relative strength of an aspect as a curve over the point of exactness (partile). Orbs are normally given as a range in degrees along the ecliptic (zodiac) and may take into account whether the aspect is forming (applying) or dispersing (separating), the former generally con-

---

10. The *Old Farmer's Almanac* published out of Dublin, New Hampshire, has been using a combination of a "secret" formula presumably based on the exact time of a lunation (full or new Moon) and also data based on predicted solar activity. The accuracy of the extended weather forecasts is claimed to be 89 percent, though studies point to rates well below 50 percent.

11. A call to the editor's office of the *Old Farmer's Almanac* inquiring as to how weather forecasts were made elicited an astonishingly hostile response and very little information. From what I could gather they were using averages over 30-year periods for long-range forecasts that incorporated any variations induced by the sunspot cycle for their forecasting section, and the secret formula for the monthly almanac pages—but this is a guess on my part.

sidered, based on observations over many centuries, much stronger than the latter (Dean and Mathers 1977, 284). Aspects are measured in celestial longitude (which could be seen as a measurement on the x-axis of a graph) and are considered to be stronger if the bodies forming them are simultaneously at the same celestial latitude, which would be their location north or south of the ecliptic and therefore a kind of y-axis measurement. Declination, distance north or south of the celestial equator, is also noteworthy as planets having the same declination will rise and set at the same points on the horizon and have the same diurnal arc (i.e., spend the same amount of time above the horizon). Planets with a declination of zero will be located on the celestial equator (and will rise due east and set due west), and the declination of two planets equidistant from the solstices (the solstice will mark the midpoint between the planets) will be the same or very close, this being the basis of the doctrine of symmetry in Hellenistic astrology.

The astrological aspects could be thought of as specific phases in a cycle, or in terms of harmonics, which is how Kepler handled them. Division of a cycle by a specific integer determines the number of waves that occur within that cycle length. For example, division of 360 by four produces four waves spaced 90 degrees apart. The subject of aspects seen as harmonics, in the context of Platonism, or in terms of wave theory, is complex and involves many details that have been discussed elsewhere.[12] Here, I wish to limit the definition of aspects to specific angular separations (Ptolemaic and lower harmonics) along the ecliptic (celestial longitude). Other factors such as distance north or south of the ecliptic (celestial latitude) and distance north or south of the equator (declination) serve to amplify or weaken aspects.

The astrological doctrine of aspects is not so far-fetched. There are many examples in nature where pure geometry forms a basis for certain types of physical phenomena. A magnetic field propagates at exactly 90 degrees to an electrical current. Crystal symmetry, where an extension of the geometric arrangement of atoms is the basis of axis geometry,

---

12. The geometry of the aspects was seen as validation of Plato's forms by many Renaissance astrologers (Dean and Mathers 1977, 277–370).

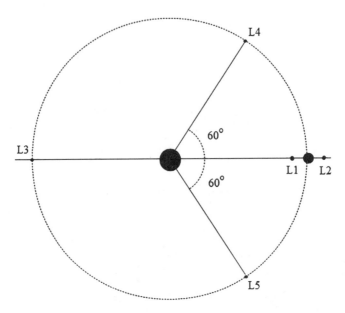

Figure 9. Around the central body (e.g., Sun, planet) five Lagrange points relative to an orbiting body are depicted. Points L1 and L2 are on opposite sides of the orbiting body (opposition) and opposite point L3. Points L4 and L5 are 60 degrees (sextile) from the axis established by points L1, L2, the orbiting body, the central body, and L3. Relative to the central body points L4 and L5 are spaced 120 degrees apart (trine).

mostly involves angles (or rotations) of 180, 120, 90, and 60 degrees. An example of pure geometry in physics is found in the angle and strength relationships of a radiation pattern, something used in antenna design, or in phase difference between oscillating waves. In these cases, angles of 180 and 90 degrees have special properties. Another example, this one on a solar system level, are the five Lagrange points between two bodies that are named for the mathematician and astronomer Giuseppe Luigi Lagrange (1736–1813). These "balance" points locate where the combined gravitational forces of two large masses (the Sun and a planet) balance the centrifugal forces on a smaller mass (e.g., satellite or asteroid). There are five Lagrange points at angles of 180 and 60 degrees (see diagram) that are used when placing a satellite in a stationary orbit. The stable Trojan asteroids on Jupiter's orbit and equidistant from the planet are located at Lagrange points.

5

A Signal from Saturn

Saturn has a higher degree of its quality in cooling and slightly drying because it is the most distant, so it seems, from the warmth of the Sun and, at the same time, from the exhalation of moisture around the earth. But powers for this star and for the remainder establish themselves through careful observation of their figurations [aspects] to the Sun and to the Moon.

PTOLEMY, *TETRABIBLOS*. SCHMIDT,
TRANS. 1996. BOOK I:4

The belief still to be found in all countries that the planets and the moon do affect the weather never had any scientific basis whatever; it is only a remnant of the many superstitions generated and fostered by that other greater superstition, astrology.

ELGIN, ILLINOIS DAIRY REPORT, 1914

In traditional farmer's almanacs, long-range weather forecasts were based on the aspects between the planets, Sun, and Moon, geo-centrically, and were associated with meteorological phenomena including heat, cold, rain, winds, storms, and the like. This is basic astrometeorology. The question *Does any of this really work?* is obvious, but what's not so simple is how to go about testing a historic qualitative methodology like this one in some meaningful way. The larger part of this problem is that astrological weather forecasting is based on a synthesis of many factors, some increasing in power and others decreasing, all relating to each other by phase angles that continuously change, and all in a context that never repeats in exactly the same way. It is the ability to work with discovered patterns and, from them, make a series of accurate estimates that constitutes the interpretive methodology central to astrometeorology. Isolating a single factor to test, which is exactly how modern reductionist science works, presents serious problems in study design. This is because weather is a system in motion and, like other systems, it resists scientific methodologies that focus only on the parts, one at a time. But apparently, as Bradley and others have demonstrated in reputable and replicated scientific studies, this problem can be overcome to some extent with proper data selection and the right experimental methods. Because such studies have been done, we can today be reasonably confident that the Moon does have an influence on weather, some of which is in agreement with what Goad found in the mid-seventeenth century.

Johannes Kepler had some ideas about how to deal with the problems of testing astrology. In his book *Tertius Interveniens* he set himself up as "the third man in the middle" between an astrological healer, Doctor Helisaeus Roeslinus, and a critic of astrology, Doctor Phillippus Feselius, both well-known physicians of his time. Kepler had disagreements with both but says he wrote the book because Feselius's attacks might sway a ruler to prohibit astrology, which (and this is part of the subtitle of his book) would be like "throwing out the baby with the bathwater." Throughout *Tertius Interveniens* Kepler puts medicine, which he regards as frequently unreliable, on trial as much as astrology. One theme emerges from the many defenses of astrology Kepler puts up and that is the importance of the aspects:

". . . I could just as easily criticize his [Dr. Feselius] medicine and cast suspicion on it, as easily as he now with such arrogant and idiotic indiscretion can deny astrologers their experiments with aspects and completely reject them?" (Kepler 2008, T-133, 195).

This is immediately followed by a section of his weather diary which is basically a demonstration of how an astrologer can carefully record weather data at the time of a specific aspect, in this case the annual Sun-Saturn conjunction, to actually learn something about it. A comparison of his collected data with the aspects leads him to make a couple of points. One is that this is the proper way to do scientific astrometeorology. You don't make a general statement about the meteorological effects of an aspect between two planets simply based on one example because there are factors other than astrology that come into play, such as the seasons, though competent astrologers knew that already. The second point is that completely separating the Sun-Saturn conjunction from the other aspects operative at the same time is nearly impossible, and consequently any effect it produces must be regarded as general. Then he goes on to bash astrologers who define the effects of planets by their zodiacal rulerships, which he regards as fictional, or make forecasts based on the entrance of the Sun into the equinoctial and solstice signs, a Ptolemaic and Arabic technique that can result in serious errors if planetary tables are unreliable—which they were.

Kepler's challenge to the doctors got me thinking about actually testing Sun-Saturn aspects myself. In times where no instruments were available to quantify weather data, astrometeorologists had to resort to descriptive language to designate the effect of an aspect. For example, in the astrological literature the aspects between the Sun and Saturn were an established family of alignments for predicting mostly unpleasant, but more often cold, weather. The following are quotes from leading writers on astrometeorology since the Renaissance in regard to Sun–Saturn alignments:

Johannes Schöner (1477–1547) was a German mathematician, astrologer, astronomer, cartographer, geographer, and scientific instrument maker, among other things. Schöner played a role in the Copernican Revolution, as he was the one who in 1538 convinced Georg Joachim Rheticus to visit Copernicus. Rheticus later dedicated the first published work of Copernicus, a short summary of his ideas that informed the rest of Europe on the heliocentric hypothesis, to Schöner.

Sun to Saturn: Spring = rain and cold, Summer = great thunder, Autumn = cold and frosty, Winter = foggy or snowy (Schoener 1994, 11).

Gerolamo Cardano (1501–1576), a classic Renaissance man who was a doctor, mathematician, biologist, chemist, astronomer, astrologer, inventor, and more, is known as one of the founders of probability theory, introduced binomial coefficients, and was a developer of algebra. Kepler regarded him as a mediocre astrologer, however, because in his aphorisms on astrology he made claims based on single examples. Regarding the Sun and Saturn, Cardano wrote:

Whenever Saturn is joined to the Sun the heat is remitted and the cold increased, which alone may be a sufficient testimony of the truth of astrology (Cardan 1970, 88).

Leonarde Digges, discussed earlier, produced an annual almanac with instructions for doing your own weather forecasting with the aspects of the year.

The conjunction, quadrature, or opposition of Saturn with the Sunne, chiefly in colde signs; snow, dark weather, haile, rayne, thunder and cold days (Digges 1605, 9).

Johannes Kepler described this aspect as follows:

. . . [T]he effect of this [Sun-Saturn] conjunction is quite general and gives nature at least an opportunity to cause turbulence in the air . . . this purifies the air, brings freezes, snow and rain. [Astrologers] . . . observe when Saturn stands opposite the Sun in the summertime, when no other planet is aspecting the Sun, and observe that the weather is cool and rainy (Kepler 2008, T-45, 106; Kepler 2008, T-135, 199).

John Goad was more specific:

[Saturn and Sun] produce cold and frost and misty weather, clouds and dark air with snow (Goad 1686, 275).

Thomas Wilsford, an English writer on astrology, wrote in his 1655 book *Nature's Secrets:*

Saturn and Sun, in conjunction, square, or opposition do cause generally rain, hail, and cold weather, both before and after, especially in the water signs, or in Sagittarius or Capricorn, and is called Apertio Potarum, or opening the Cataracts of Heaven. Particularly their effects in spring are cold showers; in summer producing much thunder and storms of hail, in autumn rain and cold, in winter snow or moist, dark, and cloudy weather, and oftentimes frost (Wilsford, 1665, 100).

Ebenezer Sibly (1752–1799), an English physician, astrologer, and occultist, was also a mason and is known for his publication of an astrological chart calculated for the Declaration of Independence not long after the event.

Saturn and the Sun in conjunction, quartile, or opposition, is Apertio Potarum, especially if it happens in a moist constellation; for then, in the spring time, it threatens dark and heavy clouds; in summer, hail, thunder, and remission of heat; in autumn, rain and cold; in winter, frost, and cloudy weather (Sibley 1798, 1026).

Alfred John Pearce (1840–1923) is another English astrologer who wrote a comprehensive textbook on the subject that is surprisingly uncontaminated with occultism or theosophy. In his section on astro-meteorology he offered a bit more in the way of meteorological explanations and included dated observations of weather events:

Saturn's action, when configured with the Sun, is to condense aqueous vapour, to lower the temperature of the air, and to excite tempests. When the atmosphere happens to be quiescent under Saturn's ascendancy, it is often dark and foggy. When Saturn crosses the equator, the atmosphere is greatly disturbed and such effects last for several months (Pearce 1970, 360).

The above statements in regard to Sun and Saturn are not completely consistent, but a common theme of coolness is apparent in phrases such as "cold increased," "remission of heat," and "bitter frost." Below is a table organizing the various descriptive words of the above authors, which shows that all do agree on an association of cold with the Sun-Saturn aspects. Doing reductionist science on a complex system such as the weather presents many problems, the greatest of which is to isolate the variable to be tested. In my study, described in the next section, only temperature will be investigated, although the table makes it obvious that this is not the only factor thought by these classical astrometeorologists to be brought about by the Sun and Saturn's geocentric alignments.

HISTORICAL DESCRIPTORS FOR SUN-SATURN ASPECTS

Source	Cold	Frost	Snow	Hail	Rain	Thunder	Fog	Dark Clouds
Schöner	X	X	X		X	X	X	
Digges	X			X	X	X		X
Kepler	X	X	X		X			
Goad	X	X			X			
Wilsford	X	X	X	X	X	X		X
Sibley	X	X	X	X	X	X		X
Pearce	X						X	X

In the final sections of *Tertius Interveniens,* Kepler brought his arguments in favor of astrology to a climax by presenting a seventeen-year diary documenting the weather around the time of Sun-Saturn conjunctions (Kepler 2007, T-134, 196). This section appears below in its entirety and is an example of how he thought about the simultaneity of multiple aspects. Note that Kepler refers to zodiacal signs in his diary, but he only used them for positioning planets, not as factors for interpretation. Also notice that the conjunctions, which average 378 days between them, occur only once a year and each time later in the year, and that his diary did not account for the full Saturn cycle of 29 years.

As concerns the aphorism that the conjunction of Saturn and the Sun in Capricorn and Aquarius should cause cold weather—on which Kepler's weather diary location, as noted on page 169 the astrologers rely, and about which Dr. Feselius says that they are stuck in it in a graceless manner—I want to make a whole philosophical process out of it. First I shall compile the weather data of this conjunction, as far as my observations go.

1592, July 9 (New Style)—I had not yet begun to list data with Cancer. Chytraeus, however, writes that the whole summer, especially around this time, was cold and wintry.

1593, July 24, in the beginning of Leo. There was a myriad of aspects. Sun, Venus, Saturn were in conjunction; Mars in sextile to Jupiter and in addition, Mercury was separating from an opposition to Jupiter and moving to a trine with Mars. The 20th, 21st, and 22nd much rain, hale, changeable. The 23rd cloudy, the 24th there was fog a day or more in a row; overcast, then warm. This [was recorded] in Tuebingen.

1594, August 7th and 8th, there was much rain around Raab; I lost my data for this year.

1595, 21st & 25th of August, at the end of Leo, in Graetz in Steiermark: thunder the whole night, hail stones, one day before and after sultry weather; cloudy.

1596, 4th of September in Virgo; cold rain.

1597, 18th of September. Again a great myriad of aspects. Saturn, Sun and Mercury made three conjunctions, and all three running in square to Mars. Then after several days there came rainy weather on the 13th, very cold air, became cold and overcast the 14th, 15th & 16th, 17th somewhat warmer, often drizzled, the 18th cold rainy air, Sun pale, 19th pleasant, 20th April weather the whole day, etc.

1598, 1st October in Libra. It rained heavily, also the whole week before, then at the same time a conjunction of Mars and Mercury, along with a lengthy sextile of Mars and Venus.

1599, 13th and 14th October in Libra. On the 12th, rain, cold. The 13th overcast, cold. The 14th cold, Sunshine. From that time on the Sun and Moon appeared red through a heavy, smoke-like, low-lying material, in such a way that the high mountain tops protruded from it, as from fog. This became a general condition.

1600, the 24th and 25th of October in the beginning of Scorpio, in Prague. The 24th, rain and Sunshine. The 25th, cold wind, freezing; the freeze lasted almost to the end of the month.

1601, the 5th & 6th of November, conjunction of Saturn, Sun and Mercury, the 1st wintry cold, 2nd strong wind, 3rd & 4th snow, 5th and 6th rain.

1602, the 17th of November at the end of Scorpio. The 16th fog, overcast. 17th fog, cold, then pleasant. 18th wintry cold, pleasant, because of a cold wind.

1603, the 29th of November, in Sagittarius. Then a transit of Sun to Jupiter to Saturn, with Venus present. Until the 27th it was mild; then wind began, on the 28th there was a freeze, from a southeast wind. Afternoon a thaw, the 29th it froze again, wind and rain in the evening, similarly on the 30th.

1604, the 8th and 9th of December; the 7th 8th and 9th cold air, bringing a freeze. At the same time a sextile of Jupiter and Venus, therefore on the 10th and 11th replaced with fog.

1605, the 20th and 21st of December at the end of Sagittarius. The 19th, 20th 21st and 22nd there was cold air, a deep freeze and pleasant weather. Before and after because of aspects of Mercury it became mild and wet.

1606, end of December and beginning of January 1607, in the beginning of Capricorn a conjunction of Saturn and Sun, also in sextile with Mars. The 30th and 31st of December heavy rain. On January 1st and 2nd, snow and rain heavy.

1608, the 12th of January an even greater myriad of aspects, then Saturn, Sun and Mercury running in sextile with Mars. On the 11th it began to thaw after a long cold period, hail, west wind, on the 12th and 13th snow flurries and strong west wind blew down walls.

1609, 22nd and 23rd of January in the beginning of Aquarius. Before it was a trine of Jupiter and Mercury, after it a semisextile of Saturn and Venus. The 19th rain, the 20th overcast and colder, the 21st frozen and snow, the 22nd snow and cold, the 23rd cold air and pleasant, the 24th change to rain (Kepler 2008, T-134, 196–198).

While Kepler used this qualitative data as evidence of astrology for the doctor he was arguing with, he also speculated on why there is such an effect. He thought the conjunction actually causes atmospheric turbulence through a kind of warming, which purifies the air in winter, bringing freezes and snow made from existing moisture. In summer the aspect brings rains. He was clear that Saturn was not operating alone—he stated that no single aspect rules alone over the weather and every listing in his record notes the connections with other planets, or other simultaneous aspects. Kepler is careful to say very little about what Sun-Saturn conjunctions do, except that they appear to cause turbulence, but this varies depending on the season. He then chastised the astrologers' opinions, which he says are based not on data, and he attacks their methodology, that is astrological charts calculated for the passage of the Sun over the equinox and solstice points (called ingresses), which would require extremely precise tables of planetary positions. Finally, he states that Sun-Saturn conjunctions are not really that potent, unless they share the same position in celestial longitude and latitude. The frustration Kepler must have felt about testing astrology, a nearly impossible project, along with his disgust for the incompetence of other astrologers, including Cardano who he describes as full of ill-conceived notions, frame his thoughts on the subject, but he never questioned its efficacy (Goad 1686, 2).

A half century later John Goad presented sections of his weather diary in his book *The Astro-Meteorologica* in order to illustrate the effects of the individual planetary aspects. For the Sun and Saturn he lists data only for the conjunctions, which occur once a year, though he reports on the other major aspects. His diary listed the day of the month followed by a qualitative description of the weather including wind direction; the portion published for the Sun-Saturn conjunction begins nine days before and ends fourteen days after the event. Here, below, is his daily record for the Sun-Saturn conjunction of October 25, 1660, (in bold) which runs from October 16 to November 8 (Goad 1686, 277).

16. Close m. p. coasting showr some places S.W.

17. Rain a.l. fiar, somet, overcast. Nly.

18. Fair, some clouds. N.W.

19. Fair, fr. Overc. 10m Nly. Mist below. N.W.

20. Fr. Fog N W, at 0.E. clear p.m. N. E.

21. Frost, black thick clouds in S. Sun occ., clear and fair. E.N.

22. Frost, clear, some wind. N.E.

23. Cloudy, windy, Nly, fiar 9 m. N.

24. Fr. Fair, windy. S.W.

25. Fr. Cold, windy, cloudy; frequent clouds in S.SW. NE.

26. Fr. Clouds curdled, close day. W.

27. Dry, cold wdy, Hail and R. 1 p. a shower 3 p.

28. Rain after mids. Cloudy.

29. Fr. Curdled clouds. N.

30. Fr. Fair; Venus seen half an hour after Sun.

31. Fr. Mist below, about Horizon; some rain, close and most even. W.

1. Close, cloudy, windy, dry yet threatening. W.

2. Fr. Venus seen half an hour after Sun rising. N.W.

3. Mist, some clouds even incling to moisture. S.W.

4. Close and cloudy. W.

5. Fog below, fleecy clouds. S.W.

6. Fair, windy. N.

7. Open, windy, storm of rain 11 m. S.E.

8. Fr. And fair; freeze hard at n. W.

These two diaries are historic examples of early attempts to constrain the alleged effects of planets on the weather; attempts to make a science out of astrometeorology. Not being convinced that these obviously learned and scientific astrologers were purposely perpetuating some ancient and remarkably stable fantasies of the imagination, I decided to do my own investigation. This project began when I was a teaching assistant for a course on evolution at the University of Massachusetts and working on my doctorate. Each semester our class would take an overnight trip to Harvard Forest, a research facility of Harvard University located in central Massachusetts, where a wide range of environmental data, including weather records, are collected. With these records readily available, I came up with a way to test the astrological hypothesis that there is a correlation, as shown in daily temperature records, between cooling trends and the geocentric alignments of the Sun and the planet Saturn, at least in central Massachusetts. The Shaler daily time series from Harvard Forest that I used contains the daily minimum, maximum, and mean temperature and the precipitation for each day of the thirty-nine-year period from 1964 to 2002 (Boose and Gould 1999). I downloaded the timeseries from the Harvard Forest website, opened it in a spreadsheet software called Open Office Calc, and began my investigation, something I had no previous experience with.

My pilot study was to test all the Ptolemaic Sun-Earth-Saturn alignments used in traditional almanac weather forecasting in the manner established by Kepler and Goad over the course of the Sun-Saturn synodic cycle. (Recall that a synodic cycle is one in which two bodies are perceived to be moving relative to each other, the best-known example being the lunar synodic cycle of new Moon, first quarter, full Moon, and third quarter.) The average rate of the Sun's motion, as seen geocentrically, is 59 minutes and 8 seconds, as arc is often measured, or 0.9855 degrees per day. Saturn's motion varies widely due to retrogradation, but its average forward daily motion in longitude is 0.0335 degrees per day, roughly 1 degree per month or 2 minutes of arc each day. It therefore takes the Sun extra time, and extra distance, to catch up with Saturn each year in order to form the various aspects as its approximately 378-day average synodic cycle unfolds (it varies from 376 to 383 days due to orbital ellipticity). Since the synodic cycle of Sun and Saturn is longer than the solar year, conjunctions and the other aspects move roughly two weeks ahead each year, making a complete cycle through the zodiac in about 29 years, which is how long it takes Saturn to orbit around the Sun. In the 39-year Shaler time series from Harvard Forest there are 38 (n = 38) of each type of aspect, about 1.3 cycles of Saturn.

I began working with the data in the time series by taking the date of each aspect (the twenty-four-hour period within which the exact aspect formed), beginning with the conjunction, as the mid-point of a fifteen-day sample of daily mean temperatures. The day of the aspect would then be day zero, seven days before would be day minus seven days and ten days later day plus seven days. I then stacked these and calculated a mean of all thirty-eight conjunctions and the seven days preceding and following, basically a composite snapshot of mean temperatures closely centered on the aspect. I next graphed the results and continued with the same methodology for the other Ptolemaic aspects in the Sun-Saturn synodic cycle in this order: sextile (60 degrees), square (90 degrees), trine (120 degrees), opposition (180 degrees), trine (240 degrees), square (270 degrees), and sextile (300 degrees; see figure). Notice that 90 degrees is the first square that forms as Earth moves ahead in its orbit, which appears to

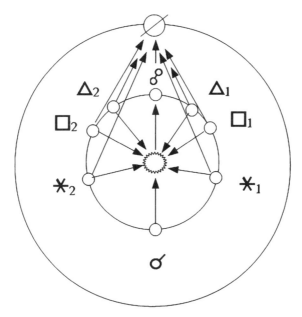

Figure 10. Here, depicted from an overhead view, the synodic cycle of
the Sun and Saturn is shown with the aspects unfolding counterclockwise.
The Sun is at center, the Earth is shown on its orbit around the Sun, and
Saturn is shown on its more distant orbit. The positions of Earth along
its orbit each have arrows that point to both the Sun and Saturn, the
angle formed between these vectors being the Ptolemaic aspects. The
conjunction symbol, at the six o'clock position, shows the syzygy of Saturn-
Sun-Earth. Moving counter-clockwise, the next symbol (asterisk) is that
of the opening sextile. From this geocentric position, Saturn and the Sun
are seen as 60 degrees apart. Next is the square where the arc opening
between Sun and Saturn is 90 degrees and then the trine of 120 degrees.
At the opposition is the other syzygy of Saturn-Earth-Sun. This sequence
of precisely defined phases is recapitulated in reverse over the second
part of the synodic cycle. (Saturn is shown stationary here, but over the
course of one Earth year Saturn will have moved ahead a distance of about
12 degrees in its orbit around the Sun.)

a terrestrial observer as the Sun moving in advance of Saturn as measured in celestial longitude. A second square occurs when the Sun
and Saturn are spaced at 270 degrees, these two aspects being like
the Moon's first quarter (waxing) and third quarter (waning). This
distinction is applied to the two sextiles and trines also.

As can be seen in the figure above, the Sun-Saturn conjunction places Saturn behind the Sun. The alignment is a syzygy, Earth-Sun-Saturn, from a geocentric perspective, of course (but it is also a syzygy heliocentrically as well). This is the aspect that Kepler recorded for seventeen years and commented on, and also one that Goad included his data for in *The Astro-Meteorologica,* examples of both shown previously. Kepler, while acknowledging that often cold followed the conjunction, thought it was actually a warming aspect that disturbed the atmosphere creating conditions for cold to be generated, at least in central Europe (Kepler 2008, T-137, 200).[1] Goad, however, reported weather on or near the Sun-Saturn conjunction in London to be cold, dark, or cloudy most of the time.

As I worked my way through the data, producing graphs for each of the aspects, variations were noted. For some of the aspects the range of the average temperatures was much lower than for others characterized by rapid and steep temperature changes. Some aspects showed warming trends and others pronounced temperature volatility on or near the date of the exact aspect. The sextiles, squares, and trines in the first half of the synodic cycle differed considerably from those of the second half. Of the eight aspect tests I had done, the one that best displayed lower temperatures at the time of exact aspect was the opposition, the aspect that occurs when Saturn is closest to Earth. Goad had also noted that the opposition was colder than the conjunction (Goad 1686, 285). Consider that the Sun-Saturn opposition occurs when Saturn is closest to the Earth, about 190 days after conjunction, when Saturn is at perigee, and the alignment is also a syzygy. The graph I produced for this aspect from the stacked daily mean of its thirty-eight occurrences showed an abrupt drop in temperature on minus two days and then a low the following day, and remaining low on day zero. My thinking was, if there was anything to this at all, this might be a possible confirmation of

---

1. Interestingly, in central Massachusetts, my results did show an increase in mean temperature peaking at the time of the conjunction and then dropping by over 3 degrees six days later.

the traditional rule in astrology that distinguishes between applying and separating aspects, the former thought to be stronger or more pronounced. From day zero to plus three days, as the aspect separated, the temperature rose nearly 3 degrees.

I decided at this point to focus almost entirely on the Sun-Saturn opposition and get serious about my methodology. The first order of business was to prepare the temperature time series properly. Understanding that my method was simply measuring whatever temperatures were recorded throughout the season, and that my sample was weighted with data from the spring season because the length of the dataset covered about 1.3 cycles of Saturn, I prepared the time series so as to eliminate this bias. Working with my spreadsheet software I first stacked the individual years of daily mean temperatures in consecutive columns and computed an annual daily mean for the entire period. Next, I pasted this thirty-nine-year mean top to bottom onto itself thirty-nine times in one column, and in another column I pasted the continuous list of daily temperatures. By subtracting the mean from the actual data, I produced a new time series of daily temperatures that were either above, below, or at the mean—this result is called the anomaly from the mean. Seasonality was thus canceled out and the dataset mean appeared on the graph y-axis as line zero. In addition, I calculated the standard deviation from the mean for the full dataset, which was 4.4 degrees, to get a sense of what the range of temperature variation, or amplitude, was like in that weather record. I then reworked the calculations for the aspects and obtained results that were very similar to my pilot study: the Sun-Saturn opposition again stood out from the others with a temperature drop near day zero. From here on all my work was done with daily "anomaly from the mean" temperature data. Below on page 182 is the graph for the opposition between Sun and Saturn, based on the revised time series, showing a drop below the mean (zero) near the exact syzygy.

While the results show a temperature drop roughly symmetrical to the center of the sample, amplitude is another matter. In the graph, the temperature on minus one day was just under 1.5 degrees and the total amplitude of all the temperature points in the sample, about 3 degrees,

Sun-Saturn oppositions 1964-2002

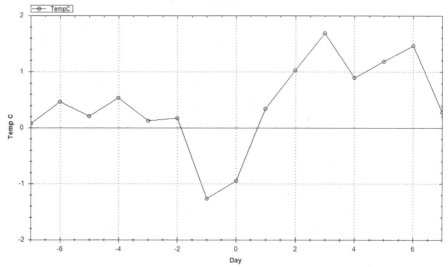

Figure 11. Sun-Saturn oppositions (n = 38) stacked, averaged, and centered on a fifteen-day period from Central Massachusetts 1964–2002. Only two dates, minus one day and day zero, fall below the mean, which is zero on the y-axis.

vary from the mean by less than one standard deviation (of the entire dataset), indicating this is a minor fluctuation. The daily temperature data itself, even after reducing it to the anomaly from the mean, is quite noisy with rapid ups and downs around the mean every few days, something that might be expected of New England weather, which is a zone where polar, southern, and maritime air masses collide. I did note that the temperature on the exact date of the Sun-Saturn opposition was below the mean 60 percent of the time.

The next step was to see if my results for the Sun-Saturn opposition could be reproduced with controls. The first type of control used an online random number generator, which I set for a range from 1 to 365 for each year of the dataset. The randomly generated dates were then used as the date of a hypothetical aspect, and strips were taken from the time series and analyzed consistent with the methodology I was using. After assembling the control samples, each consisting of randomly generated dates, a number of very different graphs were produced, none with dip at day zero. The amplitude of the control mean temperatures was similar to that in the Sun-Saturn opposition test, however.

A second method for control data was tried, this time using randomly selected dates within the range of the full dataset. Some years had two or more dates; other years had none. A third type of control used the exact dates of the Sun-Saturn oppositions between 1964 and 2002 (day zero) and added ten days, making that day zero on the graph. In both cases the controls produced a curve different from the one produced by the actual data—they lacked any emphasis on, or close to, day zero. A fourth type of control for my study was to focus on the geocentric oppositions between the Sun and Jupiter. In this control, using positional data for the largest planet in the solar system, no drop in temperature was found. Instead, a general warming trend was shown. These four procedures became my standard method of creating controls. In this graph below only four are shown, but many more were created.

In order to see if there were internal consistencies in both the curves for the study and the controls, I first took the data for the Sun-Saturn aspect and divided it into two sections, the first and second halves of the period, chronologically, with each one containing nineteen years. A first graph was produced based on the stacked mean of aspect dates between 1964 and 1983 and a second for 1984 to 2002. Each of the two graphs had a low on minus one day and displayed roughly the same curve as shown in the full thirty-eight-year period. Applying the same procedure to each of the controls produced two entirely different curves in every case. Next, I took both actual and control data and again divided them into two sets, but this time made up of alternating years. Again, the Sun-Saturn oppositions showed a matching pattern; the controls didn't. One way to put numbers on my results (the goal of any reductionist science project) was to calculate the correlation coefficient (also known as r), which describes the relationship between two samples in terms of positive similarity (+1) and negative similarity (-1), with 0 indicating no similarity. No significant positive correlation was found between the sample and the controls. The correlation between the two parts of the split Sun-Saturn sample was high, especially when the range was constrained to five days

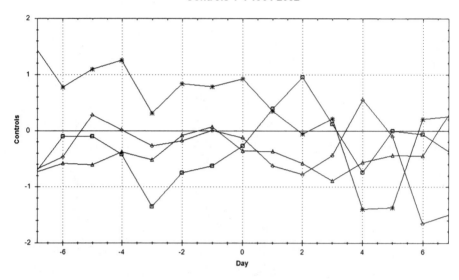

Figure 12. Control data based on randomly generated dates. These are from dates in all years (shown as squares and triangles), dates drawn randomly from all years (diamonds), and dates ten days following Sun-Saturn oppositions (stars).

before and after day zero. This was not the case with the split controls nor was it with the controls by alternating years. Overall, the actual sample had far more consistency and higher correlation coefficients, exactly what the controls lacked (Scofield 2013).

As already noted multiple times, historical writers on astrometeorology emphatically stated that the effect of one planetary pair cannot be isolated from the constantly changing configurations of the other planets, Sun, and Moon relative to the Earth. The weather is a chaotic system, and any single part that is isolated and studied out of context is not likely to say much about that of which it is a part. Given this admonition from the astrologers, and the realities of testing a complex system, which alone suggests separating the noise from the signal would not be easy, I considered some other possible approaches to the problem. My thinking about possible causes for the "effect" that I was attempting to isolate from the background noise was that gravity might be the force modulating the atmosphere at these times—atmospheric tides. It is the case that Saturn at opposition to the Sun occurs when Saturn is closest to Earth. Remember that complex and chaotic systems, climate and weather, are known to be responsive to weak signals, known as the butterfly effect, so it seemed to me that the additional mass of the Moon or possibly another planet might show in the data as an amplifier of the signal. In the thirty-nine-year database there were six Sun-Saturn conjunctions and eight Sun-Saturn oppositions that occurred within two days of a full or new Moon. On average the Moon moves about 13.2 degrees of celestial longitude per day, so the sample was limited to dates with a maximum Moon-Saturn distance of 26.4 degrees before or after the Sun-Saturn conjunction or opposition. Using these fourteen events, and using twenty-one-day samples centered on the Sun-Saturn aspect at day zero, I stacked them and computed the average. This small sample produced an interesting graph that showed temperatures mostly below the mean for the entire period and dipping to a low on day zero. The amplitude of the data is about 2.5 degrees.

Planetary positions are normally measured in the ecliptic coordinate system, specifically in celestial longitude along the zodiac, which is the astronomical positioning examined so far in this study. Celestial longitude could then be seen as an x-axis coordinate. But planets are also located in space by a vertical, or y-axis, component, and in the case of the ecliptic coordinate system, this would be celestial latitude. Because the planets orbit in a plane around the Sun, differences in celestial latitude between planets are small. If one uses the equatorial coordinate system, however, planetary positions are given in right ascension and declination, which are also analogous to longitude and latitude, though the first is specified in units of time as this coordinate system clocks the rotation of Earth. When two planets have the same declination, they are equidistant from the celestial equator and on the same side of it (north or south). In astrology they are said to be in parallel. (The contraparallel is formed when two planets are equidistant from the equator but on opposite sides.) The planets that orbit close to the plane of the Earth's orbit around the Sun will have their lowest declination when they are near the equinoxes, these being the intersection of the ecliptic and the celestial equator; the one in spring

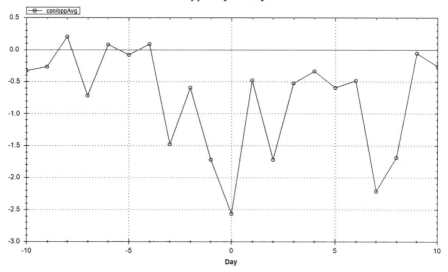

Figure 13. The six Sun-Saturn conjunctions and eight Sun-Saturn oppositions (n = 14) that occurred within two days of a full or new Moon in the Shaler 1964–2002 dataset. Nearly all data points are below the mean.

in the northern hemisphere is called the vernal point. The passage of the Sun through these points was considered significant by Ptolemy, Al-Kindi, and most scientific astrologers in history, and their weather forecasting techniques were based on astrological charts calculated for solar equinox passages when the Sun's declination is zero and therefore simultaneously on both the ecliptic and equator. (These were the charts Kepler was skeptical of because existing tables could not be relied on to deliver the accuracy required.) Given that declination holds an important place in traditional astrometeorology, I thought it worth investigating.

I decided to first stack temperature samples centered on dates when the Sun and Saturn occupied the same declination, that is when they were in parallel, and also when they were near the equinoxes. Only seven events of this type occurred when the Sun and Saturn were within a range of 3 degrees from either equinox during the years of my dataset. These seven events, when graphed, show a low on day zero that is statistically significant, well over one standard deviation (calculated

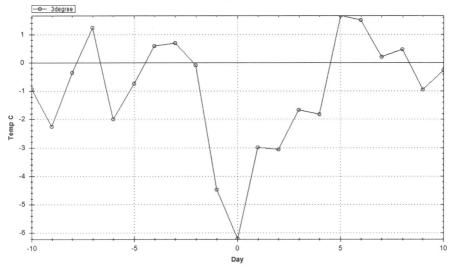

Figure 14. Sun–Saturn parallel in declination within a range of 3 degrees from either equinox (n = 7). The low on day zero is well over one standard deviation (calculated for each individual day) from the mean.

for each individual day) from the mean. Although sample size is low, the effects are impressive.[2]

One of the data points used in the above graph was for March 18, 1967, when the Sun and Saturn were both located within three degrees celestial longitude of the vernal equinox and at exactly the same declination, one degree south of the ecliptic. A graph for this single event shows a very high amplitude with temperatures dropping almost eighteen degrees to a statistically significant low on day zero. Given this extreme result, I examined both the surface and the five hundred millibar troposphere data on a daily basis from minus six days to plus six days using maps from the online NCEP/NCAR Reanalysis, a project of the National Oceanic

---

2. The three central days (minus one, zero, and plus one), which in the data amounts to twenty-one points, can be shown, using an analysis of variance (ANOVA), to be significantly different from the other points (grouped in threes) in the figure. In the controls that were generated for this test none produced a segment that was this far from the norm.

and Atmospheric Administration's Physical Sciences Laboratory. The surface maps show a compact cold cell between Greenland and Baffin Island that began to move south on minus day six and broke down by day zero with its cold air then spilling south into the northeastern United States. The five hundred millibar maps show this cell, much enlarged, and of course at a high altitude, stationary over Baffin Island and Greenland and then moving south abruptly on minus day one. This cell then moved over the northeastern United States on the next day, day zero. Given the configuration of land mass north of New England, this is a typical pattern of air mass movements, more active near the equinoxes in spring and autumn when the temperature gradient between lower and higher latitudes is steep. One example is interesting, but a large number of these events would need to be documented with reanalysis maps in motion in order to take the study further. Unfortunately, such data is limited by the years of instrumental records of this sort.

All of my work so far was focused on one dataset from central Massachusetts. Obviously, weather varies considerably, over longer periods of time and by region, so I next compiled a composite time series of ten weather stations in the mid-Atlantic states and New England for the years 1971–2000. (Thirty-year periods starting with the first year of a decade, like this one, are used routinely in weather and climate studies and are called climate normals.) The anomaly from the mean was prepared for each station and data for the Sun-Saturn oppositions in that period were stacked and averaged. The averages for the twelve stations were then averaged and graphed. The result is a figure almost exactly like the one generated at the beginning of the study showing a drop in temperatures near the center of the twenty-one-day period, but with the lowest point on minus day one. This suggests that the Saturn effect I isolated in data from central Massachusetts is operative over a much larger region.

THE AMHERST COLLEGE
WEATHER STATION DATA

Having worked the Harvard Forest Shaler time series as far as I could, I next expanded my study with a longer daily time series from Amherst

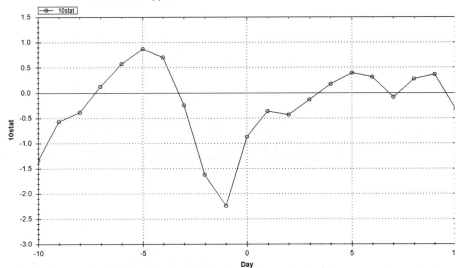

Figure 15. Composite of records from ten northeastern United States weather stations for 1971–2000. These stations are Burlington, Vermont; Cooperstown, New York; Franklin, Pennsylvania; Groton, Connecticut; Hanover, New Hampshire; New Brunswick, New Jersey; Farmington, Maine; West Point, New York; Amherst, Massachusetts; and State College, Pennsylvania.

College, one of the longest in the United States.[3] The location, about twenty miles southwest of Harvard Forest, experiences the same seasonal weather and climate patterns. While the weather records (precipitation) start in 1835, continuous (more or less) maximum and minimum temperatures were first recorded around 1893, which is where I began my new dataset. I prepared the data in several formats. First, a daily mean for each calendar day was calculated for thirty-year periods producing four thirty-year datasets covering 1893 to 2012. I also built

---

3. These were the Amherst College Weather Station Records, 1835–1924 and 1948–present. Missing years supplied by records from the nearby Massachusetts Agricultural Experiment Station also in Amherst. Climatological data was recorded by Professor Ebenezer Strong Snell (Class of 1822) and his daughter Sabra Snell from 1835–1902; and by Dr. Philip Ives (Class of 1932) from 1984–present. Early daily records for precipitation and regular daily temperature data only becomes consistent in the later part of the nineteenth century.

thirty-year datasets in conformity with the standard climate normals (recall these begin in the first year of each decade). Working first with the climate normal of 1971 to 2000, I replicated the tests I had done with the Harvard Forest data. The results were the same with both the opposition and opening square having a drop in temperature near minus one day and day zero.

Next, the Sun-Saturn opposition test was expanded to include the entire 120-year dataset, a total of 116 oppositions (n = 116). Here a low temperature on minus one day is still present, but the amplitude is low. One thing I found was that the Sun-Saturn oppositions that occurred later in the twentieth century, roughly after 1960, showed more amplitude than those that occurred before. I next separated the Sun-Saturn oppositions in the sample that occurred with 15 degrees (roughly a little more than one day) of a lunation (new or full moon), a total of twenty, which were then graphed (see above). The two lows in this sample are at minus one day and day zero. Sun-Saturn oppositions within a day or so of a new or full Moon may be a good way to separate the signal from the noise, but the sample size would need to be much larger and instrumental records are too short. More controls were generated from the 120-year time series and tested, yielding results similar to those of the Harvard Forest controls. Random dates did not produce consistently shaped patterns with a low near day zero, and the data split in two produced completely different patterns with low or negative coefficient correlation. One observation that is consistent with most tests of what might be called the Saturn effect is that the lowest data point frequently occurs before day zero. This, the applying aspect, was traditionally regarded as being more potent than the "separation" of the aspect.[4]

---

4. One interesting finding when using longer sections of data surrounding the date of opposition was steep drops in temperature focused on minus fifteen days and plus fifteen days. These points mark distances of 15 degrees of celestial longitude before and after the exact opposition. John Nelson, RCA's solar storm forecaster, reported that 15-degree increments between the planets (heliocentrically) were found to correlate with storms on the sun and consequently radio transmission disruptions. The 15-degree arc is used by many astrologers in their work today. Note that 15 degrees is one-quarter the sextile of 60 degrees, one-eighth the trine of 120 degrees, and one-twelfth the opposition.

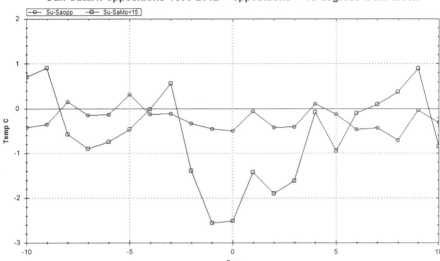

Figure 16. All Sun-Saturn oppositions (n = 116) in Amherst weather data from 1893–2012 (circle) and all oppositions that occurred when the Moon was within 15 degrees either side of the opposition (n = 20, squares). Note that in both samples most data points are below the mean.

One control I used was the substitution of Jupiter for Saturn, that is charting Sun-Jupiter oppositions. None were found with a low on day zero. Unlike nearly all the Saturn tests, which show the majority of data points below the mean, the Jupiter tests displayed temperatures above the mean and often steadily rising. When Sun-Jupiter oppositions were tested against precipitation, either average or total, a spike on day zero was found, and not just in the northeastern United States, but also in England. In figure 17 below, both temperature and precipitation (total) for 120 years of Amherst data are graphed.

DATA FROM EUROPE

Having found that the Sun-Saturn opposition registers as a discernable signal in the data from the northeastern United States, I next turned to Europe and downloaded two long datasets of daily temperatures, one

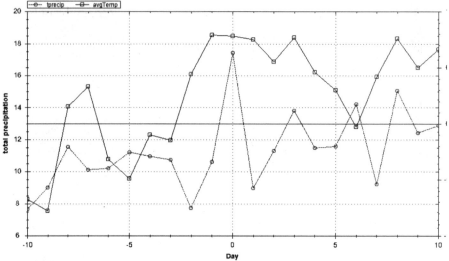

Figure 17. Sun-Jupiter oppositions for 120 years (n = 110) from Amherst College dataset. Temperatures C on right y-axis (squares) during the twenty-one-day stacked sample are above the mean from minus two days through plus six days. Total precipitation in inches on left y-axis (circles) show a spike on day zero.

from England and the other from the Czech Republic. I began with the Hadley Centre Central England Temperature (HadCET) time series, the longest instrumental record of temperature in the world (a source of hard data points that Goad, had he lived for another century and a half, might have used to do his study properly). The HadCET dataset of daily temperatures are representative of a roughly triangular area of the United Kingdom enclosed by Lancashire, London, and Bristol (Parker, Legg, and Folland 1992).[5] In preparing the time series for my studies I began with the climate normal from 1971 to 2000 and found that the standard devi-

---

5. The mean daily temperature data begins in 1772; the mean maximum and minimum daily and monthly data begin in 1878. These data were updated to 1991 (Parker et al. 1992), when they calculated the daily series. Both series are now kept up to date by the Climate Data Monitoring section of the Hadley Centre, Met Office. Since 1974 the data have been adjusted to allow for urban warming.

ation for this period (a measure of variability) was 2.63, half that of New England, which is expected given the difference in the geography of both regions (New England is subjected to westerly continental winds, and the climate of England being moderated by proximity to the sea). I then ran tests for all the Ptolemaic aspects to look for any correspondences with low temperatures and noticed that the Sun-Saturn conjunction showed an increase in temperature, peaking at day zero, while the opposition showed lows at minus four days and plus four days, symmetrical to day zero. Of the other aspects, only the opening sextile (which displayed a drop from minus five days to a low on day zero, followed by an immediate rise to the high of the sample on plus three days) and the opening square showed any marked activity near day zero.

In general, my tests of the Ptolemaic aspects and the parallel were not strongly suggestive of a correlation between Saturn and temperatures. I decided to take a closer look at the Sun-Saturn conjunctions that occurred within one month of the equinoxes, a total of ten conjunctions between 1971 and 2000. In this sample a steep temperature drop occurred on minus four days to a low on plus two days. The entire period was below the mean. Next, I turned to the full dataset (1772–2008) and selected the seventy-seven Sun-Saturn conjunctions that occurred within one month of the equinoxes (see graph below). This data showed a ten-day drop in temperatures reaching a low on day zero, with a second low on plus nine days. The same tests were done with the oppositions of Sun and Saturn. Unlike the central Massachusetts and northeastern United States weather data, my findings are that the opposition in England does not show a pronounced correlation with cold temperatures, but the conjunction, especially when it occurs near the equinoxes, does show a downward trend reaching a low point near day zero.

Next, I moved my focus to the 225-year dataset for Prague, Czech Republic, and prepared the data (Klein 2002).[6] This location was selected as a study site for two reasons: first because it is a very long

---

6. This is the blended series from the location labeled "Czech Republic, Praha" (location ID 24). The higher standard deviation for the period 1971–2000 is probably an effect of global warming.

daily temperature series, almost as long as the HadCET, and secondly because Johannes Kepler spent many years in Prague where he studied the weather and noted correlations with planetary aspects. (Recall that it was Kepler's comments about evidence for astrometeorology to be found in his descriptions of the effects of Sun-Saturn conjunctions that originally got me started on this project.) The Prague data temperature variations were much like the HadCET data—temperatures did not fluctuate much and the standard deviation for the entire dataset is 2.47. With long datasets, the effects of climate change can be seen and the calculation of means will need to consider the length of the test. As was shared earlier, normally meteorologists use a climate normal, which, as previously noted, is what I have done, and did here, when sampling a longer time series. This figure turns out to be especially appropriate when studying Saturn, which has a sidereal cycle of 29.5 years, ensuring that its position throughout the zodiac is represented evenly.

I next tested the entire Prague time series of 226 years for Sun-Saturn oppositions. The results showed very little amplitude and only one steep drop from plus two and three days to plus six days. I then organized the oppositions from 1775 to 2005 by meteorological seasons (i.e., March, April, and May is spring; June, July, and August is summer, etc.) and also by individual months, which came out to fifteen to twenty oppositions per month. In this test, organization by civil calendar, rather than zodiacal sign, which is based on the equinoxes and solstices, was chosen as an arbitrary way of breaking the data into sections of the year. The graphs for April and October, the months after the equinoxes, showed the most robust indications of volatility and movement toward lower temperatures on day zero, which was expected given the steeper temperature gradients between northern and equatorial latitudes at those times of the year. The oppositions that occurred within 3 degrees of the equinoxes between 1775 and 2004, a sample of only six, were next averaged and graphed. An abrupt temperature drop of about 3.5 degrees from minus one day to plus one day, over one standard deviation, was found in this sample. A look at the Sun-Saturn conjunctions occurring within thirty days of the equinoxes ($n = 73$) was tried next and compared with the HadCET data discussed above.

The result produced a curve with a low near day zero but with very little amplitude (see graph below). One interesting observation is that for twelve days the data from the two stations moved in parallel, the next eight days they moved inversely. Many more tests were done, some showing little or no correlation, others suggesting they be investigated more thoroughly in the future.

Kepler, who worked in Prague for twelve years and always lived in the same general region of central Europe, stated that Sun-Saturn conjunctions are not necessarily productive of cold and actually have warming effects. In his book *Tertius Interveniens,* after criticizing other astrologers who claimed (based on zodiacal rulerships) that Sun-Saturn conjunctions bring cold, Kepler reported that he observed these conjunctions to cause turbulence in the air and to bring warmth. His explanation, at least part of it, is that cold conditions depend on the amount of aqueous matter (moisture) in the lower world (below the surface of Earth), which, when warmed, evaporates (a cooling process) and falls again as rain in summer

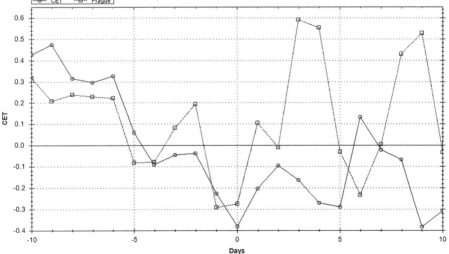

Figure 18. Sun-Saturn conjunctions that occurred within thirty days of the equinoxes. Hadley Centre Central England Temperature (HadCET) data: 1772–2008 (shown as circles, n = 77) and Prague, Czech Republic, 1775–2004 (squares, n = 73).

or snow in winter, both of these precipitation events being cooling factors. So, in terms of the meteorological science current in his time, which was basically Aristotelian, cold should follow a warming.

> When an aspect of Saturn causes Nature to perspire wind or fog, then Saturn causes this lower world to become warm . . . although later on, the released wind brings the most severe cold with the help of the landscape [environment] (Kepler (2008) T-133, 134, 137, 196, T-137 194–201).

I decided to test this observation. Using the Prague time series, I first tested all Sun-Saturn conjunctions between 1775 and 2004. A small but consistent trend in temperature above the mean was found as shown in the graph below. I then selected only the Sun-Saturn conjunctions within five days of the equinoxes and solstices, a total of twenty. While those occurring near the equinoxes showed higher temperatures roughly centered on day zero, the conjunctions that occurred near both solstices showed the same but with a much higher amplitude. Kepler did say Sun and Saturn conjunctions having the same celestial latitude were more significant than those not so close to each other.[7] This would be the case for conjunctions near the solstices, as Saturn's nodes (the intersection of Saturn's orbit with Earth's) were, and still are, located near these points. The resulting graph below seems to have confirmed Kepler's finding.

DATA FROM AROUND THE WORLD

At this point in my exploration of daily temperature data I had found a few correlations with Saturn-Sun aspects in the northeastern United States, England, and Central Europe. I wanted to know if there were effects in other parts of the world as well, including the southern hemisphere. First I ran temperature studies on Sun-Saturn oppositions

---

7. Kepler wrote: "Also, the conjunctions are not the strongest among the aspects, unless they are corporals [conjunct in both latitude and longitude]" (Kepler 2008, T-134, 198).

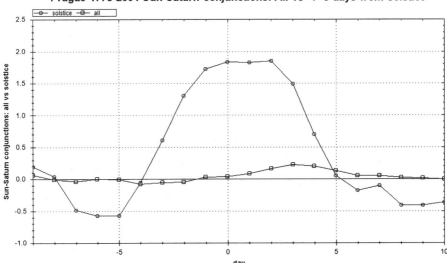

Figure 19. Sun-Saturn conjunctions 1775–2004 from the Prague time series. All conjunctions (shown as squares, n = 221) versus conjunctions plus or minus five days from summer and winter solstice (circles, n = 20). A three-day running mean was used for these curves.

and conjunctions for twelve selected locations at higher northern latitudes around the Earth.[8] What I found was that the opposition appears to correlate more consistently with lows near day zero in the eastern North American datasets and that of Greenland. Cold temperatures occurred several days before day zero in Alaska and western Canada, however, which is where some polar and arctic air masses form and begin to move south and east across North America, with the lowest temperatures reaching the Northeast on or just before day zero. I also found that cold temperature correlations with the opposition were weaker or non-existent in northern Europe and Asia. This difference may be attributable to land mass configurations.

The southern hemisphere presented problems for my investigation.

8. In North America, time series were aquired for Fairbanks, Alaska; Yellowknife, Saskatoon, Ottawa, Halifax, and Igualuit, Baffin Island, Canada; Mount Washington Observatory, New Hampshire; and Tasiilaq, Greenland. In Europe and Asia, I obtained datasets for Oslo, Norway, and Moscow, Novosibirsk, and Jakutsk, Russia.

First is the fact that there is limited land mass in the southern hemisphere at latitudes near or greater than 40 degrees south (the reciprocal of the latitude of New York City). The second is that the only regions of this category, the southern tip of South America and New Zealand's South Island, have relatively limited daily weather time series, and many of these are not complete. Third, nearly all the stations that record weather data are near the coast, and the climate is therefore maritime and less stable than a continental climate. In spite of these constraints, I managed to get data for six southern hemisphere stations, three in New Zealand, two in Argentina, and one in Chile.[9] I replaced the missing data (roughly 10 percent, most of it during holidays) with a mean calculated from the entire time series. The individual datasets were noisy, but a composite of the three South American stations showed, for the Sun-Saturn opposition, a nearly 1.5-degree temperature drop following plus one day. This was not found in the composite of the New Zealand stations, however. A graph of all six southern hemisphere stations combined showed a cooling trend lasting for several days following the alignment, this being most obvious in the South American data alone. In general, I found no strong signals centered on day zero and, having only this limited data for the southern hemisphere, ended my investigation.

DISCUSSION AND CONCLUSIONS

The tests performed in this study isolated single components from mostly noisy datasets. The opinion of those who have worked with astrometeorology, including Kepler and Goad, is that this cannot be done; meteorological correlations with planets are thought to be a result of many interactive factors. Having been warned that a reductionist approach to the complexities of the weather system has its limits, I went ahead and attempted to isolate the geocentric aspects of the

9. Data came from the NOAA Satellite and Information Service, National Environmental Satellite, Data, and Information Service. The locations are Christchurch, Invercargill, and Nelson, New Zealand; Neuquen Aeroport and Rio Gallegos Aeroport, Argentina; and Puerto Montt, Chile.

Sun and Saturn from other planetary activity. After that I added the Moon and also tested those aspects that occurred near the equinoxes or solstices. A few things were discovered, in particular that the opposition aspect of Sun and Saturn (and combinations with the Moon) frequently corresponds to cold weather in the northeastern United States. Of interest to me was that this syzygy occurs when the Earth is closest to Saturn in its annual orbit around the Sun and therefore Saturn's gravitational effects, though minuscule, are strongest then. When the Sun-Saturn opposition occurred about the same time as a full or new Moon, the effects were shown to be amplified. Another finding is that the aspect appears to be stronger when near the equinoxes, points where Saturn and the Sun are close in declination, and declination is low.[10] To test the oppositions that simultaneously meet these criteria, a much, much longer dataset would be required to reach a reasonable sample size, but none exist.

When a planet is at a low declination, it is near the celestial equator and therefore at right angles to Earth's poles. As my investigation found that both the proximity of Saturn to Earth (geocentric opposition) and its positioning at right angles to the poles (low declination) correlated with a stronger down-trending temperature signal, I hypothesized that atmospheric tides could possibly be the mechanism behind the phenomena. Recall that, in some studies, gravitational tides from the Moon on Earth's atmosphere were proposed to have an influence on certain types of weather. It follows that a tidal force on the atmosphere from the Sun, Moon, and planetary bodies when they are at low declination (near the celestial equator) should cause a bulging at the terrestrial equator and consequent compression at the poles, which may then force cold polar air southward (see diagram below). This movement of polar air masses to the south will necessarily be

---

10. This follows from the fact that the planets orbit the sun in a plane, the plane of the ecliptic so named because eclipses can only occur on it. But Earth's axis is tilted by some 23.45 degrees from this plane, causing the extension of the equator, that is the celestial equator, to cut the ecliptic at the equinoxes. Declination is measured above and below the equator, which means that a planet with low declination must be very near or on one of the equinoxes.

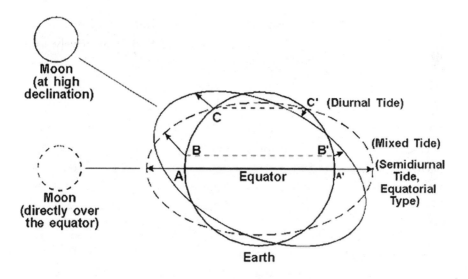

Figure 20. In this diagram, Earth's equator is shown horizontally with the north pole at 90 degrees. When the Moon is directly over the terrestrial equator, the resulting tidal bulge causes a flattening at the poles. In contrast, high lunar declination produces a tidal bulge away from the equator and flattening far from the poles.

conditioned by the size and features of land mass, more of which exists in the Northern Hemisphere. My tests of Sun-Saturn alignments suggest that air mass movement south means that temperature lows near day zero, the focal point of the alignment, will not occur at all locations at the same time and that cold air was found to be more pronounced and consistent in some locations but not so in others. It was also found that maritime conditions, as shown in certain Northern Hemisphere datasets and especially those in the Southern Hemisphere where there is little land mass, tended to distort the pattern.

The Moon's gravitational effect on Earth is quite strong, and the Sun has its own tidal pull just less than half that of the Moon. But how strong of a tidal pull does Saturn have? A rough back-of-the-envelope calculation of the gravitational forces on Earth's atmosphere (not the solid Earth) shows the force of the Sun and Moon being much greater than any of the planets. The average gravitational force of the planet

Saturn on the Earth's atmosphere is weaker by four and six orders of magnitude compared to that of the Moon and Sun, respectively, and three and five orders of magnitude when it is at perigee. Since many of the tests done in this study were of alignments occurring near the equinoxes, my hypothetical explanation is that tidal forces raising the atmosphere at the equator would lower the height of polar air and this would push colder air toward lower latitudes. But Saturn's gravitational force would appear to be far too small to have such an effect by itself.

If there is a gravitational factor behind the cooling trend at the time of Sun-Saturn oppositions in the northeastern United States, then it follows that a similar trend should also occur at the time of Moon-Saturn oppositions, or Sun-Moon oppositions, such as the full Moon. Several tests suggested Moon-Saturn oppositions sometimes correlate with cold temperatures, but not so with Sun-Moon oppositions by themselves. It is the syzygy of three points, Moon-Sun-Saturn, that was found to correlate best with cooling near day zero in the graphs. If there is such a tidal effect from a Sun-Saturn syzygy to Earth's atmosphere, then perhaps some amplification of the signal by resonance, such as that proposed by Seymour (see Chapter 2), may be an explanation. Tides in the Sun's atmosphere driven by a planet's gravitational field that are greatly amplified by additional planets have been proposed, and it is possible that the tests performed in this study may be evidence of a similar terrestrial effect. The methodology developed here, that is the use of a focal point day (aspect day zero) and a sufficient sample size, allows the aspect under study to be separated from the changing planetary context in which it occurs, and in many cases a consistent signal is detected visually in a graph. Control samples do not reveal such a regular signal. The evidence for a tidal effect from Saturn produced in this study is intriguing but not conclusive, and the fact that Jupiter did not correlate in the same way complicates an explanation. Further testing will be needed to take the system-like qualities of weather into account if anything practical is to be accomplished in studies of astrometeorology. Should it be conclusively found in further correlative studies, or by direct measurement of an atmospheric tide or other transient phenomena, that there is indeed a planetary effect on the Earth's atmosphere, the implications are substantial for the meteorological researchers and climate modelers. Further studies of other traditional astrometeorological techniques, such as those reported by Goad, might be attempted and newer satellite data may be useful for further studies, as would longer time series and cross-comparisons in tracking air mass movements.

In conclusion, my study of a possible effect on the Earth's weather from a planet nearly a thousand million miles away raises a number of issues, especially the unconsidered-yet-complete dismissal of the Western astrometeorological tradition. With more sophisticated and sustained studies the phenomena measured by classical planetary weather forecasting might be placed in the same category as orbital variations, solar cycles, and cosmic ray flux levels: geocosmic influences on the Earth system. More than just local weather may be investigated. Longer range datasets, when available, would allow for investigating recurrent alignments of planets, especially conjunctions and oppositions of the larger planets, which may turn out to be significant factors in climate studies, as I will touch on below (Scofield 2016).

CLIMATE

Climate refers to the weather over long periods of time in a certain area. The climate system is the long-term interactions of the atmosphere, hydrosphere, cryosphere, biosphere, and the land surface. These components of the climate system respond in certain ways as a unit to both internal and external forcings, some of which are quite regular and predictable. Climate cycles are observable patterns, some being correlations between repeating astronomical signals and conditions on the ground. It hasn't been that long, just a few decades, that an understanding of orbital (Milankovich) cycles has informed climatology and paleoclimatology in profound ways, providing an integrating structure that was previously lacking. The success of correlations with the three main orbital cycles and the patterns of the ice ages and other climate phenomena led to a search for other, shorter, cycles that many think may be related more directly with the Sun, these being discussed in Chapter 2. One mysterious cycle that has been proposed and estimated to be about 1,500 years caught my attention some years ago, and I studied what was known about it. I even came up with a few ideas of my own on what may be driving it, if it is a cycle at all.

The planets Saturn, Uranus, and Neptune conjoin from time to time. These triple conjunctions occur at predictable intervals, one being about 180 years, which is the orbit of the Sun around the barycenter of the solar system. The conjunctions can also be grouped in three separate series where the spacing between them is 1,542 years. Ice core records do show a consistent approximately 1,500-year pulse of temperature changes toward the end of the previous ice age, about 20,000 to 40,000 years before the present, that are not apparent before or after. These pulses, which show as regular temperature fluctuations, are known as Dansgaard-Oeschger events, and they occurred as the ice age was changing from a cold to a warm regime, a time when the climate system was unstable and fragile and more susceptible to weak signals. My hypothesis was that the conjunctions of the gas giants have an effect on solar activity that in turn modulates Earth's climate. I didn't see such an effect as a cycle, more as a punctuation that, when conditions were right, could appear to be a regular beat of about 1,500 years in the climate record. My plan to test the hypothesis required some long datasets of solar activity, one of these being an 11,500-year dataset of solar variability reconstructed from tree ring radiocarbon concentrations that has a resolution of a decade at best (Solanki et al. 2004). With this data, expressed as estimated sunspot numbers, I constructed some graphs and inserted the dates of Saturn, Uranus, and Neptune heliocentric conjunctions. What I saw in the graphs were correlations with reversal points in temperature, not always in the same direction, with some of these spaced about 1,500 years apart. As the resolution of the reconstructed solar and ice core data is not at all precise, I next looked at more recent climate history (Scofield 2016).

The second part of my study examined the reconstructed climate record, with decadal or better resolution, from the present to 750 years ago. The points where sunspot numbers were low coincided with times that European climate was cold, these being named by climatologists as minimums, the Maunder Minimum being the best known and often associated with the general label "Little Ice Age." I then calculated all the major outer planet conjunctions and oppositions during this period and placed them in the graph. In 1306 there was a Saturn-Uranus-Neptune conjunction that also included Jupiter. It fell at the low point (which is a reversal point) of what's called the Wolf Minimum, a time when the previously warmer climate of the Medieval Maximum had turned cold and wet, leading to crop failures

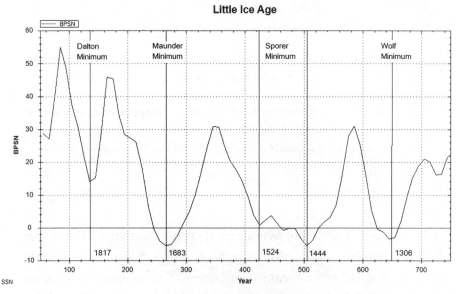

Figure 21. Little Ice Age cool periods (BPSN) indicated by low sunspot numbers, shown over the years before present (i.e., 1950). Names of minimums (cold periods) are indicated at the top, and vertical lines indicate dates of outer planet conjunctions (1306 is Jupiter-Saturn-Uranus-Neptune, 1444 is Jupiter-Saturn-Uranus, 1524 is Jupiter-Saturn-Neptune, 1683 is Jupiter-Saturn opposite Neptune, and 1817 is Jupiter-Uranus-Neptune. Holocene reconstruction of sunspot number is based on dendrochronologically dated radiocarbon concentrations. This 11,500-year dataset is an indicator of solar variability and has decadal resolution (Solanki et al. 2004).

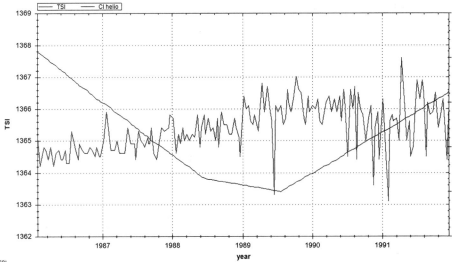

Figure 22. Total Solar Irradiance (TSI) data from the Earth Radiation Budget Satellite (jagged lines) plotted against total distance (angular separation) between Saturn, Uranus, and Neptune. The lowest figure for combined inter-planetary distance as measured in celestial longitude coincides with a sudden drop in TSI in June of 1989. Using the above data, a twenty-point moving average of TSI correlated with the cyclic index displays a weak negative correlation (r = -0.356).

a decade later. The Sporer Minimum appears close to the 1524 Jupiter-Saturn-Neptune conjunction and the Dalton Minimum around the 1817 Jupiter-Uranus-Neptune conjunction. The Maunder Minimum, generally given as lasting from 1645 to 1715 is more complicated, but Saturn was opposite a conjunction of Uranus and Neptune in 1648, Jupiter and Saturn were opposite Neptune in 1683, and these three were conjunct in 1702.

None of this proves anything, especially without instrumental records (which don't exist and never will), but it suggests to me that heliocentric conjunctions of the larger planets may play a role in climate change along with volcanic dust, ocean currents, and orbital solar variations. Planetary alignments may stimulate solar activity (higher insolation, uptick in coronal mass ejections, flares, solar wind, etc.) that could intensify a general trend in climate, reverse a trend, or perhaps negate it temporarily. Due to the regular motions of the planets such correlations may appear to be a cycle, but they are more likely a punctuation that, if the climate conditions are unstable, could cause a directional change or amplify a change already operating. It may also be the case that close outer planet alignments initiate changes in climate that take decades to be of consequence. The most recent conjunction of Saturn, Uranus, and Neptune occurred in 1989, and total solar irradiance did increase as the conjunction reached its focus (see fig. 22). Given that astrologers have long pointed to correlations between planetary alignments and weather, and also geological activity, investigating these claims in a way that respects the complexities of testing a system could prove fruitful. Such studies would not be unlike those presently studying solar cycles and the gravitational forces of the planets on Earth's orbit that modulate incoming solar radiation.

PART 2

The Decline of Astrology in Thought and History

6

The Downfall of Astrology

"As revived in the twelfth century, astrology was to last until finally destroyed by the cosmology of Copernicus."

Charles Homer Haskins,
The Renaissance of the 12th Century

"There was no scientific revolution in the sense of a new world order, pioneered by a few brave men who pursued a titanic struggle to overthrow the old world. There was instead a complex interplay of ideas as competing cosmological narratives vied with each other, and the principal protagonists remained deeply engaged with ancient and medieval concepts of the world."

Nicholas Campion,
A History of Western Astrology

In Chapter 4 the history of astrometeorology was traced from its origins in Mesopotamia to the seventeenth-century efforts of Goad to test it by comparing weather records with planetary aspects. His attempt at this nearly impossible task was not followed up by others, and the effort to modernize at least this part of natural astrology was abandoned. But the subject, which survived unchanged in the popular almanacs, did not disappear, though by the late seventeenth century few in the experimental science community found it of interest. Astrometeorology's fall as a respected body of knowledge among the intellectual elite followed that of judicial astrology. To better understand how and why this happened it is necessary to consider the decline and marginalization of all branches of astrology, especially judicial. When the term *astrology* is used in this section, it will refer to both natural (weather, agricultural, plagues, medical) and judicial (natal, interrogatory, elections). Those who developed and practiced what eventually came to be called science will be referred to as natural philosophers, which is what they called themselves. While science, from the Latin *scientia,* was in use as a term for knowledge, *scientist,* as a social label, only began in 1834.[1]

What was it that happened between roughly 1500 and 1700 in Western Europe that resulted in the near extinguishing of a subject that, while always controversial to some extent, had such an impressive historical pedigree? The common answer is that an event called the Scientific Revolution, usually explained as the triumph of rationalism over superstition, proved astrology to be nonsense. Standard accounts of this revolution generally point to several main events. First, Copernicus's description of a heliocentric cosmos challenged, and then triumphed over, Ptolemy's geocentric model. Tycho Brahe (1546–1601) was an astrological reformer who became the leading producer of accurate astronomical data in his time. His observation of a comet cutting through the presumably solid planetary spheres and Galileo's discovery of sunspots demolished Aristotle's concentric spheres model of the cosmos and the notion of a perfect and changeless Sun. Kepler's discovery of elliptical, not perfectly circular, planetary orbits best fit the observational data collected by Brahe; Ptolemy's model was far less precise. While the correlation between these events and the decline of astrology seems obvious, several decades ago I found this explanation unsatisfying and was prompted to search for a more complete answer. It soon became apparent to me that astrology's change in status during this historical period was not a simple matter and that only multiple strands of different kinds of history, when woven together, could tell a plausible story of its decline.

---

1. William Whewell introduced the term *scientist* as a label for professionals in the knowledge-making business.

Steven Shapin, in his book *The Scientific Revolution,* elaborates on what he calls the historian's dilemma, emphasizing the shaping factor of the storyteller's perspective, mine being astrology:

> Since in my view there is no essence of the Scientific Revolution, a multiplicity of stories can legitimately be told, each aiming to draw attention to some real feature of that past culture. This means that selection is a necessary feature of any historical story, and there can be no such thing as a definitive or exhaustive history, however much space the historian takes to write about any passage of the past. What we select inevitably represents our interests . . ." (Shapin 1996, 10)[2]

There were multiple revolutions in knowledge during the Renaissance in astronomy, physics, chemistry, mathematics, medicine, anatomy, and many other fields, each of which occurred in its own time frame. The most important feature in my weaving of the strands of history, because it is most relevant to astrology, is the revolution in astronomy, also called the Copernican Revolution. Still, this chain of events, insights, and discoveries should be seen as only one of many other historical trends and events. For example, there was the influx of ancient writings into Europe, a trend that intensified in the late Middle Ages and the early Renaissance. This flood of new information, much of it full of astrological symbolism embedded in philosophical and metaphysical doctrines, but also complete astrological textbooks, challenged established Aristotelian-Christian explanations of the world. These new ideas stimulated intellectuals to reconsider established doctrines regarding human life and the nature of the cosmos, well before the height of the Scientific Revolution in the early seventeenth century.

A major source of trouble for astrology had long been its conflict with organized religion and the issue of God's role in one's personal life. Another conflict was with humanism, an intellectual movement

---

2. If one considers the substantial contributions of natural philosophers from ancient times through the late Middle Ages, the Scientific Revolution might be seen more as an evolution, moving at a much quicker pace than previously, however.

of the time that emphasized human freedom, progress, and free will, the latter seen as lacking in the apparent fatalism of judicial astrology. These doctrinal and philosophical issues were spread widely due to the rise of printing in the mid-fourteenth century, which accelerated the diffusion of information throughout Europe over the following two centuries. At the same time, printing became a new revenue stream for practicing astrologers who were previously working within a patronage system. They soon began to publish their own almanacs, compete with each other, and make extreme political predictions with results that brought sharp responses from critics and lowered the public's opinion of astrology. Economic and social factors, including the breakdown of the feudal system in the wake of the Black Death and its recurrences, the rapid growth of capitalism and the rise of a middle class, the effects of climate change, and the depletion of resources shaped the period in profound ways and, most importantly, changed the type of knowledge that was valued. In all this astrology got lost and, beaten and battered, slipped into the background.

The decline of astrology, compactly summarized above, is generally brushed over or completely ignored in histories of the Renaissance and accounts of the rise of modern science. When it is considered, it is usually mixed with natural magic and alchemy, subjects that used parts of astrology for symbolic thinking and ritual timing. The great historian of the "wretched" occult sciences, Lynn Thorndike, confused matters in his monumental six-volume work *History of Magic and Experimental Science* by mixing astrology with occult practices, a juxtaposition that uses the word *magic* far too loosely, and is not quite right, in my opinion, because it focuses too much on the practitioner, not the subject matter. While magic, alchemy, and experimental science are activities that act on nature, this manipulative quality doesn't apply so well to astrology.

Given the potential for serious misunderstandings here, some clarifying definitions are necessary. To review, astrology was developed by expert skywatchers who cataloged correlations between sky motions and Earth events based on centuries of careful empirical observations. With the addition of Greek geometry these observations were organized into a more formal methodology, and a time-slice technique (the astrological chart) was developed as an analytical focus (a seed moment). The resulting set of interpretive methodologies established a tradition based on observation and interpretation that slowly advanced over the centuries. Exactly how the planets influenced Earth was debatable, but this was not a major concern of those who practiced the subject, so this part needs to be treated separately. The cultures within which astrology evolved also supported or accepted religion and magic, both concerned with the supernatural, and there was some mixing of these subjects at their boundaries, and also in regard to cosmology. However, at its core, astrology began as, and remained, a study of correlations that required precise astronomical observations and real mathematical calculations, not numerology or magical rituals, to generate data from which to make forecasts or to describe a person or situation. Electional astrology involved calculating planetary positions ahead in time to locate moments that, based on previous experience, might be fortuitous for the initiation of certain activities. Only this branch of the subject could be said to be somewhat like magic, but it involved nothing supernatural and no rituals except that of doing spherical trigonometry precisely. Astrology stands on its own as the subject that attempts, by means of methodical observation, record keeping, application of mathematics, and pattern recognition, to interpret correlations between the motions and positions of solar system bodies and conditions on Earth. Astrology was so rigorous that, up to the early seventeenth century, those practicing it were called by either of three interchangeable names: astronomer, astrologer, or mathematician.

Magic is much older than astrology and is found in most early cultures. Some anthropologists have described it as a pre-religion or an animistic proto-scientific practice based on belief of action at a distance. The word *magic,* which derives from the Magi of the ancient Persia or possibly an even earlier Near Eastern source, has many names and forms: sympathetic, white, black, demonic, high, low. It was a well-known practice in the cultures of Greece and Rome but was not highly regarded and often considered fraudulent. There is little agreement on how to define magic except that it is not a socially-organized religion, does involve rituals and divination, and is manipulative.[3] Christianity (having its own authorized magic rituals of changing wine into blood, wafers into flesh, and talking to a supernatural deity) has generally defined non-Church magic as an evil practice that relies on demons, and many people, like witches, paid the ultimate price for practicing unauthorized rituals.

During the Renaissance a high form of magic, what came to be called *magia naturalis,* was practiced by many leading intellectuals. The stimulus for magic during this time came from ideas found in the *Corpus Hermeticum,* these being the recovered texts of Hermeticism, Neoplatonism, and Stoic metaphysics that spoke of an interconnected cosmos in which parts influence other parts and higher spirits can be called upon to bring changes in human life. The *Picatrix,* a lengthy book of magical formulas, is one example of such texts. Natural magic was then a practice that attempted to communicate with spiritual entities and to use knowledge of occult cosmic connections for manipulative purposes: healing, success, prediction, altering a perceived fate, and so on. In many cases, those practicing magic would use parts of astrology, such as zodiac signs or individual planets, as a symbolic language and a kind of cosmic taxonomy with which to organize their participatory actions and presumably make them more effective. This practice, what Frances Yates calls "astral magic," merely borrows from astrology, it does not contribute to it (Yates 1964, 60).[4] The great writers on natural magic of the Renaissance strained themselves to reconcile these ideas with Christian theology, this requiring distinctions between natural

---

3. In modern times, magic has been divided into natural magic, which involves the creative use of physical substances that are believed to have powers and communication with spirits, and ritual or ceremonial magic, which is more of a group activity, one form being religious services. A third kind of magic might be the sanctioned magic of religion, such as the ritual of the Eucharist. And today there is also trickster magic, for example the kind of performance entertainment frequently found on Las Vegas stages.

4. Perhaps the best example of scholarly Renaissance natural magic is to be found in *Three Books Concerning Occult Philosophy* by Heinrich Cornelius Agrippa von Nettesheim (1486–1535). This classic on natural magic influenced John Dee and Giordano Bruno and is still in print.

and demonic magic that were not well understood, unfortunately, by the public, government, and ecclesiastical authorities.

Alchemy originated in Hellenistic Egypt out of an older metallurgy tradition. Early alchemists worked with the refinement of metals, which they saw as a kind of sacred process, no doubt influenced by ancient notions of a living and feminine Earth. Alchemy was partly a philosophical pursuit and partly experiments in transforming matter, described in the context of four-element theory. It was taken up by the Arabs, who advanced the field by studying the components, processing, and treatment of a variety of substances through distillation, sublimation, and crystallization. During this period the theory of matter transformation was developed, such as turning lead into gold by a specially created substance called the Philosopher's Stone. Experimentation was a major part of an alchemist's work, and results of experiments were often written in cryptic forms to keep such knowledge out of the wrong hands. By the time of the Renaissance, alchemy was a practice that also produced medicines like those found in the influential and innovative works of Paracelsus (Philippus Aureolus Theophrastus Bombastus von Hohenheim, 1493–1541). Alchemy continued to be studied and practiced in the seventeenth century by scientists like Robert Boyle and Isaac Newton, though eventually most of the subject that proved useful morphed into chemistry. Some concepts in alchemy were based on astrological taxonomy, including the planetary rulership of metals, and the timing of operations might employ electional astrology. But other than these uses, alchemy was not astrology and shouldn't be confused with it. While the same person could be involved in astrology, magic, and alchemy, astrology stood completely on its own. In many respects, the role of astrology in those subjects was analogous to the role of mathematics in physics, astronomy, chemistry, and other sciences today—it integrated them in some ways but was a stand-alone subject in itself.[5]

I suspect that one reason for which astrology hasn't been treated thoroughly and honestly in most histories of the Scientific Revolution

---

5. Thorndike (1955) goes as far as saying astrology was the first universal law, one that applied mathematics to natural phenomena and was only replaced by Newton's laws.

is that authors either don't understand it or they find it infuriating, or both. It's fairly easy to compare the angel magic of John Dee to a church ritual where higher powers are called upon, or the laboratory chemistry and symbolism of alchemy as either a kind of meditation or a complicated get-rich-quick delusion. Such shortcuts to judgment ignore the fact that these subjects are highly subjective, personal, and participatory, and are therefore probably better evaluated by psychologists than historians of science. I argue that astrology is different. It was, and is, a group enterprise in the building of an empirical body of knowledge, much of it best expressed using a symbolic language, and organized by rules or laws that have proved themselves in practice. There was always a general unanimity in regard to most of its methodologies and its procedures were standardized and consistent. It was also more readily available to the public in comparison to those of magic and alchemy, which were often mysterious or purposely vague. Ancient astrological texts like those of Ptolemy or Valens, those of the Middle Ages, such as Abu Ma'shar and Bonatti, and those of the Renaissance, such as Schoener and Lilly, are not exercises in unrestrained imagination like magic or religion, they are practical manuals on math and method and what they interpreted to be fundamental laws of the cosmos. These documents, and others like them, are what I would call "pure astrology" and are representative of the mainstream of astrological knowledge. Historians, for the most part, have generally failed to discriminate between pure astrology and the use of astrology in other disciplines, especially magic and alchemy, and they have in almost all cases failed to consult with the leading modern astrologers (not to be confused with those who are expert at marketing themselves or lack appropriate credentials) who have much to offer in understanding these past methodologies.[6] There

---

6. The 1999 discovery by astronomer Dr. Anthony Misch of an astrological chart drawn by Kepler is an example of this disconnect and its perpetuation. The astronomer and others in his field he called on had no idea what certain numbers meant in this chart and could only speculate, which got them nowhere except in terms of media exposure, and which portrayed them as brave researchers delving into the cryptic scribblings of a Renaissance genius. Most knowledgeable astrologers knew exactly what was in the chart—an early example of what are called secondary progressions, a fractal-like method of analyzing the data in an astrological chart—but they were not consulted. See Whitehouse (1999) for more on the topic.

are rare exceptions to this bias, however, one being the two-volume social history of astrology by Nicholas Campion, a professor of cultural astronomy who has had firsthand experience with the subject and treats it with the respect it deserves.

Given these issues, my approach to a better understanding of what happened to astrology will be to organize the historical material relevant to this great transition into general categories—intellectual, ideological, religious, economic, and social—which are then called on to produce the strands that together tell the story. This approach should, at the very least, provide a stable framework on which an understanding of a very complicated, multi-level situation like the decline of astrology might be based, the bare minimum necessary for any intelligent discussion of the problem. But first I'll begin this long haul into history with a review and critique of the common perspective of the Scientific Revolution as a triumph over a static and superstitious past, a blast of rationality leading the way to modern times.

THE PROGRESSIVE EXPLANATION

Georg Joachim Rheticus (1514–1574), a professor at Wittenberg University, spent two years traveling in Europe visiting some of the best educated and most knowledgeable scholars of the time who, like he, were also experts in those three nearly inseparable subjects: astronomy, astrology, and mathematics. During his travels he learned of Copernicus, who reportedly held some radical ideas about the Sun and planets, so he headed west to Poland to meet him in person. Their meeting in 1539 led Rheticus to spend the next two years managing the announcement, completion, and publication of *On the Revolutions of the Celestial Spheres* (*De revolutionibus orbium coelestium,* or more simply, *De Rev*), the book that many believe launched the Scientific (or Copernican) Revolution. Copernicus's description of a Sun-centered cosmos is thought to have been powerful enough to set off changes that, from the modern perspective, appear to have blown away the magical beliefs of the Renaissance and led in a straight line to the Age of Reason and ultimately our present scientifically informed civilization.

The interpretation of the Scientific Revolution as a singular progressive historical event, a time when sensory information, not the
speculative authority of Plato or Aristotle, became a basis for reality is
widespread. The common view usually starts with Copernicus's *De Rev*
in 1543 and ends with Newton's *Principia* in 1687 and is found in many
textbooks and histories of the period. (The period designated as the
Scientific Revolution is also sometimes given as roughly 1500–1700 or
1450–1650.) The event has long been considered a major turning point
in Western European history and is described as the shift to modernity after two earlier historical periods called ancient and medieval. A
similar perspective in regard to social and intellectual change was also
held by many of the founders of modern science themselves, and they
expressed it by the use of the word *new* in titles of publications such as
those by Bacon (*New Organon, New Atlantis*), Kepler (*New Astronomy*),
and Galileo (*Two New Sciences*). They were right—they lived in a time
when ancient authorities were being rejected by many leading intellectuals and an entirely new way of making knowledge was being birthed,
one that could only be contained within a new worldview.

An explanation of the decline of astrology is neatly found in this
progressive explanation, which states that those who supported the
astrological worldview, taken as superstitious nonsense, were found to
be wrong, and their downfall took astrology with them.[7] When astrology does come up in regard to the revolution in astronomy a common
version of the story goes as follows: the Aristotelian model of the universe had provided a causal explanation for astrological planetary effects
in (1) the notion of changeless, circular, distinct, and concentric crys-
talline spheres, each containing specific celestial phenomena, including
the planets, and (2) in the idea of motion having a natural direction
toward the Earth's center. Assuming a loosely interpreted Aristotelian
view, it would seem logical that, if the planets exerted any insensible,

7. Butterfield (1957), in particular, has articulated the view that the Scientific Revolution
destroyed astrology, and when it is mentioned in his book it is either associated with
witchcraft or labeled a fraud. This view is also found in Thorndike (1923) and Thomas
(1971), among others. For a summary account of the downfall of Aristotle and Ptolemy
see Butterfield (1957, 67–88) and Thomas (1971, 349–350, 643).

and therefore hidden, powers, these would naturally move downward through the spheres in some manner and ultimately be felt on Earth. But observations of sky phenomena during the Renaissance, some using new kinds of instruments, exposed problems with Aristotle's model. To many observers, the comet of 1577 (Brahe's comet) appeared to cut through the supposedly solid spheres of the celestial realm. The supernovae of 1572 and 1604, along with Galileo's reports of sunspots, were further blows to another traditional Aristotelian tenet, the idea of a perfect and changeless universe. Also, the perfection of the circular universe was undermined when Kepler showed that the planets orbited the Sun in ellipses. The above events conspired toward the distrust of Aristotle and his eventual downfall as an authority on cosmology. One could then make the assumption that the failure to find another causal explanation for astrology led to its own collapse.[8]

Copernicus's heliocentric cosmos also contradicted the geocentric model articulated by Ptolemy, the great scientist of Roman times. Following early news of the hypothesis and then the publication of *De Rev* in 1543, Ptolemy's complicated Earth-centered astronomical system, set out in his major work, *Almagest,* lost credibility as mathematical and visual evidence for the heliocentric model steadily accumulated, especially from Brahe's data, Kepler's laws, and Galileo's telescope. Ptolemy, who was also an astrological authority for his highly regarded textbook of astrology, *Tetrabiblos,* became the other great ancient authority to fail, and again, one could assume that he brought astrology down with him as the timing was more or less coincident.[9] There is thus a common assumption that the Copernican Revolution dethroned both Aristotle and Ptolemy, causing the fall of astrology. It is true that the former rose in status at roughly the same time the latter authorities declined.

There are some problems with this account of astrology's rejection, however. First, it doesn't consider that astrology, completely dependent on astronomy and mathematics, is mostly a practical discipline shaped

---

8. Kepler did argue for another explanation, one of harmonics and resonance, but this assumed the Earth was a living entity, and that notion was discarded in the seventeenth century.

9. Field (1984, 225) has argued that this factor in the decline cannot be underestimated.

and driven by experience and not theory.[10] Neither causal nor metaphysical explanations were necessary to make forecasts, analyze a nativity, or answer questions, and few astrologers cared to say anything more than that astrology worked by hidden forces. The authors of textbooks on astrology barely, if they did at all, discussed mechanisms or explanations for the subject. From the astrologer's point of view, heliocentricity, or any other solar system theory for that matter, was not a threatening theoretical problem. In astrology, there is far more concern with the exact location of a planet in the sky and far less with theories explaining why it is there when it is. What practicing astrologers have always wanted are good tables of planetary positions; if the Copernican model was able to produce more accurate ones, then so much the better. The fact that astrologers were among the earliest supporters of the heliocentric model and did much to educate the public on the subject, because it made computational sense, is significant.[11] Again, astrology was, for the most part, a practical discipline based on experience and tradition, more like medicine, engineering, or even agriculture. Very few astrological authorities offered explanations, proposed theories, described mechanisms, or speculated on its purpose—these issues were left to the philosophers and theologians.

One could also argue that it was the concern of astrologers (most of whom were out of necessity also mathematicians) for more accurate tables of planetary positions that actually drove the astronomical and mathematical innovations that propelled the Scientific Revolution in astronomy. It is a fact that data collection methods and mathematical

10. In this sense astrology was not a branch of natural philosophy, nor was it astronomy, which was regarded as a branch of mathematics. Astrology is so unique that it should probably be placed in its own category. Most natural philosophers of the period knew a great deal about all three subjects, as they were interrelated.

11. John Dee appears to have been interested in the heliocentric hypothesis, at least in its consistency with ideas of solar supremacy inherent in Hermeticism (which was also true for Copernicus himself). Dee's pupil, Thomas Digges, son of almanac maker Leonarde Digges, was an advocate for Copernicism, as he plainly argued for it in his father's astrological almanacs (Digges 1605). Most importantly, it was also regarded as an important methodology for calculating planetary positions, which is all that astrologers require from astronomy. Most astrologers were, by nature, pragmatists, and the loss of a mechanism did not affect the practice of the subject in any significant way (Goad 1686, 18).

modeling of data, the foundations of modern astronomy and science in general, were advanced by Brahe and Kepler, both of whom practiced astrology. Both also complained about the inaccuracies of the planetary tables (ephemerides) of their time and both were motivated to do something about it. Brahe built large instruments that enabled him to capture the positions of astronomical bodies to a half second of arc and Kepler labored to produce the *Rudolphine Tables,* the first very accurate tables based on his improvements of the heliocentric model. This theme of astrology's powerful influence on astronomy and mathematics is slowly getting more attention as graduate students and scholars run out of safer topics to pursue and find, in astrology, a mostly untapped and yet important strand in the history of science.

Another fact already alluded to is that certain founders of experimental science, the same ones who were eager to overthrow Aristotle and Ptolemy, either practiced, endorsed, reformed, or defended astrology during the seventeenth century. Copernicus certainly knew something about astrology as he studied it in Italy and was a physician; medicine in those times was informed by astrology (it was called iatromathematics). Rheticus, who brought the work of Copernicus to European intellectuals, was a highly respected reform-minded astrologer in addition to his work in astronomy, mathematics, and cartography. As already noted, Brahe's effort to better chart the sky was a response to the inadequacy of existing tables available for making horoscopes. His observations showed him that the Alfonsine Tables, produced in the thirteenth century, were inaccurate by a wide measure, and the more recent Prutenic Tables based on Copernicus's work weren't that much better, though a bit easier in regard to calculations. Brahe understood astrology well and knew its limitations. He made astrological reports for his patrons and published an astrological prognostication on the 1572 supernovae and the comet of 1577. For this historical time period, both of these reports were the equivalent of a scientific paper.[12]

Francis Bacon called for a reform of the subject, what he called

___

12. See Kusuwawa (1995, 173) on Copernicus's credibility as an astrologer. Also see Campion (2009) on Rheticus as an astrologer and Christianson (1979) on Brahe's German Treatise on the Comet of 1577.

a "sane astrology." While he thought popular judicial astrology was infected with superstition, he didn't abandon that branch of the subject entirely and instead advocated an experimental program that included the planetary aspects and orbital variations, similar to what Kepler was doing at about the same time. We also know that Kepler was a serious astrologer who was consulted by highly placed patrons like the military leader Wallenstein and Rudolf II, the Holy Roman Emperor (Kepler succeeded Brahe as Imperial Mathematician). He produced annual almanacs and ephemerides, wrote two books on astrology, and included the subject in his major work, *Harmonics of the World*. Galileo practiced astrology for his wealthy patrons, worked on the horoscopes of his daughters and friends, taught medical astrology, and apparently got into trouble with the church for the fatalism of some of his readings. But he didn't write any books on astrology, and his association with the subject has been successfully sanitized by historians (Bacon 1970, 462–465).[13]

The fact that these men, each of whom made significant contributions to the Scientific Revolution in astronomy, either practiced, proposed reforms to, or were knowledgeable about and open to large parts of astrology, and didn't renounce it as a subject in itself, raises questions about exactly how and why the downfall took place. These questions, until very recently, have been more or less avoided. Many accounts by reputable twentieth-century historians display extreme bias, an example being Herbert Butterfield's influential book, *The Origins of Modern Science*, in which astrology is consistently placed adjacent to the words *witchcraft* or *fraud*, or it is completely omitted. It is true that Brahe, Bacon, Kepler, Galileo, and others made critical comments about astrology, but these are routinely cherry-picked by writers of history so as to be misunderstood when out of context and appear to be more reputation protection actions than modern scholarship. For the most part, the criticisms from the more scientific astrologers of the time were directed toward other less technical astrologers who they argued were ignorant of the subtleties of human free will, sloppy in their work, and lowered the standards of the field by excessive self-promotion and outrageous predictions (Kepler 2008;

13. See also Campion and Kollerstrom (2003) on Galileo's Astrology.

Campion 2009, vol. 2, 134). So, other than the ignorance and bad behavior of some astrologers of the time, we could say that it was the needs of astrology that drove much of the astronomical part of the Scientific Revolution and that many of the important contributors practiced it or understood it well, defended it in principle, and did not think it was a superstition or fraud.

The idea that astrology was disproven during the Scientific Revolution is not supported by evidence. Consider that the subject stood on firm, long-lasting, and respectable mathematical foundations. From the time of Hipparcus to the sixteenth century, both astronomy and mathematics were developed to serve several practical purposes: astrology, navigation, mapping, and calendars. Much of astronomy's subject matter had to do with locating stars and planets against abstract frameworks of the celestial sphere (celestial mechanics). The planetary positions that the astrologers needed had to be calculated using trigonometry, specifically spherical trigonometry. But by the mid-seventeenth century the evolution of trigonometry, spurred in part (if not largely) by the need to calculate accurate astrology charts, was more or less complete. From this point on, astronomy and mathematics changed focus in order to solve newer problems in physics. The instrumentation (telescopes) and mathematical modeling (Kepler's and Newton's laws) that developed during this century accelerated astronomical knowledge and led to the general form of the science as we know it today. But these developments had almost no effect on astrology, which persisted in a practical form that had changed very little over the preceding centuries. If it wasn't the case that astrology was simply disproven, then either it was a subject that could not be easily modernized and benefit from progress in instrumentation, or its functions and general domain became irrelevant to the new type of scientist in the seventeenth century—or both. There were, as already noted and will be discussed in a later chapter, attempts to reform astrology as a discipline, but in the end these were unsuccessful.[14] Given the above,

---

14. In the later part of the seventeenth century attempts at creating a heliocentric astrology failed. These attempts seem desperate, as if it was a case of too little too late, and they were not adopted by other astrologers probably for the reason that they didn't work very well in practice. Childrey (1652) and Hunt (1696) are examples.

a case could then be made that the more competent astrologers became discouraged and moved on to other challenging projects in astronomy, with less baggage and better payoff, and so astrology was left to the less qualified practitioners who eventually tainted its reputation. The result was that the subject, still intact in its traditional form, was sidelined as what's called the Scientific Revolution proceeded full steam ahead. Why astrology couldn't be modernized and why it became irrelevant to the mainstream are great questions that I've already commented on and will also be considered in the next sections.

The progressive argument usually assumes the criteria of modern scientific theory and the data produced by its own methods to be necessary and sufficient to account for the historical failure of astrology to become a respectable science like astronomy, physics, chemistry, or others. From the modern perspective and its tacit assumptions of what is real and what is false, the rejection of astrology during the seventeenth century by the evolving experimental scientific community (which included astrologers) was due to the lack of a certain species of proof, along with the lack of a satisfactory mechanistic explanation for exactly how the planets affect the Earth. In other words, if the alleged planetary effects central to astrology were not measurable using the instruments or methodologies of the time, or those of today, then there is no other option—astrology must be declared absolutely false in an objective sense and is therefore nothing more than a belief. This assumption is found in quite a few histories of the Scientific Revolution.[15] After four decades of

---

15. Many influential historians and science writers have dismissed astrology as either irrelevant or so wrong-headed that it can hardly be taken seriously. Dampier (1943, 89) states that astrology was a dangerous foe to modern science and the subject is not even mentioned in Crombie (1959) and Burtt (1954). Butterfield (1957, 47) offers a backhanded putdown of the subject and Shumaker (1972, 54) rejects it outright. In the latter's discussion, astrology is said to be dishonest and impossible, which, as presented, it seems to be, certainly in the context of reductionist scientific materialism. In reaching this conclusion Shumaker sets up straw men, cuts them down, and refers to a limited and inadequate collection of modern sources. But this scarcely scrapes the surface of this deep-seated bias. See Vaughan's introductory chapter to Kepler's *Tertius Interveniens* (2007), in which she exposes the shallowness and superstitions of many other respected scholars and science writers who have made bold but ignorant pronouncements on the subject of astrology.

this situation, involving what I believe to be careful and
tion on my part and that of many others, this dismissal
ed on the (1) fall of Aristotle and Ptolemy and the power
sm, (2) misunderstandings of what certain Renaissance
ologers really thought about the subject, and (3) rejection based on lack of proof (an objection that is also applied to climate
change by some of a particular political persuasion), doesn't seem to me
to be a closed case. The situation was, and is, more complicated.

I see three basic kinds of natural knowledge, that is knowledge about
nature. Aside from the common knowledge useful for living (which is
collective knowledge that is widely known and verifiable, like rocks fall
when dropped and rain is wet), all else is either subjective, like faith-based
knowledge or the products of philosophical speculation, or knowledge
produced collectively using a logical methodology (science). For myself,
I take the scientific method (critical and systematic thinking, observation, description, measurement, experiment, replication, and modeling)
to be the best, and most democratic, way to produce a comprehensive
knowledge of nature that does not include any speculative or supernatural explanations. In the case of astrology, a couple millennia of collective
empirical observations are behind its knowledge claims. What is basic
knowledge to astrology was not dreamt by a prophet or thought up by a
philosopher. In Chapter 5 I made the case that, at least in regard to astrometeorology, there exists some evidence that at least parts of it are as real
in our time as they were perceived to be in centuries past. A reading of
Chapters 1 and 2 should convince anyone that the cosmic environment
does influence Earth and its inhabitants in profound but complex ways,
and that scientific findings are steadily closing in on what astrologers
have been getting at for millennia. The fact that traditional astrological
methodologies chart photoperiod, locate phase angles, and track diurnal
motions is evidence that the subject was long ago developing intelligent
ways of measuring phenomena that are either the same or closely parallel
to what is being done today in certain scientific subdisciplines.

So, if astrology is not something made up, a belief system, or a mass
hallucination, then it should be treated as a kind of collective or democratic knowledge that probably just needs a good checkup and some

replacement parts. Still, astrology, just the word alone, causes a kind of cognitive dissonance in the minds of many science popularizers and the self-appointed pitbulls of science who call themselves skeptics. The former appear to have status protection issues; the latter are driven by a faith in scientism, the belief that only reductionist science produces real knowledge. Scientism, to some social scientists and historians of science, is not a rational mindset but more a quasi-religious species of fundamentalism.

Uncritical assumptions about astrology, especially the confusion of astral magic with pure astrology, are persistent and cross academic boundaries. Even the Wikipedia page on astrology is closely guarded by what appears to be a legion of volunteer defenders of scientism, and any evidence-based contributions posted that support astrology will be replaced, often in a matter of minutes. The willingness of Bill Nye the Science Guy and Neil deGrasse Tyson to flaunt what amounts to an embarrassingly minimal understanding of astrology is obvious only to the very small minority who are not ignorant of the subject. Astrology has thus been branded by academics, wiki-geeks, and even some celebrities (who aren't really doing their own thinking) as irrational nonsense, completely non-scientific, misinformed, and, at best, a waste of time. Just discussing it in respectable company is to risk becoming tainted, and the connection between one's academic reputation and keeping food on the table has kept many scholars from taking on the job of assessing the situation fairly.[16] There also appears to be a widespread, infectious subconscious fear that human free will is compromised by the determinism of birth circumstances (which it obviously is by genetics and enculturation). The result of all this, a collective decision that (1) the "patient" should be declared dead because its continuing existence is a big problem and (2) attacking and ignoring it are the best options, can be compared to the war on cannabis. There, the herb was classified in such a way that it wasn't allowed to be studied, and so it remained illegal and open to evidence-free claims for decades.

---

16. In 1989 I approached a history of science professor at the University of Massachusetts and inquired as to the possibility of developing a graduate program for myself focusing on the history of astrology. He immediately dismissed the suggestion and said it would be the "kiss of death" to his career and mine.

7

THE RENAISSANCE MIND
AND ITS ROOTS

We may think of the stars as letters perpetually being inscribed on the heavens, or inscribed once for all, and yet moving as they pursue the other tasks allotted to them. Upon these main tasks will follow the quality of signifying, just as the one principle underlying any living unit enables us to reason from member to member, so that, for example, we may judge of character and even of perils and safeguards by indications in the eyes or in some other part of the body. If these parts of us are members of a whole, so are we; in different ways, the one law applies.

PLOTINUS, ENNEADS: SECOND ENNEAD, THIRD
TRACTATE (ARE THE STARS CAUSES?), CHAPTER 7.
TRANSLATED BY STEPHEN MACKENNA
AND B. S. PAGE. 1921.

What people believe to be true and real is conditioned by their culture's general conceptions of reality, personal life history, and their capacity to think critically. People's worldviews are largely imprinted during early childhood by the social world they were born into; change is possible only with considerable learning or some form of forceful mental re-programming. The "truth" is well-defined in tightly controlled societies with a single ideology or religion (North Korea, Saudi Arabia), or it can be limited to sub-groups within a larger pluralistic culture (United States). Beliefs about reality shape perception, experience, and destiny. The Aztecs were defeated by the Spanish in part because their leaders believed Cortés was the fulfillment of a prophecy, not because they lacked sufficient warriors or opportunities to completely destroy the invaders. More than a few conservative Christians in twenty-first-century American politics regarded Donald Trump as a tool of God, and they fully expect Biblical Armageddon to occur in the Middle East. Yet these people live normal lives and they vote, seeking to base international policies on completely unsubstantiated beliefs.

In this chapter I will focus on developing a general understanding of the worldviews and consciousness of those who lived during the sixteenth and seventeenth centuries, a time when Christian theology and ancient natural philosophy and metaphysics were thrust together in a variety of ways. One thing that will be apparent over the following pages is that theological and philosophical doctrines supportive of astrology were ubiquitous during the Renaissance.[1] Another is that, with a few exceptions, most practicing astrologers did not do theology, philosophy, or metaphysics and, in this regard, they collectively contributed very little to those subjects during the Renaissance. However, the legitimacy of their existence as astrologers was conditioned by the belief systems of the age. Perhaps by perusing these historical mindsets another kind of understanding, or another story, of astrology's decline may become apparent.

For millennia a large part of the Latin West's cultural belief system was, and still is, the Christian religion. In medieval Europe Christianity united communities and regions by giving people a common set of beliefs about human life and the natural world, and it was the background landscape on which Western knowledge was organized. The religion gave rules to live by, a set of morals and explanations of life's

---

1. This chapter attempts to provide a basic foundation to the many thought systems that converged before and during the Renaissance. It does not get into depth, however, which would consume far too many pages. Interested readers can discover these details from numerous sources, including the original documents themselves.

mysteries, including answers to questions such as what kind of beings we are and what happens after we die. Of course, this religion was used by some to justify destructive behaviors, but it mostly served as a cultural integrator and linked diverse parts of Europe. The official document of the religion, the Bible, is composed of ancient Judaic writings that include an account of the Hebrew mythology of creation, lines of genetic descent, floods and other disasters, and the survival of the people who followed the god of Israel, a desert god named Yahweh. An appendix to this older collection of writings is the "New Testament" which describes the life of Jesus of Nazareth, believed to have been an actual incarnation of this god and a demonstrator of life after death. In general, Christianity and its Hebrew foundations provide for people a worldview that regards humans as a special creation that is obligated to their creator, and a set of rules (ten commandments plus two added later) that keep society more or less stable, the latter being what religion does best. But Christian theology based entirely on the Bible is very thin on metaphysics and cosmology, and it doesn't offer much in the way of insight into the natural world either. This is why the ideas of pagan Greek and Roman philosophers were utilized by theologians and other Church intellectuals to develop a more complete Christian worldview.

The Christian theology of the early Middle Ages was largely shaped by a few writers. One, Augustine of Hippo (354–430), the most important of the Church fathers, lived in Northern Africa near Carthage but taught for a time in Milan. Augustine was well-educated in Greek and Roman philosophy and sought to both distinguish and synthesize his knowledge of these rational traditions, particularly Neoplatonism, with faith. In his autobiographical book *Confessions,* Augustine explains how, as an isolated individual, he found a way to transcend the material world and find unity in God. His position, a metaphysical dualism, was that body and soul are completely different things; humans are a composite of two distinct qualities, the former being material substance and composed of the four elements, the latter not confined to space and superior to the former. Another species of dualism is found in his work *The City of God Against the Pagans,* in which history is interpreted as a struggle between God (channeled through the Church) and the Devil, who lives in the form of pagan gods, secular society, and government. Augustine, who was exposed to astrology in his early life (but didn't actually study or practice it), was passionately against it and believed that human destiny was in God's hands, not under the light beams of the planets. People had free choice to choose the good, though only God knows what those choices will be. He cited the differences between twins as evidence of astrology's falsity and thought that accurate astrological predictions came about either by chance or due to help from demons, which implies that astrology is a kind of divination, not a natural science. Regarding the Magi's star, he said it was a special creation of God, not an alignment of planets. His attack on the subject, not entirely original, was, and still is, a major resource for Christians.

Another influence on early Medieval thought was Boethius (477–524), a Roman philosopher who translated and commented on Plato and Aristotle and was also influenced by Neoplatonism and Stoicism. He effectively preserved knowledge of the great ancient philosophers and interpreted their ideas relative to Christian doctrines. His popular work, the *Consolation of Philosophy,* is a dialogue between a man (Boethius) in a prison cell waiting for execution and a woman that personifies philosophy. She explains to him how God is good and happiness is available to all, no matter what the predicament. Boethius also wrote essays using reason and logic that cemented and fortified the theological foundations of Christianity in the Latin West. In this regard he is considered to be one of the early founders of Medieval Scholasticism, the program of learning using logic and dialectic that flourished from the eleventh through fourteenth centuries. Efforts to prove the existence of God, to understand the immortality of the soul, and to solve the problem of free will, among other theological issues, were debated for centuries using Augustine and the limited number of writings of Greek and Roman philosophers that were passed down through the writings of Boethius. Then, mostly during the twelfth century, the bulk of Aristotle's works were recovered and translated, a surge of information that contrasted the differences between natural philosophy and a theology that emphasized the life of the soul. Church intellectuals responded by producing a highly complex synthesis of faith and reason, what might be called a Christianized Neoplatonic-Aristotelianism, and it was taught using the rational methodology of Scholasticism. This was the worldview of the educated Latin West that was in place prior to the Black Death and the early Renaissance.

PLATO AND ARISTOTLE

In the early European Middle Ages, not all that much was known about Plato, though many of his ideas were transmitted through Augustine's version of Neoplatonism. Today Plato is best known for his theory of forms, which stated that nature is constantly striving toward reproducing ideal structures (perfect geometric figures would be one example) that exist in a realm lying somewhere behind the world of everyday appearances. Influenced by Pythagoras (c. 570–c. 495 BCE), who lived a century and a half earlier, Plato viewed mathematics as more real than the physical world; he viewed numbers as having the intangible property of remaining the same regardless of what material substances they may count or measure. Plato's philosophy is known as rationalism, in reference to the mind as the means by which reality is apprehended, as opposed to sensory input alone. His metaphysical writings, particularly the *Timaeus,* which was the only Platonic dialogue that was known in early Medieval Europe, described the natural world as a living entity with intelligence and its own kind of soul (Plato 1961). Plato saw the cosmos as the creation of a divine craftsman who brought order to a pre-existing chaos. The world is alive (anima mundi) and is intelligent, which means it has a soul, and its order is displayed by, among other things, the motions of the stars and planets. The ideas of a higher being producing a living world, and the dualism of the sensory-material world and spiritual-eternal world of forms, were modified and transmitted through Neoplatonism, Augustine, and Boethius to the Middle Ages. But during and after the twelfth century, Christian thought had to also consider the philosophy of Plato's greatest student who was less interested in things other-worldly.

The enduring authority of Aristotle in both philosophy and science, from Hellenistic times to the seventeenth century, is really unmatched in history. His surviving writings, mostly lecture notes, were the dominant source of knowledge about the natural world in the Latin West for centuries and a major contribution to the evolution of the scientific method. Aristotle, so impressive a thinker as to be referred to as simply The Philosopher, offered a way of reasoning (logic) and a scientific method (observation and classification based on sensory perception) that he described in great detail, far more than any other ancient philosopher, and these (along with Euclid's method of proofs in geometry) were, and continue to be, cornerstones of natural science. Aristotle's cosmology was also influential. He described the cosmos as eternal but also finite; as circular with levels arranged in a hierarchy. It was composed of the central Earth surrounded by the Moon, Sun, planets, and stars, all themselves spherical and all moving around Earth in the perfect geometrical shape—circles.

Aristotle thought the objects of the heavens were fundamentally different from those of Earth and therefore not subject to the terrestrial laws of nature. Further, he said that the cosmos is most dense in the terrestrial center and that matter diminishes as one moves away from Earth up to a point where it no longer exists. The motions of everything in the cosmos are caused by a mysterious prime mover, a celestial being of pure intelligence that is extremely remote from humans. The prime mover, however, acts on the outermost sphere, itself being a lesser kind of intelligence, and this action is then transmitted downward, in sequence and less perfectly, to the inner spheres. These spheres of intelligence are, of course, the stars and planets. Aristotle didn't write on astrology (he lived before the bulk of it spread out of Mesopotamia), but his cosmological model gave it a kind of higher-to-lower rationale where the unique energy-intelligences of the higher spheres are stepped down to eventually generate phenomena in the sublunar realm, the world located inside the sphere of the Moon where humans live. Recall that it was Aristotle's cosmic scheme that Ptolemy adapted to his explanation of the actions of the planets in astrology.

Another key characteristic of Aristotle's cosmos, and also his general philosophy, is its qualitative way of describing material substance and physical causes. He used the four elements as the key to understanding matter, these being proposed by Empedocles before him but undoubtedly having origins in earlier civilizations. The four elements were considered states of matter; everything was built on combinations of them. (Since the high heavens existed where low matter didn't, Aristotle added a fifth element he called aether.) In the scheme of this cosmology, elements could change (this idea fueled alchemy) and they had a natural motion: fire and air rise, water and earth fall. The elements themselves were produced by the actions of four primary qualities: hot, dry, cold, and wet, where fire was hot and dry, air was hot and wet, water was cold and wet, and earth was cold and dry. None of this early physics, obviously, lends itself to mathematical quantification, but it is a method (based on specified qualities in a continuum) to describe nature, and it was applied to astrology, alchemy, natural magic, and medicine. Four-element theory was adopted by the Stoics as part of their physics, and it was, and still is, used by astrologers to classify fundamental distinctions between the signs of the zodiac.

There is another component of Aristotelianism that is relevant to this discussion, one that was strongly reacted to during the Renaissance. Aristotle, who thought far more like a biologist than a mathematician, argued that nature has a tendency to change in very definite ways and that processes of becoming and stages of development are inherent in

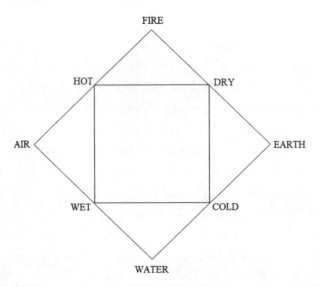

Figure 23. The four elements and their relations to the four qualities.

nature. This is called entelechy, and an example would be the growth of a seed into a mature plant. Here, where the seed apparently has an internal plan, matter and form are seen as two parts of the whole plant with the plan being the form that guides the matter. (Contrast this idea with Plato's separation of matter and form.) Aristotle saw the inner driver of the plant as its vital function. Similarly, in regard to inorganic falling bodies like a rock dropped from a window, he thought something internal must be moving them toward the center of the Earth. In both examples the inner form is also the goal, which implies a certain direction and purpose, or teleology. This view, as expressed by Ptolemy, also supported the idea that the astrological chart at birth was like a seed and that it signified a person's destiny, though it was usually conditioned by a statement suggesting that the human soul could make choices as it moved toward its final form (the stars incline, but don't compel). It was this Aristotelian teleological view, also expressed in his account of causation in which the final cause implies purpose, that later became a target for attacks launched by Bacon, Descartes, and others. Its refutation, necessary in the context of the deterministic materialism that was developed in the seventeenth century, was a factor in making astrology less appealing to those in the Western European scientific communities (Collingwood 1960, 93; Copleston 1962b; and Osler 2000).

Before we go further, it is important to get at what the term *soul* meant to ancient and Medieval philosophers and theologians. Precise definitions varied among the ancient Greek philosophers, but *soul* or *psyche* referred to interior phenomena: thought, memory, consciousness, and feeling. *Stoic pneuma* (breath) was a related concept, but seems to be more like Hindu prana or Chinese qi—the force that gives the body its life. Aristotle taught that all living things had a soul, though there was a natural hierarchy with plant souls at the bottom, animals midway, and human souls at the highest level. The soul was the force in a body that drove and guided it toward its ultimate purposes; it had *entelechy,* meaning "the process of intention in nature that leads to actualization." Plato (speaking through Socrates) described the soul as immortal and the body its prison, but wrote in *The Republic* that the soul was made up of three parts located in three regions of the body: the intellect component residing in the head, the emotive part in the heart, and desire located in the stomach. Stoics taught that the soul, which they regarded as refined pneuma, was a universal structuring force that operated in both individuals and in the continuum of the larger cosmos, making soul a kind of cosmic glue. Other philosophers described the soul as a faculty of living things that possesses desire, experiences pleasures, and is the location of the sense of morality. The distinctions among Greek philosophers regarding the soul and the mind were many, but soul and body were generally considered to be fundamentally different. This dual nature of existence and the cosmos was elaborated on in Neoplatonism, which strongly influenced Augustine and, consequently, Christian theology of the Middle Ages.

Plato's ideas were discussed and interpreted by philosophers for centuries, but Plotinus (c. 204/5–270), a Roman philosopher born in Egypt, expanded on them. His and other similar reinterpretations of Plato's thought were named Neoplatonism in the nineteenth century and defined as traditional Platonic idealism influenced by metaphysical ideas found in a wide range of sources, including Pythagoreanism, Aristotelianism, Stoicism, Judaism, and Gnosticism. Plotinus may have also been influenced by Indian philosophy, as he was a student of the Alexandrian philosopher Ammonius Saccas, who is thought to have had connections with India. Neoplatonism was the last major school of ancient pagan philosophy, a product of the era of Hellenistic syncretism during the first several centuries CE that was centered in Alexandria, Egypt, the cultural and intellectual center of the Mediterranean region. While the details and subtleties that Plotinus developed in his reinterpretation of Plato are well beyond a short summary, it could be said that the central doctrine was both quasi pantheistic (God's emanations penetrate all of nature and beyond) and monotheistic (a single One that is able to emanate a divine intellect or mind, such as Nous, also called the Demiurge) that is capable of objectivity. The intellect then produces, through self-contemplation, the Platonic forms. These emanations from the One and highest source, through intellect, give rise to the world soul, which contemplates its own existence and both manifests and inhabits the lowest level of the material cosmos. The immortal human soul was thought to descend from the divine realms, passing angels and the planets along the way (and being influenced by them) to dwell in the world of matter (i.e., couple with the body). The goal of an enlightened human was to transcend the material world and rise back up to merge in the oneness of God. In Neoplatonism the cosmos was seen as an organic unity in which each part has a sympathetic relation to the rest, an idea probably sourced from earlier Stoic cosmology. By the Middle Ages this vertical structure of the cosmos took the form of what's called the Great Chain of Being, a hierarchy of God, angels, humans (at the halfway

Figure 24. The hierarchical cosmos known as the Great Chain of Being. At the bottom are stones, followed by plants, animals, people, angels, and the stars. The Sun illuminates the entrance to the house of God. From Ramon Llull's *Ladder of Ascent and Descent of the Mind*, 1305.

point between spirit and matter and partaking of both), animals, plants, and minerals. It was depicted in many woodcuts as in fig. 24.

Because in Platonism there is an integrative world soul, and in Neoplatonism there is a hierarchy of beings with spiritual energy flowing up and down, it follows from both that the planets (which can be seen with one's eyes to exist at a higher level) must have some kind of link to the lower, denser levels. The problem is exactly what the link is and what its precise influences are. In responding to this problem, Plotinus offered highly detailed explanations that kept the effects of astrology, which he appears to have regarded as obvious, from eclipsing God in any way (Plotinus 1921). His position was that the planets were living, ensouled higher beings that were not evil, located well above the sublunary realm of nature and people, somewhat indifferent toward humans, and are only one part of the larger cosmos (see quote at the beginning of this chapter). They had an effect on some things (impulse, desire, temperament, the body) but in general were more like signs or instruments of God than causes in themselves. Planetary influences were like a celestial wind blowing on a ship at sea (the material body) that is responded to by the steersman (the soul), who is free to make navigational decisions. Plotinus was critical of certain methods used by astrologers but focused more on issues such as the limits of fate, the exercise of free will, and the multiple causes that lay behind any event. His position on astrology was held by many Christian philosophers and theologians from the Middle Ages to the Renaissance.

There were other subtle modifications of how astrology fit into the scheme of things that are found in the writings of other Neoplatonists. Porphyry (c. 234–c. 305), a student of Plotinus who edited and published his teacher's writings (*Enneads*), was a writer on philosophy, a vegetarian (like Pythagoras), and also a critic of Christianity. He also wrote on astrology, though only portions of his introduction to Ptolemy's *Tetrabiblios* have survived. A mathematical system of diurnal cycle division (the houses or domiciles) bearing his name is still used by some astrologers today, though it was known before his time. Porphyry was far more knowledgeable of, and favorable toward, astrology than Plotinus, and he argued that because the soul in its descent into matter chose its fate, knowledge of it (through astrology) was useful for the soul's return to greater heights. Here astrology becomes the key to free will and a tool for the transcendence of fate. A third early Neoplatonist, Iamblichus (c. 240–325 CE), regarded astrology as a true natural science concerned with the lower bodily life, not the intellectual or soul life. He also thought transcendence of one's fate was possible mainly through theurgy, or communication with the gods.

Building on earlier Greek philosophy, Plotinus's Neoplatonism attempted to explain in great detail how an ideal, intangible world that is imperceptible to our senses produces, and then ensouls, the material world and the bodies that live in it. This split view is consistent, in general, with Plato's separation between the world of the senses and the world of the mind and the forms, the latter being where he believed real knowledge was to be found. Using only their minds and not much actual data, Plotinus and other Neoplatonists developed Platonic soul-body dualism into a holistic vision of a dynamic and massive cosmic system, with the spiritual equivalent of arteries and veins flowing through and connecting the highest and lowest levels, and those in between. Taken in its entirety, notions of cosmic unity through interconnectedness and movement up and down a ladder of beings not only supported at least some forms of astrology, but also supported the possibility of divination, communication with higher (or lower) beings, and the influence of the imagination on the body as practiced in natural magic.[2] It was, however, Augustine's Christianized version of Neoplatonism that became the dominant theology and metaphysics in Western Europe for centuries.

While Plato and the Neoplatonists had described the soul and its place in the scheme of things, Aristotle's more naturalist legacy in Western Europe was at first limited. He was, however, well-known to the medieval world of Islam, where his works were collected, copied, and commented on (notably by Averroes in Spain), and then were later brought to Western Europe during the eleventh and twelfth

---

2. It should also be noted that the "New Age" movement of the 1970s and later, which promotes positive thinking, dreamwork, symbolism, ritual, and various kinds of natural healing, is essentially a resurfacing of Neoplatonism and other ancient ideas.

centuries. This period, sometimes called the Renaissance of the twelfth century, was when an influx of mostly scientific and mathematical texts (including plenty of astrology) coming from the Arab world via Spain, Sicily, southern Italy, Byzantium, and possibly the Crusades filled in huge gaps in classical knowledge. This material, especially Aristotle's natural science and cosmology, was a massive information dump of carefully worked out and very down-to-earth philosophical thought that was devoid of theology. It was immediately a problem for the religious authorities, and several condemnations of his teachings were proclaimed from Paris, the center of learning at the time.

A century after its appearance in Europe, Aristotle's philosophy of the natural world came to be adapted and synthesized with the traditional Christian ideas built a few centuries earlier on the writings of Augustine and Boethius. Astrology, seen as a natural science, but one which the Church was previously hostile toward, became a part of this new synthesis due to its inclusion in the works of Aristotle-influenced Arab astrologers such as Abu Ma'shar. Islamic philosophers had reconciled God and astrology by identifying Aristotle's Prime Mover with God and letting the planets be the means by which God signals and communicates with the world. Astrology was then a useful way to know God's plans and work with the planets, especially in regard to the body and how to heal it. The Dominicans Albertus Magnus (c. 1193–1280) and his student Thomas Aquinas (1225–1274) did more or less the same thing for Christianity. Albertus, from Bavaria, thought astrology was an important body of knowledge and a means to understand and improve free will. He even wrote a major work on it called *Speculum Astronomiae* ("The Mirror of Astrology"), in which he argued against its critics and pointed to some parts of judicial astrology, notably elections, as evidence of the existence of free will. He regarded the stars and planets as inanimate vehicles through which God spoke. Aquinas, from Sicily, thought natural astrology was a legitimate science but argued that the influence of the planets on people was limited to the body and feelings. Planetary influence could always be overridden by the use of free will, a property of the mind or soul and therefore not corporeal. Like Augustine, he thought that exact predictions for people were only possible with the help of demons.[3]

In England, more or less contemporaneous with Albertus and Aquinas in Paris, Roger Bacon (1214/1220–1292) lectured on Aristotle at Oxford after having spent a decade being a Franciscan friar. Like Albertus and Aquinas, he blended science and philosophy with faith. Unlike them he is also considered to be a precursor of seventeenth-century experimental natural philosophy in his bringing together of observation, measurement, and mathematics. Bacon had an interest in astrology; he wrote on it and included it in his proposed university science curriculum reforms. His attitude toward it was much like that of Albertus's: it was acceptable as a natural science as long as it didn't get in the way of free will; but Roger Bacon also saw astrology as a kind of environmental science in need of experimental verification (Thorndike 1923, vol. 2, 655–683; Hand 2014, 66–75).

These three Christian philosophers accepted Aristotle's cosmology in which divine power and control was transmitted downward via the spheres of the planets to the sublunar realm of nature and humans. This transfer applied only to an individual's corporeal state, which made astrology a useful source of knowledge about the conditions of the natural life of the body and how to self-govern it, its medical applications being especially valuable. Astrology could then be seen as practical knowledge that allowed for adjusting and improving a life. Predictions about one's fate were possibilities, not inevitabilities, that were contingent on other factors, which is exactly what Ptolemy had said. What is very clear is that the metaphysical system constructed in the Middle Ages by these friars and others, though primarily Aristotelian, had a place for astrology. It's also clear that these systems of thought were built by philosophers and theologians, not by astrologers. At the same time,

---

3. Albertus Magnus's *Speculum Astronomiae* was a defense of astrology that was referred to, or criticized, for centuries. He regarded astrology as a Christian form of knowledge. See also Aristotle, *De Anima* iii, 4, 9, and Aquinas, *Summa Theologica*, Part 1, Question 115, Article 4. Whether the heavenly bodies are the cause of human actions?

other Christian theologians were skeptical of the subject and argued that most, if not all, of judicial astrology should be completely banished.

My point here is that, following the rediscovery and intense study of ancient texts on Aristotelian natural philosophy during the twelfth century, astrology had a major cultural resurgence because it came to be interpreted by many as being loosely compatible with official Christian-Neoplatonic thought. It was considered to be a natural science, part of the book of nature that, since the time of Augustine, was seen as complementary to the book of scripture. Astrology was also supported by the cosmology of the great and authoritative philosopher, Aristotle. However, most clergy thought the judicial part of it had to be kept in check due to the issue of free will, something that the immortal, non-material, God-created human soul was believed to possess. Augustine saw free will as the source of both good and evil in the world, this being demonstrated in the Adam and Eve myth, where breaking the rules and eating the apple of knowledge was the result of a free choice. Without free will there would then be no guilt or original sin, and without original sin there would be no devil, no need for God's salvation, and ultimately, no need for priests and the church itself. This chain of consequences in theological thinking, and its implications for astrology, cannot be overemphasized.

Extreme astrological fatalism (which most astrologers did not subscribe to) was easily avoided by regarding the planets as able to influence only the body, not the soul, though the planets could have an indirect effect on the latter by stimulating feelings and thoughts. The remedy for this "leakage" between body and mind was to turn to God through prayer for intervention. Another issue important to the clerical monopoly was that prayer had to be understood as having the power to bypass or overrule the planets. So, with these doctrines in place, the influence of the planets, limited to the physical body, weather, or other natural phenomena, was then not off limits for practice or study. Questions and elections, which imply fate and free will, respectively, were a more complicated problem and required a deep understanding of astrology, which most theologians and clergy did not have, so these practices were generally rejected.

THE MULTIPLE STRANDS OF
RENAISSANCE THOUGHT

As noted, the synthesis of the Aristotelian philosophy, astrology, and Christian theological ideas produced by Albertus, Aquinas, and Bacon was stimulated by translations of previously unknown works of Aristotle and astrology into Latin, and also Arab commentaries on them. This process of assimilating new information took place over the course of the twelfth and thirteenth centuries. At this point astrology had become more embedded in the thought of the literate (which amounted to a very small percentage of the populace) than it had been previously. The fourteenth century brought disaster. First the climate cooled, disrupting agriculture, and at mid-century the Black Death ravaged Europe—and then flared up periodically for the next three centuries. This disruption should not be overlooked as a contributor to the expansion of astrology in the centuries preceding the Renaissance and during most of it. The effects of this plague were so extreme and raised so many questions about fate and fortune that astrology was often turned to for answers the Church didn't have. But eventually populations recovered and trade in the Mediterranean expanded. By the fifteenth century the importation of the writings of ancient authors increased, especially those of Plato, Plotinus, and Aristotle, but also texts on Stoicism, Hermeticism, and other exotic traditions like Gnosticism and Kabbalah. These writings, many of which were Greek and came by way of Byzantium (which had fallen to the Turks in 1453), were quickly translated and distributed widely, the process being accelerated with the advent of printing in mid-century. This new and readily available material, much of it pagan, had an even more powerful and disruptive impact on fifteenth-century European intellectuals educated in Aristotelian Christianity than did the recovery of Aristotle and astrology in the twelfth century.

The influx of pagan learning resulted in *the* Renaissance, a European cultural development roughly dated from the fifteenth to the seventeenth century. This was a time when artists, writers, statesmen, theologians, and philosophers (again, a small percentage of the population) explored alternatives to the late Medieval status quo. Two major themes of the

Renaissance were humanism and naturalism. Humanism, the progressive view that focused on the dignity and freedom of the individual, had an early beginning in the fourteenth century with the Italian scholar and poet Petrarch (1304–1374). He thought humans should cultivate their God-given intellectual and creative abilities, something that the ancient Greek and Roman writings offered. Revived classical texts were used by humanists to strengthen arguments for the value of the individual and to downplay the role of the supernatural in human life. The more spiritually inclined found ideas in rediscovered theological texts that placed the immortal individual soul at the midpoint of matter and spirit in a perfectly ordered cosmos ruled by God. Secular humanists saw humans as intelligent and beautiful in their own right, not here just for the glory of God. Both stressed the special nature and value of the human individual and the power of the human mind. Another theme found in the pagan classics was the view of nature as a continuum that has its own laws, a condition that supported astrology and, later, experimental science. Armed with intellectual sharpness (the result of centuries of scholastic critical thinking and dialectical exercise) and possessing ancient manuscripts full of lost ideas, Renaissance thinkers sought new answers to social, philosophical, and metaphysical questions. The abundance of wide-ranging and eclectic worldviews that were produced by the intellectual elite was one of the trends that made the late fifteenth and sixteenth centuries such a unique period in Western history.

A collection of texts that had a particularly strong influence on Renaissance thinking was the *Corpus Hermeticum,* supposedly the writings of the ancient Egyptian god Thoth, known in Greek as Hermes Trismegistus. The Hermetica, which includes the eighteen chapters of the *Corpus Hermeticum,* the *Asclepius, The Emerald Tablet,* and several other texts, is a genre of metaphysical and mystic writings that originated in the Greek-speaking, Roman-ruled world of Alexandrian Egypt sometime between 100 and 300 CE. Most of these writings are in the form of dialogues between a teacher, Hermes Trismegistus (thrice-great Hermes), and a disciple. The spiritual or philosophical contents of the texts are themselves a synthesis of a number of spiritual wisdom traditions extant during that period, a time when the rational traditions of Greek philosophy had ceased to innovate and when much cultural mixing was taking place. At that time a need for a deeper and more emotionally satisfying understanding of reality may have driven a search for solutions using intuition and belief rather than cold logic. In the Renaissance, the Hermetic texts were appealing in large part because they were thought to be of great antiquity; a window into the original and most pure of theologies (*prisca theologica*). It was thought that, since Plato incorporated Pythagorean ideas into his philosophy and Pythagoras allegedly got his ideas from Egypt, then Egypt must be the source of all wisdom. At the head of that wisdom tradition was the ancient Egyptian god Thoth, god of time-measurement, writing, astronomy, astrology, medicine, and alchemy and presumed contemporary of Moses. The wide-ranging spirituality of the Hermetica and related texts was also taken as a universal theology, a cosmic vision that, in ways, encapsulated the teachings of Neoplatonic-influenced Christianity. For these reasons the humanist scholar, philosopher, and priest Marsilio Ficino (1443–1499) was ordered by Cosmo de' Medici, his patron, to translate the Hermetic Corpus prior to translating the dialogues of Plato (Thorndike 1923, vol. I, 287; Yates 1964, 20).

The philosophical and spiritual writings that made up the Hermetic Corpus were attractive to fifteenth- and sixteenth-century European intellectuals as they offered a detailed cosmology and a program for individual mystical contemplation and enlightenment (no priest or Pope needed). In the Hermetic synthesis are elements of Platonism, specifically in regard to God being the All, or One, that exists beyond the material cosmos, which He created. Stoic organicism (cosmic sympathy and the living cosmos) is also found in the synthesis. Another component of Hermeticism, in this case coming from Neoplatonism, were ideas about angels, demons, and the soul's return to its origin in God. In addition, a wide range of Near Eastern doctrines, including the uses of magic, alchemy, and astrology, is mixed into these writings. But because these texts had multiple authors, there is no precise consistency to be found in them, and, as with the Bible, any number of interpretations can be made in regard to specific doctrines.[4]

Hermeticism involved the seeking of knowledge through the use

---

4. See Schumaker (1972) and Yates (1964) for details on Hermeticism.

of alternatives to reason and faith. For example, the intuitive recognition of resemblances and correspondences in nature, which explained the world as fundamentally interconnected and having living properties, was a method of apprehending reality that bypassed reason and did not require faith. The orderly Hermetic cosmos was generally structured as follows: a high god, basically pure mind or intellect, exists over and beyond a creator god that makes the visible universe. Below this god are the planets, with the Sun and Jupiter most dominant, the thirty-six decans of the zodiac (10-degree sections that were used by the Egyptians), and the daemons (forces) that gave the planets their power. Below this were humans, composite beings having a duality of mind and body, mind being a piece of God that seeks return to its heavenly source after death. Below humans, which are (of course) a special case, is the rest of creation. Some key concepts of Hermeticism include vertical microcosm-macrocosm correspondence (as above, so below), the anima mundi, the Sun as the center of the material cosmos, and the use of magic as a means of acting on the world. This cosmology, supportive of planetary influences, did much to promote astral magic, that is the use of astrological symbolism in magical rituals, which in turn did much to discredit serious astrology during the Renaissance.

The notion of the macrocosm being reflected in the microcosm has roots in Greek philosophy as far back as Pythagoras and Plato, and it held a special place in Stoic metaphysics. It is also found in a mysterious and influential alchemical manuscript called *The Emerald Tablet,* ascribed to Hermes Trismegistus (but probably from the sixth to eighth centuries), that contains the line: "That which is below is like that which is above and that which is above is like that which is below." The idea here, seen as a natural law, is that the structures and patterns of the whole cosmos are replicated in its parts, especially in humans. In a modern sense, this appears to be a doctrine of self-similarity, as seen in fractals.[5] One expression of this in astrology are

_____

5. A possible way of describing, in modern terms, these ancient ideas of parallel symbolisms and microcosm/macrocosm doctrine is to consider fractals, patterns that retain their form at all scales. This phenomenon is called self-similarity. Of course, discussion of fractals is normally limited to the domain of mathematics, where they have

the correspondences between the planets of the solar system and the parts of the human body system: Sun equates to the heart, Moon to the stomach and breasts, Saturn the bones, Mars the muscle, and so on. A similar set of linkages involve zodiacal signs as displayed in a figure called the Zodiac Man. This figure, which probably originated during Hellenistic times, has never been out of print and can still be found in almanacs. Both systems of correspondence were used in medical astrology for diagnosis and treatment.

Another expression of the microcosm/macrocosm doctrine in Hermeticism is cosmic communication, the notion of contacting or drawing down higher beings through the use of magic words or by replicating them symbolically with objects such as talismans. This presumed property of nature was the basis of sympathetic magic, a set of techniques that allow humans to participate with the cosmos. In many respects Hermeticism was more than an eclectic philosophy; it could be a kind of personal religion, a factor that put it in conflict with the church. It called for an intuitive, mystical, and imaginative approach, or personal program, through which an individual could seek illumination and ascend toward the divine source of existence—without the aid of any religious tradition, priestly authority, or institution. An enlightened human seeker could become like God (or at least like a pope) through the use of a special kind of knowledge, that is the relations of the parts of the larger system. And Hermeticism was practical. Certain documents, like the *Picatrix,* a massive collection of rituals employing a simplistic version of astrology that aimed to satisfy whatever needs one may have, were circulated among some of the learned elite, including Ficino and Pico (both will be discussed later). In a fragile and uncertain world in which an army could seize a town, or a plague decimate the population, believing that control of the external environment was possible if one made a talisman out of a certain stone during the rising of a specific planet was very appealing.

It was probably the elements of Stoicism blended into the Hermetic

---

been studied since the seventeenth century, but it is the case that fractals appear in nature over a wide range of scales, examples being DNA molecules, blood vessels, crystals, ocean waves, mountain ranges, and the rings of Saturn. Today, what's called fractal cosmology attempts to explain the universe as a series of self-similar structures.

writings that were most important for astrology. In Chapter 3 Stoicism was addressed in regard to its influence on Hellenistic and Roman astrology. Recall that it was a highly respected synthetic philosophical system founded in 300 BCE in Greece by Zeno of Citium that included a complete metaphysics and natural philosophy, in addition to logic and its better-known ethical philosophy (Hahm 1977, 136; Sandbach 1975). The Stoics taught that the cosmos—the orderly, physical universe—was a dynamic continuum, a massive, living, intelligent being having an eternal cycle of creation and conflagration. Stoic cosmology was a monistic pantheism (one god everywhere), a materialistic vitalism, or what is sometimes called cosmobiology, and from a more contemporary perspective, a closed system that is God in its totality. Stoic natural philosophy influenced Neoplatonism, but by itself was not compatible with the Christian view on nature, which saw God as a non-material entity, separate from the world that He created out of nothing. Stoicism also differed from Christianity in its doctrine of causal determinism (fate) and in its emphasis on self-control and self-reliance, not on prayer.

The Stoic natural philosopher Posidonius has been associated with the concept of sympathy as an explanation for the coordination and

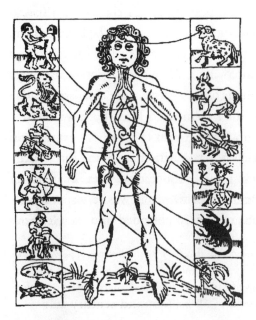

Figure 25. From a woodcut in a 1702 almanac (probably reproduces an earlier image).

interconnectedness of natural phenomena. Cosmic sympathy assumes correspondences among the various components of the living Earth, celestial phenomena, and the entire Stoic cosmos. It is in the doctrine of sympathy that the notion of two-way communication between the parts and the whole is based (there are similarities between this concept and modern systems theories, (e.g., feedback loops)). In ancient astrological theory, all phenomena were classifiable by the symbolic language of the planets. Each planet had a domain and was said to rule or resonate with a specific set of phenomena within that domain. An alignment or positioning of a specific planet was thought to produce a kind of vibration that was able to affect, through sympathy, all the objects, organisms, and systems of the living Earth that were receptive to that specific planetary energy. For example, all things that resonate within the domain of Mars (i.e., outgoing force, war, anger, wounds, blood, weapons, iron, the color red, etc.) would be affected or modulated by that planet's motions and configurations with other planets. Venus would affect distinctly different things (ingoing force, attraction, art, beauty, copper, sensuality, love, etc.), and so on for the other planets and the Sun and Moon. This planetary taxonomy, along with the doctrine of sympathy, and also its opposite, antipathy, served as a pathway for human participation with nature, and it is the basis of astral magic. Astrological symbolism alone can then be used for more than interpretation and analysis of a birth or of natural phenomena; it can be used as a code or tool for furthering the ends of those who desire to participate and act on nature—and without the requirement of tedious trigonometrical calculations.

An extension of cosmic sympathy was the doctrine of signatures, a Medieval system of classification with ancient roots in plant medicine that is also found in Hermeticism and is the basis of much of natural magic. Ancient writers on botany and medicine, such as Pedianos Dioskorides (c. 40–90), the author of *De Materia Medica,* a well-read summary of ancient herbalism, drew connections between how plants looked and how they acted on the body when used as medicine. These correlations were extended to include planetary and zodiacal rulerships to facilitate the use of herbs in astrological remedies and magic. An example is Thessalus of Tralles, a second-century author of

De Virtutibus Herbarum who produced a botanical and astrological work, which claimed links to ancient Egypt and the astrologer-priest Nechepso (one of the alleged founders of astrology), that offered inspired medical prescriptions for bodily problems (Piperakis 2016). This herbal manual and collection of remedies based on cosmic sympathy was a predecessor of Renaissance astrological herbals (e.g., Culpepper's seventeenth-century herbal) and shows how the history of Western plant-based medicine has roots in Hermeticism and astrology.

The Stoics saw nature as composed of diverse parts, interconnected in complex ways, within a dynamic continuum (the universe) where the same laws operate everywhere. These connections were describable through the qualitative language of astrology, and the empirical justification for this taxonomy of quasi-organic linkages was the actual astrological effect of the planets, which most seemed to agree was real. Stoic physics (the laws of the dynamic continuum and causal determinism) then provided a naturalistic explanation for astrology that was not based on any duality of soul and body. For astrology, this description was also more or less compatible with the hierarchical models of Aristotle and Ptolemy—but it had conflicts with the Christian worldview.

Fate was a feature of the Stoic cosmos and was regarded as an effect of causation, a natural tendency in nature. Stoic doctrines of determinism (called causal determinism) were universal, that is to say everything that has happened, or will happen, has a prior cause. It follows that if every event has an antecedent cause (or a swarm of them), then clearly a deterministic force must be operating. If a certain cause always tends to produce a certain kind of effect, then event prediction is possible, this being how medical diagnosis works. Stoicism teaches that while fate is an ongoing and complex process in which multiple strands of causation move forward in time, within these constraints there is some room to move. This philosophical position, called compatibilism, is one where determinism and free will are mutually compatible. The Stoics thought that individual choices are shaped by the personality, which is fixed at birth or during early childhood. People then make decisions according to their natural tendencies in the context of the causal web they exist within. Being conscious and aware of external circumstances, and following one's true nature in a virtuous way, preserves a measure of free will, but because of the existence of greater, and impossible to manage, causal forces, individuals were advised to focus on the elements of their life that were truly within their power and not be distracted by what was out of their control. The second head of the Stoic school, Cleanthes, is reported to have said, "Fate guides the willing, but drags the unwilling," implying that fighting what is beyond personal control could be counterproductive (Epictetus, Enchiridion, 52, Seneca, Epistles, cvii, 11).

Stoic ideas on fate entered the Renaissance through the writings of ancient astrologers. In Hellenistic and Roman astrology, a personal fate would be seen in the pattern of planets at one's birth. Ideally, individuals could use astrology to understand their fate, accept it, and then work with and improve it. This connection between Stoic acceptance of fate and natal astrology was made clear in the "Astronomicon," an early first-century CE Roman poem, and also in the writing of Vettius Valens and Ptolemy. Manillius, an author about whom little is known, argued that knowledge of one's fate obtained from the details found in astrology does not diminish a person's achievements in any way, and that with it, humans have the ability to delve more deeply into their own fate and better understand it (Manillius 1977, book 4).[6] Astrology then provided useful information about initial conditions and personal tendencies that could add to self-understanding and be applied to conscious decision-making. This appears to be a realistic and sophisticated view of human life and also a prescription calling for personal knowledge, a clear awareness of personal circumstances, and intelligent decision-making to relieve the pressures of relentless causal determinism. There is no denial of free will here, just a realistic acceptance of its limitations. Stoic doctrines were demanding and required self-knowledge and personal insight (astrology was especially helpful in this regard), but they were ultimately incompatible with Christian-Augustinian ideas on human free will being the source of good and evil, prayer as a direct connection to God, and God as all-knowing (which includes having

---

6. Ptolemy and Valens (see Valens, *Anthologies,* Book IX) are also considered to have been proponents of Stoic philosophy.

knowledge of future choices people will make using their free will!).
Astrology had plenty of critics, including some Stoics, but it was generally not criticized on scientific or theoretical grounds (as an explanation
for how the physical world works); it was criticized on philosophical,
moral, and religious ones. This topic will be covered in another chapter
and will include a summary of anti-astrology arguments.

Two other metaphysical or spiritual traditions were also significant strands in the landscape of Renaissance thought. Gnosticism was
a form of early Jewish Christianity. Very roughly, as there are different
accounts, the cosmology of this tradition begins with God, the One,
or Monad, out of which emanates the aeons, male/female pairs, out of
which comes the Demiurge, which then creates the physical world, a
denigrated place where humans live. Gnosticism taught that humans
had a spark of God within them that could be accessed by gnosis, or
personal knowledge of God, not just faith, and this spark would later
reunite them with God. In contrast to Hermeticism, which encourages
learning about the material world, Gnosticism encourages escape from
it. Elements of Gnosticism survive today in the Middle East as a small
population known as Mandaeans.

Kabbalah, which is a form of esoteric Judaism largely based on a
text called the Zohar, also influenced Renaissance thought as certain
aspects of it were seen to be compatible with Christianity. Its earliest origins are debatable, but it did appear in Spain and in southern
France around the twelfth or thirteenth century. The central feature of
Kabbalah is a structure called the Tree of Life, a symbolic map consisting of ten Sephirot that depict a series of stages and their interconnections from the first emanation of God (Keter) to the densest level of the
creation (Malkuth). Another feature is the use of numbers not as quan-
tity, but as symbols of broad concepts (essentially numerology), and also
the power of numbers, words, and letters to access different levels of the
cosmos. Knowledge of the linkages between stages, which can be activated with the right combination of words, numbers, or letters, made
Kabbalah also a kind of magic, a way to make contact with angels.

Jewish Kabbalah had a powerful influence on Pico della Mirandola,
the humanist who launched a brutal attack on astrology at the close

of the fifteenth century. He, along with others, found in Kabbalah ideas compatible with Christian thought (in that context it is sometimes called Christian Kabbalah or is spelled Cabala). Gnosticism and Kabbalah were more or less in line with the general trend of Renaissance thought: idealist metaphysics, the centrality of humans in a multi-level cosmos, the free will of the individual, the possibility of cosmic participation by directly accessing higher levels of being or even merging with God. Along with the above esoteric philosophies, Neoplatonism, Stoicism, and Hermeticism were combined in various ways during the Renaissance by the leading thinkers, including some proto-scientists. It is difficult to say whether someone was a Hermeticist, a Neoplatonist, a Stoic, or anything else because the doctrines tended to overlap to a large extent (Yates 1964, 250–251). Individuals drew freely from each of these spiritual programs, but most also attempted to reconcile them with Christianity, the dominant ideology. Given the theological constraints and the power of the Church during the Middle Ages, the rediscovery of these strange and exotic philosophies and practices was as disruptive as it was enlightening, a difficult mixing that appears to have induced a kind of collective cognitive dissonance among the intelligentsia that lasted for roughly two centuries.

A number of nature philosophies emerged from the various philosophical streams that converged in the fifteenth and sixteenth centuries in Italy and other parts of Europe. Many of the intellectual elite of the time, including Marsilio Ficino (1433–1499), Gerolamo Cardano (1501–1576), John Dee (1527–1608/9), and Giordano Bruno (1548–1600), held views that considered nature as a unity, governed by its own laws, alive in a sense (anima mundi) and also astrologically-structured, holistic, and organic—basically a system. The rediscovered ancient organismic doctrines, with roots going back to the pre-Socratic philosophers of nature and elaborated in Stoic physics, provided them with ideas about a living Earth that follows universal laws, the vertical model of spirit and matter, and the self-similarity of the macrocosm and microcosm (shades of Koestler's holons). These ideas, force-fitted in some cases with Christian theology, stimulated the formulation of many original and transitional natural philosophies that fed into the rise of experimental science in the seventeenth century. Although there were significant variations among them, there were enough similarities that the labels Renaissance naturalism or Renaissance natural philosophy serve to group them in a practical way, and I will use these from here on.[7] The inherent conflicts between Christianity and ancient variants of organicism, however, were extremely difficult to reconcile, and Renaissance natural philosophy could not completely escape the thought police of the church. We will return to this issue in another chapter when Pico della Mirandola, the humanist who vehemently attacked astrology, will be discussed.

7. Shapin (1996, 43–44) uses Renaissance naturalism in contrast to the new mechanical philosophies. Copleston (1962b, 248; 1985) finds it impossible to label this trend in thinking under one label—except Renaissance natural philosophy.

8

THE EVOLUTION OF THE DOMINANT IDEOLOGY

The truth is that we can always find previous world views lacking if we judge them in our terms. The price paid, however, is that what we actually learn about them is severely limited before inquiry even begins.

MORRIS BERMAN,
THE REENCHANTMENT OF THE WORLD

Within the context of an all-encompassing ideology, world view or conceptual system, the boundaries between fact and fiction are usually quite clear. However, within another global world view these boundaries are perceived differently. The existence of facts that are entirely objective is then hard to prove and we are left only with interpretations of reality, whatever that is.

Many historians have attempted to explain the phenomenal rise of experimental natural philosophy in the seventeenth century by focusing on individuals, ideas, events, technology, and cultural trends. Thomas Kuhn, in *The Structure of Scientific Revolutions,* pointed to basic historical changes that he called paradigm shifts as the operative force behind the radical developments in science during the seventeenth century. These shifts, when they occur, mark deep changes in the practices and concepts, essentially the ontology (the basement layer in the understanding of being or reality), underlying a scientific discipline (Nescolarde-Selva et al. 2017; Berger and Luckmann 1967). Sociologists similarly consider historical changes in terms of the observable differences between societies and their definitions of reality, this subject matter falling into the domain called the sociology of knowledge. *Ideology* is the term often used by philosophers when talking about a worldview, which itself contains an ontology and set of values, religion being the classic example. Staying with this term, ideology is something that has evolved along with human society; myths as explanatory narratives were among the earliest structural features of civilization. Societies, historical and modern, generally have a dominant ideology that contains common features of other subordinate ideologies that co-exist in a culture.

While descriptive terms for the basic assumptions of a culture at any given time in its history may vary, there is a kind of group consciousness that emerges from countless individual communications, distinctive to each particular group, culture, or society that allows for the transmission of beliefs from generation to generation. This field of consciousness, which functions as a framework that determines what is known and establishes values for both the group and individuals, could also be defined as a special type of system in that it is self-organizing, synchronized, becomes more complex, and can exist over long periods of time without much modification. But at times, due to interruptions, new information, internal conflicts, or confrontations with very real events (e.g., political power shifts, plagues, weather events), the collective field of consciousness enters a state of crisis and is forced to change. In Western culture one of the biggest shifts in fundamental assumptions and beliefs occurred during the Renaissance when a sea change of thought was initiated by intellectuals who were processing the rediscovered thinking of the ancient world. Over the course of roughly three centuries these changes in thinking impacted cultural institutions and eventually trickled down to the public.

The French historian and philosopher Michel Foucault, a self-described archaeologist of consciousness, argued in his book *The Order of Things* that the rules of scientific (or other) knowledge will change when a historical episteme changes. For Foucault, an episteme is the epistemological field, or roughly a worldview, a thought system, the unconscious sum of general assumptions, the dominant ideology, or perhaps a cultural paradigm, within which knowledge conditions history and controls collective perception and values. He described the general mindset of sixteenth-century Renaissance naturalism as one based on an episteme of kinships, resemblances, and affinities (Foucault 1970, chapt. 3). To know anything in that world view required knowing how things were related, similar, or analogous. Foucault viewed the perceived connections between things in any episteme as a closed loop, an all-constraining preconceptual grid that determined what was possible and true within it. His description of Renaissance naturalism appears to apply mostly to the worldviews of Stoicism, Neoplatonism, and Hermeticism, views that include cosmic sympathy, the doctrine of signatures and macro-microcosm interchangeability, and other natural laws that supported astrology, magic, and alchemy. I would add that this kind of knowing is essentially built on pattern perception.

In maddeningly dense prose, Foucault argued that this grid of consciousness, or Renaissance naturalism, had to break down in order for the modern world to come into being and, beginning around 1600, it did. What happened was that in the early seventeenth century Bacon and Descartes introduced a way of using thought to project order on the world by focusing on distinctions and measurements. This new thought (Foucault calls this collective mindset the "Classical episteme") included strong critiques of resemblance as a way to understand and describe the world. These critiques led first to changes in how knowledge should be created and, consequently, a major reprioritizing of things. They also favored, that is selected for, a very different kind of thinker. Previously it was the ability to perceive connections, patterns, meaning, and significance that was central to knowing; later it became the ability to perceive distinctions and think in mathematical terms, these two epistemes being mutually exclusive. A reshuffling of priority is then what happens in those rare occasions when a change of episteme occurs.

Foucault's model suggests that what is true in one episteme may not be in another and that truth may be a relative thing, an issue that falls down the slippery slope of postmodern relativism. Assuming the post-modernist position that a society's definitions of reality are conceptual constructs built on ideology and language, it is easy to see how astrology as a subject could be regarded as true in one century and false in the next. In the Renaissance episteme, astrology had acceptable evidence. It described the interconnectedness of the world through a kind of qualita-tive cosmic taxonomy, much of it expressed using a framework of geom-etry, in which a host of widely differing objects, organisms, and processes were thought to operate within organized and distinctive domains. Observable effects, such as weather, a person's fate, or bodily health, in the context of these domains was sufficient proof of the validity of astrol-ogy for most people, including the intellectual elite, of the time. In the Classical episteme of the seventeenth century, that taxonomy and its evi-dence made no sense and was useless, so astrology had to fall out of favor. The implication here is then, if reality is something that is relative to a cultural mindset, as in postmodernism, there is no need to look any fur-ther for why astrology declined. I think it is more complicated than this. How ideas in a culture evolve and take on new forms and reshape defini-tions of reality is a worthwhile project for historians and sociologists, but in all cases of worldview transformation, most commonsense knowledge remains real. My argument is that astrology does measure things, in its peculiar way, that are quite real and transcendent of any episteme, just as gravity does, and people can see and experience it firsthand, but there was simply no way to put precise numbers or sharp boundaries on such things in the classical episteme. That worldview demands quantification through reduction as a prerequisite to acceptable knowledge.

One criticism of Foucault is that he based his socio-archaeological model on the assumption that Neoplatonic/Hermetic thought was most typical of the Renaissance. But the logical language of Aristotelian Scholasticism was still a dominant framework of thought well into the seventeenth century, and there were probably many who could manage more than one world thought system. "Truth" during the Renaissance was in constant flux as various combinations of ideologies circulated among the small percentage of the population who were well-educated. Consider that today many people are able to live with multiple and even conflicting views such as the forced blending of Christianity and capitalism. Whether those who are drawn into this particular mix suffer from cognitive dissonance is not always apparent, and most live normal lives within the larger framework of the dominant ideology. Maybe episteme is an over-simplification of a historical period's domain of knowledge taken to be a generalized cultural phenomena, but I think it is a useful concept in beginning to sort out why experimental natural philosophy succeeded while astrology declined.

A major driving force in the thought crisis of the Renaissance that led to the Scientific Revolution was how to process and organize the flood of new information that had been injected into the prevailing religious worldview. There was the need to understand how rediscovered ancient philosophies and theologies that granted personal knowledge and spiritual power to the individual should, or should not, be mixed with Christianity. The constant fighting over beliefs and doctrines related to human freedom, seen in the Humanism of Pico della Mirandola and the Protestant Reformation set off by Luther, are evidence of a profound cultural conflict between this new emphasis on the individual and the constraints imposed by the traditional authoritative religious institutions. Natural philosophy was also a part of this conceptual crisis. A loose mixing of newly invented versions of natural philosophy and mystical theology provoked a reaction from the more critical thinkers who wanted to understand nature on its own terms. And then, in the early seventeenth century, a solution to this collective mental chaos came about in the form of a method that solved multiple problems in one stroke. This was a new kind of natural philosophy, one that used quantifiable data, experiment, and mathematical modeling. It could be used to read the book of nature (God's creation) without resorting to theology and, from it, make practical knowledge, all the while respecting the actual book of Christian scripture, the Bible.

The decline of natural magic, alchemy, and astrology in the seventeenth century was coincident with the rise of the new experimental method. Given that many natural philosophers were also involved in these ancient subjects, a case can be made that elements of these traditions played formative roles in the Scientific Revolution. On the surface, the astrologer's need for better data spurred on astronomy, and alchemists experimenting with substances pioneered chemistry and the experimental method; but deeper transfers have been noted. Both astrology and alchemy contained the notion that nature had laws that could be discovered. In her book *Giordano Bruno and the Hermetic Tradition,* Frances Yates argued that the natural magic of Renaissance naturalism encouraged a manipulative orientation toward the world. The concept of cosmic sympathy, for example, does suggest that knowledge of how the planets worked (laws of nature) might be used to influence the things and processes under each planetary domain. If this was so, then the magician could take the appropriate formulas for talismans and rituals and make contact with higher realms, learn more about nature, and maybe even reach God—goals of Neoplatonism and other ancient mystical philosophies. Within the framework of these purposeful, and also experimental, activities a new kind of practitioner arose—the Renaissance magus. This category included personages like Ficino, Pico, Heinrich Cornelius Agrippa, Dee, Bruno, Tommaso Campanella, and Robert Fludd, all of whom were among the most learned men of their time.[1] They were deeply involved with metaphysical, philosophical, scientific, and technical knowledge, more or less in the context of Neoplatonism, Hermeticism, and Cabalism. But they also had an experimental attitude toward their work, and through the use of sympathetic (natural) magic and some knowledge of alchemy and astrology, they set out to manipulate nature.

For centuries Scholasticism had contemplated the world from a university tower using the logic, critical thinking, argumentation, analysis, and classification methodologies of Aristotle, all in the service of Christianity. Here, maintained by generations of schoolmen, was to be found the most detailed and authoritative knowledge produced since antiquity. In contrast with this cerebral and elitist learning was the Renaissance magus. The magus was comfortable with the hands-on world of technology, that is willing to get hands dirty working at the forge or in the fields and woods picking herbs. He believed that nature had laws and

---

1. They also had many opponents, however, which underscores the point that not everyone at the time was in agreement with Renaissance naturalism.

he intended to use these to act on nature. Yates argued that the role of the magus was a matrix for the modern scientist—a role in which practitioners use tools and equipment to experiment, measure, and harness nature and thereby discover its laws. During the Renaissance, natural philosophy, which the universities included in the scholastic curriculum, came to include the skill sets of the technical trades and epistemological framework of the Hermetic tradition, both being active ways of addressing the natural world and controlling the environment. The sixteenth-century Renaissance magus could then be seen as a prototype for the experimental scientist of the seventeenth century. Yates's argument makes sense considering that alchemy is an experimental activity that attempts to reduce substances into individual components and requires tools and apparatus, these being the prerequisites for laboratory work. Astrology requires accurate astronomical data, hence the need for observation, equipment to do so, and the quest for better mathematical models of solar system dynamics. Take away the Hermetic, Stoic, and Neoplatonic metaphysical philosophy that was allied with alchemy and astrology and they look a lot like chemistry and astronomy (Yates 1964, 155–156; Berman 1981, 99).

The Renaissance magus was not representative of all philosophers of nature. Some less inclined toward magic chose to explore and experiment with certain assumptions of Renaissance naturalism in other ways, including the components within the cosmic system that were thought to communicate in some way with each other. (Today we might see these as internal feedback loops in a self-organizing system.) It does appear that the forces or processes connecting the parts of the Earth and cosmos, explained by sympathy, became targets for investigation by natural philosophers like William Gilbert (1544–1603) and Kepler. Both attempted to map out these connections through close observation, experiments, and mathematical modeling. In doing so, both replaced the loosely defined vital energies and sympathies of the anima mundi with more or less mechanical ones in their work, although they may not have intended to do so. Gilbert, after many experiments, concluded that the invisible force that moves a compass was actually a property of the Earth itself; that it had a magnetic field. Kepler projected Gilbert's Earth magnetism to the other solar system bodies, proposing that magnetic links between the Sun and the planets kept the latter in their orbits. By the seventeenth century, natural philosophers were moving away from organicist metaphysics and theology and replacing qualitative descriptions with more mechanical explanations that had the potential to be measured quantitatively using mathematics.

Recall that Kepler thought the Earth to be alive, the proof of this for him being the efficacy of astrology, and he argued that the planetary aspects, the angular separations (phase angles) between the planets viewed geocentrically, produced a kind of mathematical harmony to which the Earth resonated (Kepler 1997, book IV; Kepler 1987). For him, Earth had to be alive if it could hear this harmony, and the fact that both weather patterns and human behavior corresponded with the changing aspects was proof enough for him. But the description of his proposed resonance phenomena, which he attempted in *Harmonics of the World,* was of necessity incomplete as units for astrological effect were elusive. In contrast, Kepler's work on the mathematical "laws" of the planetary orbits, which he regarded as a sign of intelligence behind nature, was precise. These were a wonderful discovery for Kepler, who called himself a Lutheran astrologer, and was very much in line with the trend in natural philosophy away from metaphysics and toward a search for concrete evidence of God's laws in nature. The fact is, however, that his third law of planetary motion ultimately led (in Newton's reworking of it) to the confirmation of a mechanical view of planetary motions that ignored any mysterious forces associated with astrology.[2] In ways like this the bizarre organic cosmologies of Renaissance naturalism stimulated the careful investigation and mapping of nature that became the central theme of experimental science, and then these cosmologies were replaced with "facts"—quantitative evidence of objective phenomena in a neutral (non religious) context. And then equations became explanations.

---

2. Newton wasn't able to explain the action at distance of gravity, leaving it as a kind of occult force. This hole in his system would include the tides, a topic traditionally considered astrological and debated for much of the seventeenth century.

THE RISE OF EXPERIMENTAL SCIENCE AND THE MECHANICAL PHILOSOPHY

At the same time that ancient treatises on philosophy, metaphysics, natural magic, and other topics were being translated and discussed in Europe, there was also a recovery of ancient texts on engineering and mathematics. These stimulated an experimental attitude among scientifically inclined readers like Leonardo da Vinci and, later, Galileo. Archimedes (287–212 BCE) and Hero of Alexandria (c. 10–70 CE) were inventors and engineers whose surviving notebooks and writings described gadgets and mechanisms such as windmills, steam engines, pumps, and water-powered automata. The automata of these ancient engineers were mechanical animals or people that moved, did tricks, and appeared to be alive and, when built by Renaissance mechanics and engineers, were often publicly displayed in gardens or fountains. These hydraulic and pneumatic constructions involving gears and other moving parts could be extremely complex and, through their sequences of actions, even tell stories. At the same time, clocks, which shared a history with automata and were also on public display, became increasingly complex and able to do much more than simply tell time. Interest in these mechanical wonders was very high during the sixteenth century, and they no doubt played a huge role in the rise of the modern scientific method, which is based on the usefulness of experimentation and acceptance of a mechanical philosophy. Here was a situation where the magic of automata, in plain sight in the form of gears and springs, could replace some aspects of natural magic and also damage its reputation. Automata surely had something to do with the notions of a clockwork universe (and who was the clockmaker and what was his role?), and over the next century their visibility influenced both scientific theory and religion (Grafton 2002).[3]

The actual rise of the new experimental, mechanistic, and mathematical-modeling scientific method, what might be called the lift-off point, can be reasonably dated to the early seventeenth century.

---

3. The influence of mechanism on social theory is also apparent in Thomas Hobbes's *Leviathan* (1651).

Interestingly, during the late sixteenth century predictions of a transition to a new historical era of considerable significance were being discussed by astrologers. Their focus was the 1603 conjunction of Jupiter and Saturn, the first in eight hundred years in a fire sign. The conjunction, which occurred in Sagittarius, traditionally the sign of philosophy and knowledge in general, was joined by Mars the following year and, amazingly, was sweetened by the surprise appearance of a supernova close to it, an event that riveted the astronomical and astrological communities. Kepler measured the event precisely and determined that this new star was not a planet, nor was it an effect produced by the triple conjunction. His observations showed him that it was located well beyond the realm of the planets into the zone of the fixed stars where things (in the Aristotelian cosmos) were thought to be permanent and unchanging (Granada 2005). But he hesitated to make any specific predictions based on it other than it might be a sign from God, though exactly what God was saying he didn't know. He did predict a flood of predictions and ridiculed other astrologers who he thought were going public with forecasts based on flimsy assumptions. Kepler probably thought the Jupiter-Saturn conjunction by itself was an astrological marker of changes, however, as he had argued that a similar three-planet conjunction in 7 BCE marked the birth of Christ and, of course, everyone knew this began a new era.[4] Kepler's critical thinking, attention to details, and sense of responsibility as an astronomer-astrologer are evident in his book on the new star (*De Stella Nova in Pede Serpentarii,* 1606) and also in his debate over it with contemporary astrologers. His disagreements with them reveal a widening of the gap between progress in technicalities in astronomy and traditional practices and assumptions held by practicing astrologers.

In the years after 1600 big changes were certainly afoot in regard to the development of a new kind of natural philosophy. Early in the century Francis Bacon laid out a program that rejected Aristotle, in fact it was intended to replace Aristotle as an authority. In his *Novum Organon,* published in 1620, he criticized Aristotle's empiricism and argued that

---

4. The following Jupiter-Saturn conjunction occurred in 1623, but it was too close to the Sun to be observed carefully. That conjunction (and the one in 2020) was very close, the two planets being separated by about 5 minutes of a degree.

the method did not express any real, useful insight into things as it uses artificial constructs of logic. Bacon argued for improvements in making knowledge, specifically a more meticulous method of data collection to be followed by experiments. His basic idea was that nature must be quantified and forced to give up its secrets. This is reductionism, the analysis of parts that, along with abundant skepticism, mathematical modeling, and materialism, are the main pillars of science as we know it today. From the close observation of the parts and the accumulation of data, a general understanding, or theory, is produced through inductive reasoning. This "bottom-up" procedure, known as the Baconian method, was a major influence on natural philosophers in Europe, and this method has had far-reaching consequences. For Bacon, nature is there to be dissected, mastered, and used. This is a view that, on the one hand, has similarities to that of the Renaissance magus, but on the other it ignores notions of a living cosmos and treats nature as dead and like the inner mechanisms of automata and clocks. Because it is fundamentally manipulative, the Baconian view also conveniently facilitates capitalism, the economic system based on exploitation that grew alongside experimental science. This linkage will be returned to later.

In the first decades of the seventeenth century, as Bacon's ideas spread and Kepler worked on the orbits of the planets, a second phase of change in knowledge-making began and the transition to the new mechanical and reductionist natural philosophy accelerated. An important part of this process was the rise of skepticism, an attitude toward knowledge that was stimulated by accounts of ancient Greek skepticism found in Cicero. This attitude is apparent in Pico della Mirandola and others earlier in the Renaissance, but by the mid-sixteenth century a translation of the philosophical works of Sextus Empiricus (160–210 CE), the major source of Pyrrhonian skepticism, became available. His writings, which stressed suspension of judgment and the limits of human faculties in acquiring knowledge, were widely read by natural philosophers during the next century. The revival of skeptical thought did not offer a philosophy that explained life or nature, it merely stressed methods of critical thinking. The timing of this revival was significant—skepticism was a needed remedy for the chaos in philosophical thinking that followed the rise of Renaissance naturalism and the decline of the Church as a central authority following the Reformation.

Another major source of ideas in the birth of the new natural philosophy was atomism. Around the start of the seventeenth century this ancient naturalistic philosophy, first described by the fifth-century BCE presocratic Greek philosophers Leucippus and Democritus, and later elaborated on by Epicurus and the Roman poet Lucretius, experienced a resurgence of interest. The basic doctrine states that the universe is composed of discrete material elemental particles moving in a void. These "atoms" and the void are all there is. They are indivisible, indestructible, of infinite number, and come in different shapes that allow them to form connections with each other and thereby create the world we experience through our senses. Atomism is thus a materialist view with an ancient pedigree that reduces nature to minute pieces of matter in motion, a mechanical determinism that turns out to be very compatible with a reductionist methodology like that of Bacon. It was also acceptable to Archimedes, the great mathematician and engineer of the ancient world who was held in high regard by Galileo. Archimedes only cited one philosopher in his surviving works—Democritus. Unlike the organicist philosophies of Renaissance naturalism, atomism offered a metaphysical explanation for the emerging mechanical and materialist worldview. The only problems were that it was tainted with determinism and the atheism of Epicurus.

Like Renaissance natural philosophy in general, there were many different versions of atomism. Bacon and Galileo used classical atomism as a way to explain elemental state changes (solid to liquid to gas) and how the senses are impacted by objects. René Descartes (1596–1650) came up with his own version of atomism, one in which the world is made up of tiny pieces of screw-like matter described as swirling vortices. Gassendi (1592–1655) was more forthright in restoring and elevating atomism by Christianizing it (God creates the atoms), though he did include the animistic notion of vital heat as an explanation for living phenomena. Corpuscularianism, a replacement for alchemical explanations of matter, was a materialistic theory similar to atomism, except that the corpuscles making up matter were divisible. It was the explanation of natural phenomena used by Robert Boyle, who also first used the term "mechanical philosophy." Atomism in any of its forms is mechanistic and deterministic, that is it implies that all is matter in motion and causes of this motion can be traced and predicted. Like other metaphysical theories, it was untestable and controversial at the time and remained so until the rise of a far more precise atomic theory in the twentieth century, but it served as a bridge from a holistic, organicist metaphysics of resemblances to one of reductionist materialism.

The philosopher John Locke (1632–1704), a contemporary and a friend of Boyle, was influenced by corpuscularianism and the mechanical philosophy, though his thought was focused more on consciousness and social matters. He argued that people were not born with innate ideas; they were, like machines, subject to outside forces and therefore a tabula rasa, or blank slate. It was then the experiences of living, not an immortal soul, that made the continuity of the self apparent. Locke's influence on the modern era, from the founding fathers of the United States (all men are created equal) to behavioral psychology and the nature versus nurture debate that continues to the present, was another way in which the mechanical philosophy, which had no place for astrology with its unquantifiable hidden influences, qualitative measurements, and challenges to assumptions about free will, shaped the post-Renaissance world.

At more or less the same time as Bacon, Descartes argued that nature was just a machine that exists to be measured and used, and it was also separate from humans. Descartes was a rationalist who used methodological doubt as a tool for the discovery of truth and the apprehension of reality, not the methods of Scholasticism. In his view, the mind must be skeptical so that it can first eliminate uncertain beliefs in order to discover innate ideas and thus be capable of finding certain knowledge. Descartes argued that mathematics is the most solid platform on which to build a philosophy of nature as it allows for a kind of certainty not found in other approaches to knowledge. This practice of placing data in a mathematical context is, like the Baconian method, another cornerstone of the new science. Descartes's view of the world was also atomistic and therefore materialist and deterministic; nature was dead and predictable. But while Descartes built his case for the creation of a mathematically structured mechanical philosophy, he became stuck at the problem of mind and how it related to the objective world, which is the mind-body or consciousness-materialist problem, and he left a legacy of dualism, the separation of mind and matter in humans.

In Cartesian dualism, also called substance dualism, the mind (consciousness) is separate from the body, but controls it like it would a machine. People were a special case, however. The tenuous and problematic connection between the human mind and the body, the two substances, he explained as something that is mediated by the pineal gland, but this was not true of animals and other living things, which he considered to be machines. His dualism was a big step away from the various mixtures of body and soul in Renaissance naturalism, and it led to a split in natural philosophy that continues to the present. Descartes argued that there was only matter in motion and cognition; matter being a measurable substance to be studied by experimental science. The study of cognition, however, became the domain of modern philosophy and psychology. Descartes's dualism has persisted into the twentieth century, one example being the practice of allopathic medicine, which treats the body like it is a machine.

Meanwhile in Italy, another contemporary of Bacon and Descartes found ways to blend their ideas. Galileo (1564–1642), more than anyone else, put mathematics and experiment into one package. By rolling balls down inclined planes, and in similar gravity-physics experiments, he was able to quantify terrestrial motions with mathematical formulae and move toward a confident conclusion that nature could be decoded using mathematics and that truth in nature was to be found in mathematical facts. In this way Galileo established the archetype of the modern scientific method, at least for experimental physics. In his experiments he demonstrated clearly that nature could be mastered by taking parts of it out of context and placing it in a study situation where it could be measured quantitatively and reduced to an equation. Galileo thus brought Bacon and Descartes together in a philosophical view of nature where the certainty of quantification trumps description, resemblance, and pattern perception. It is this manipulative and reductionist approach to nature that allows for mathematical modeling and consequently precise predictions that are visible to all. The early seventeenth century was the point in history when scientific methodology in its modern form was first developed, unquestionably one of the greatest accomplishments of *Homo sapiens,* comparable in ways to fire-making, agriculture, religion, and urbanization.

Bacon, Descartes, and Galileo were the pioneers in the movement away from the thought systems of Renaissance naturalism, a process that was more or less complete by the end of the seventeenth century. Simultaneously, the astrological, alchemical, and magical traditions central to Renaissance holism were abandoned by leading natural philosophers, defeated in public philosophical and theological arguments, and replaced by a very different kind of thinker. During the eighteenth century nearly all natural philosophers came to adopt one variation or another of a basically reductionist-mechanistic-materialist (RMM) view of nature, which had proven itself over and over with its predictive power.[5] By declaring loud and clear in their writings that the reliable and verifiable knowledge found in this way (i.e., by reading the book of nature) revealed the glory of God, the architects of the new natural philosophy were able to adapt it to the still-dominant ideology—Christianity. The mechanical and experimental approach to nature, with mathematical quantification and aided by instrumentation, was at first applicable to astronomy, physics, optics, ballistics, and, a bit later, chemistry. Other disciplines, like botany, biology, geology, and medicine were more Baconian and lagged behind, relying primarily on observation, accurate documentation, classification, and organizational models. The transition from holistic knowledge, based on resemblances with the parts explicable only in the context of the whole, to knowledge based on isolated and quantified data produced by close observation and experiment that conformed to mathematics, was then complete.

Astrology, the effects of which cannot be easily reduced to measurable units, could not be quantified mathematically. Documentation or tabulation of its effects, like the astrometeorological studies of Kepler

---

5. While the history of science and its evolution since the Renaissance is complex and nuanced, reductionism, mechanism and materialism did become central themes in how science is done and understood today. My choice of the abbreviation RMM is one that I consider to be a useful generalization for the purposes of this book.

and Goad, were still subjective and qualitative as the appropriate instrumentation had either not yet been invented, was hard to come by, or lacked precision. And the astrology of people (judicial astrology), which posed far more complex problems in regard to reduction to units, was simply untenable. I would argue that this problem, inherent in a subject that attempts to map self-organizing systems, is a major reason why astrology could not be a part of the fundamental reorganization of perspective and method among makers of knowledge during what is called the Scientific Revolution. It simply does not easily lend itself to linear measurements, and you can't run experiments on it by taking it into a laboratory for multiple trials. Because astrology works with a constantly changing and never exactly repeating cosmic environment, it is not possible to reproduce a specific astrological effect—except maybe with time travel! If it, and I mean here pure astrology and not astral magic, could have been quantified or at least easily tested, it may have been able to circumvent the theological attacks on it, as it had done for centuries. It could have been seen as God's handiwork, certainly the natural part of it and parts of judicial, which is what the astrologers themselves were saying with titles like William Lilly's *Christian Astrology*. Of course, astrology did make more sense within the context of Renaissance naturalism, but because the subject itself was never deeply rooted in theory and most practitioners of astrology were exactly that, practical, it probably could have even survived in the context of physical explanations like atomism. But its disappearance from the central concerns of leading thinkers, coincident with the rejection of holistic philosophies and the rise of the experimental method and mathematical modeling, implies that theory did count for something in the minds of seventeenth-century natural philosophers and the rest of the intellectual elite. We'll now view this historical situation from a sociological and anthropological perspective.

HOLISM VERSUS REDUCTIONISM

The discussion so far—which has been a consideration of ideas as historical artifacts, contrasts between cosmologies and worldviews, and evolving ideologies—has hopefully served as a foundation for an understanding of one of the most transformative developments in world history. During the Renaissance, the definition of reality, at least among the small percentage who made knowledge, had shifted from one of a living Earth regulated by cosmic sympathy to one made of particles, the material atoms that obeyed deterministic laws. Explanations of causality from sympathy measured by qualitative symbols changed to one of bodies in motion measured in precise quantities. These are starkly differing viewpoints, and they appear to be mutually exclusive. In Stephen Pepper's model of world hypotheses, he presents organicism and mechanism as two fundamental integrative views that strive to establish a plan or baseline for reality (Pepper 1942). Pepper, an epistemologist, describes organicism as the view in which events are best understood in the context of a process. It is a view that emphasizes an organic whole, which transcends apparent contradictions found in the parts, much like the fundamental ideas behind systems theories. He describes *mechanism* as the view of the world as a machine where parts are preeminent and determine any understanding of the whole. This fundamental dichotomy in worldview is not only found in divergent historical periods of a single culture as discussed here, it is also found in the profound differences in thinking between Western European and East Asian peoples.

In *The Geography of Thought,* Richard E. Nisbett presents considerable historical, sociological, and psychological evidence that traditional East Asian thought and art tends to be field dependent, that is things are perceived in their context and are essentially holistic. In a social sense, this would emphasize the group or collective, something reflected in East Asian culture and government even today. In contrast, Western European thought and art has tended to be field independent and essentially reductionist and individualist, and, perhaps inevitably, it was in that culture that modern science's reductionist worldview originated (Nisbett 2004). Like Pepper, Nisbett seems to be suggesting that "organicism, holism, and collective" is one grouping and "mechanism, reductionism, and individual" is another. To Nisbett, these two primary perspectives offer descriptions of reality, with an example of each embedded in two different civilizations thousands of miles apart.[6] Why this is so may be due to the particular geographical, historical, and political circumstances that shape and focus each perspective.

It is possible that an explanation for the existence of the organicism/mechanism duality might be found in studies of how the sensory system and brain assemble the world. Or maybe quantum indeterminacy, in which both wave and particle exist but not at the same time, has relevance here. Staying within the context of the historical theme of this section, however, consider that the efforts of the Renaissance magus John Dee, and others like him, were directed toward a reconciliation of scientific and magical worldviews. Their perceived failure to do so could be considered evidence of the fundamental incompatibility of reductionist and holistic philosophies, at least in terms of what was known and technologically possible at that time. This ontological dualism, in which interpretation only occurs in the context of one reality filter (like Rubin's vase where either a face or vase are seen), would seem to be related to the fate of astrology, as that subject appears to be far more easily framed within a holistic view. When the dominant ideology, paradigm, or episteme shifted to mechanistic, materialist reductionism, like a change in fashion, astrology instantly became meaningless baggage to leading intellectuals of the time, and this opinion influenced others in their social class. Fashion, being a ranking strategy employed by domesticated human primates that can use both visual signals or philosophical persuasions, cannot be underestimated as a factor in historical change. The seventeenth-century paradigm shift from holism to reductionism was part of a social class reshuffling during the late Renaissance that was

---

6. Smuts (1926) established the modern definition of holism and contrasted it with mechanism in his book *Holism and Evolution*. See also Rudhyar (1936, 46). The differences between the words *holism* and *organicism* are subtle, and they are often used interchangeably. Pepper defines organicism as the "integrative organic process," though others use the word only when something shares qualities with living systems. Holism he defines as a key quality of evolution where an increase of unity occurs as parts are blended. Others define holism as the whole that is more than the sum of its parts and that determines the behavior of the parts.

stimulated not just by ideas, but also by the rise of two other factors: capitalism and Puritan culture. More on this later.

The organismic and holistic natural philosophies of the ancient world and the Renaissance assumed a particular type of consciousness where humans were perceived as being integrated with the living world and were able to participate with it, this being the justification for natural magic. Participatory consciousness is a state in which a person is not objectified from the environment, but actually joins with it in multiple ways (Berman 1981, 69–114).[7] Participation is both self and not-self, identified at the moment of experience, the sharing of a larger process. Common examples are moments of artistic creativity, absorption in acting or dance, feeling elevated and inspired by nature (finding God in nature), experiencing a strong intuition, tripping on psychedelics, and having a sexual experience. Magic and alchemy, which utilize parts of astrology, are participatory activities entirely based around interactions between humans and nature, though generally with goals of control or the management of transformations of substance. In natural magic, rituals and substances are used to amplify intention and change circumstances or fate. Alchemists attempt to accelerate processes in nature through the use of laboratory devices and structured methodologies. Prayer is a more familiar technique, used in both magic and religion, that is also based on participation.

The subject matter of parapsychology, which includes phenomena such as telepathy, extrasensory perception, psychokinesis, and the like, is very difficult to test with reductionism. However, some findings are suggestive of both deep connections and information transfer between human minds and even the external world. Divination, which means to be inspired by a god, utilizes these linkages and has been practiced in probably every traditional culture in one form or another. It usually involves rituals that allow the practitioner to select pieces of information in an apparently random manner and assemble them in such a way that allows for insight and possibly prediction. This selection of data is essentially an exercise in closing the rational mind and shifting

---

7. Owen Barfield originally used the term "participating consciousness."

into subconscious pattern recognition. Divination may not be very different from intuition, which Jung defined as perceptions coming from the unconscious. By engaging with the unconscious and using the information produced, a person doing a divination may be participating with both their inner self and the extended social world of collective communications. Such engagement, and any insights produced, may appear to be magic, but it may also be an altered state of mind, unburdened by rationality and doubt, that is able to see connections and patterns not apparent through other means.

Some astrological techniques are, more or less, participatory. The branch of astrology called interrogations, or horary astrology, uses astrological charts calculated for the moment that a question of concern is asked. This question is then interpreted using astrological symbolism, so it is like a divinatory practice but it doesn't require being in an altered state. A practitioner of electional astrology observes both self and nature (the planets) as two ends of an ongoing cosmic process (time) that could be consciously entered and even modulated to some extent using what the Stoics would regard as enlightened free will. This procedure allows a practitioner to choose precise times (based on calculations of future planetary positions) to "enter" the time stream and initiate a purposeful activity. The choice of a particular starting time, which would be based on knowledge gleaned from previous observations, is thought to be more propitious for achieving specific goals than if activity was commenced at other times. During the Renaissance the timing of an event like a coronation (that of Elizabeth I was calculated by John Dee) or inauguration, the launching of a ship, the start of a journey, the first shot fired in a war, or the vows at a wedding were frequently based on the configurations of the planets calculated ahead of time (Hand 2014).

The rejection of participating consciousness during the Scientific Revolution may not have been that radical of a step as Western culture has roots in kinds of dualism that are traceable to the Jews and the Greeks. In Judaism, Yahweh is a transcendental god that is not in nature; he is also a portable god that is transported via the Ark of the Covenant to a temporary tent-like Tabernacle that may be located anywhere. Revealed and priestly knowledge, codified in writing, is therefore divorced from place and separate from nature. This is very different from the pagan view of sacred places in nature like wild mountaintops and watery grottos inhabited by multiple gods and goddesses. In the Greek tradition, especially in Plato, a sharp distinction is made between ideal forms (of the mind, and also mathematics) and the tangible objects of the material world. Contrast this with Renaissance naturalism in which Earth is alive, resemblances are the basis of knowledge, participation is taken for granted, and spirit and matter are not so easily distinguished from each other except at the extremes of a hierarchical cosmos. Non-participation, then, did not have to be invented, and it had the added bonus of links to established religion and Platonic philosophy.

Non-participatory consciousness characterizes modern RMM science. The objective and mechanistic scientist, who regards nature as effectively dead, studies it at a distance by breaking it into parts that can be isolated and manipulated. Obviously, participation is not a factor in that activity, and to be completely objective requires total abandonment of participation. These two approaches to nature, participation and non-participation, are clearly very different things, but it appears to have been possible for a single individual to embrace both perspectives. John Dee, Kepler, and also Elias Ashmole, an alchemist, astrologer, and founding member of the Royal Society, are well-known historic examples. Were these individuals clear thinkers with very broad perspectives that reconciled opposites, did they simply have compartmentalized minds, or were they tolerant of ambiguity and paradox? It's my opinion, based on my own personal experiences, that the former (clear thinking) along with some of the latter (tolerance of paradox) is the case. Perhaps organicism and mechanism can be juggled with the right mental equipment or training. Consider that artists and musicians, who work with intuition, rough estimates, and what's called body memory, do similar things with their minds, things that are different from how most other people think.

In his book *The Reenchantment of the World,* Morris Berman's anthropological approach to understanding the historical dynamics of the Scientific Revolution is to view it as a shift from participatory consciousness to non-participatory consciousness. He focuses on the leading figures of the period and their views, Descartes being singled out as a crucial contributor because he so enthusiastically attacked medieval philosophy and argued for a new foundation in scientific thinking based on skepticism and mathematics. In Cartesian dualism the mind is a tool, doubt is a method, and what exists out there in the world can only be known for certain through the use of mathematics. For Bacon, nature is external and to be observed, but it must also be pushed around and shaken (some would say tortured—think of laboratory animals) in order for knowledge to be generated, this being a more active expression of Aristotle's empiricism. With Descartes and Bacon we see two different types of non-participatory consciousness, the important point being that, in both cases, humans are clearly separate from nature.

What appears to have happened in the early seventeenth century was the amazing realization that rationalism (Plato and Descartes) and empiricism (Aristotle and Bacon) are not incompatible pathways to knowledge and that these views could be synthesized into an objective and reductionist philosophy. (This is apparently a lot easier than synthesizing organicism and mechanism.) As pointed out earlier, with Galileo this all comes together in his mathematical modeling of motion through carefully designed experiments in which the natural behaviors of objects are isolated. One result of this shift is that asking *How do things work?* (typically expressed in equations) becomes the goal of science, and *Why is it so?* is no longer asked. A second result is that teleology, a central tenet of Aristotle (i.e., entelechy), is completely denied and the world (i.e., the book of nature) becomes an acceptable object for humans to study and use. This approach was compatible with church theology, which viewed nature not as God nor as having a soul (anima mundi), but as the creation of God. Understanding "how it works" does not say anything about God's intentions, which are to be found in the book of scripture, the Bible; it is just a way of measuring (and initially appreciating, though later dominating) God's creation. Nature is then objectified and is more or less a laboratory for curious humans who, the Bible says, were given dominion over it (Bono 1995, 193–198).

A few decades after Galileo's work, Newton, when he wasn't doing alchemy or theology, launched a fully articulated philosophy of nature based on this synthesis of rationalism and empiricism with laws that unified terrestrial and extraterrestrial motions. With confirmation in observable evidence and agreement among those who could understand his work, this new natural philosophy appeared to generate universal and certain knowledge. Weak points in his natural philosophy, such as action at a distance and a precise definition of gravity, he handled by stating that a force like gravity need not be explained if it can be measured. Again, equations became explanations, and how it works, or quantification, replaced any investigation into why it is so. In many respects his ideas were not a problem for the Church—God remains as the creator and a presence necessary to sustain the world system. Newton's objectivist, reductionist, mechanistic scientific model became the dominant worldview of many intellectuals for the next two centuries and, while critiqued and challenged by some reactionary movements, notably Romanticism, was not significantly challenged until the advent of twentieth-century relativity and quantum mechanics—and also general systems theories. While these conflicts have engaged philosophers and historians of science, the average working scientist and certainly most engineers remain fundamentally reductionist, materialist, and mechanistic in their thinking, and this trickles down to the well-read and scientifically literate public who, like most people, are more interested in certainty than nuance.

A reaction to what appeared to be universal and certain knowledge claims of the Newtonian synthesis came in the thought of the British empiricists—John Locke, Bishop Berkeley, and David Hume. These philosophers focused on the fact that direct knowledge of an objective world is impossible—all we can work with is what our senses bring to our minds. For all we know, what's outside of us could be anything, including God playing tricks. Hume's extreme skepticism of ever having certain knowledge, which leads to relativism, was countered in the latter part of the eighteenth century by Immanuel Kant in Germany. He argued that sensory information was shaped by a priori (pre-existing) forms and structures of the human mind that allow for the discovery of order in nature. Since he proposed this reality-making cognitive apparatus to be universal and permanent, that is embedded in the architecture of the human brain, the new science could still claim a kind of certainty in regard to our concepts of the world. But go deeper into Kant's convoluted (and still discussed and debated) arguments and you find that his model, in which the mind imposes causality on sensory data, complicates the essential objectivist mechanistic determinism of the Newtonian synthesis, and so we are still left with something short of certainty.[8] In modern times Werner Heisenberg summed up this situation by saying, "Natural science does not simply describe and explain nature; it is a part of the interplay between nature and ourselves; it describes nature as exposed to our method of questioning," an observation that puts claims of objectivity in jeopardy (Heisenberg, 1958).

During the latter part of the seventeenth and through the eighteenth century a wave of social unrest and progressivism in intellectual matters and the arts transformed Europe, particularly in regard to self-determination in thought and action. Confidence in the incredible power of rationality, the reliability of sense data, and the method of the new science led to what is called the Enlightenment. This movement weakened religion and produced a cultural change that has lasted to the present; freedom of thought, democracy, the idea of the blank slate, and modern science are among the most obvious inheritances of that time. The rise of enlightened reason and, with it, the dominance of the themes central to modern RMM science generated reactions, however, notably the emphasis on inspiration and subjectivity central to the movement called Romanticism during the late eighteenth and through much of the nineteenth century. In this cultural movement the elevation of emotions, feelings, beliefs, and the arts was to compensate for the dehumanizing effects of the Industrial Revolution that was made possible by the new science. The philosophical thinking of Friedrich Wilhelm Joseph Schelling, and the ecological thought of Alexander von Humboldt and others, also expressed these themes, while actual elements of Renaissance naturalism, presumably defeated, reappeared in the holistic science of Johann Wolfgang von Goethe and later Rudolf Steiner. Theosophy in

---

8. Hume regarded causality as something that exists in the mind as it associates events over time and builds expectations of sequence. Kant argued that causality is not a mental habit, it is the result of a priori mental categories that are universal to people.

the late nineteenth century, and the New Age movement of the late twentieth century, were also resurgences of Renaissance naturalism and its ideas and practices, much of it barely changed during the intervening centuries. The organicism of Henri Bergson and the process philosophy of Alfred North Whitehead, which focuses on change as opposed to stasis, are examples of philosophical critiques of mechanism and objectivism. Modern ecologists, nature philosophers, and historians such as Fritjof Capra, Carolyn Merchant, Rupert Sheldrake, and many others have addressed the disenchantment inherent in the still-dominant RMM scientific paradigm, which is assertive, controlling, anti-ecological, anti-feminist, and hostile to nature and human participation with it.[9]

RMM science, formalized in the seventeenth century by Descartes, Bacon, Galileo, and Newton, became more sharply focused and widely accepted during the mid-nineteenth century, the time when the designation scientist replaced "natural philosopher." Even observational scientific subjects like botany, biology, and geology adopted the assumptions and methods of the physicists.[10] A standardized procedure consisting of a hypothesis to guide experimentation and analysis evolved and then, in the twentieth century, a pure form of RMM science, paired with technology, accelerated advances in material life at a staggering rate. Although philosophers and historians grapple with knowledge issues and social scientists study its cultural effects, for governments and the vast majority of people RMM science is the signature of modern times and the guarantor of solutions to practical problems and continued

---

9. From a very broad perspective, RMM science has been steadily countered by a wide range of reactions to its objectivity and control. Romanticism could be seen as just a sophisticated expression of the need to believe and participate with the world. Likewise, the late nineteenth and twentieth centuries saw a rise of various fundamentalist religions which favored instinct and emotion, and featured an anti-science stance. This uncomfortable juxtaposition of competing ideologies may be evidence of larger cultural cognitive dissonance. Meanwhile, astrology, with its natural laws of planetary forces guided by pure geometry that can be engaged with by a knowledgeable practitioner, hovers between the two and yet is rejected by both.

10. Darwin's theory of evolution by natural selection, a concept without much proof and certainly not contained within a mathematical formula, was not physics, but it was atheistic, statistical, and not teleological. Evolutionary theory became more "rigorous" later in its merging with Mendelian genetics, called Neodarwinism.

progress. But this dominance, certainly deserved given all the material benefits, comes with costs. One of these is a persistent dualism that is exposed by the continuing reactions to it noted earlier. The great dualistic paradox at the heart of modern RMM science seems to come down to the relation between the objective (closed to participation) and subjective (open to participation), a very deep dichotomy indeed (Rudhyar 1967).[11] As already noted, the common assertion that modern science is objective is problematic and has been attacked by postmodern critics. They argue, based on Thomas Kuhn's thesis, that scientific revolutions have a strong subjective element in them, that science is a culturally conditioned kind of knowledge (Kuhn 1962). (This, of course, leads to relativism and a knowledge free-for-all.) It's hard to see how objectivity is even possible given that the acquisition of knowledge of nature is not only indirect (because we can only work with the data our senses give us), but that this knowledge is conditioned by any number of ontological assumptions. An independent and objective "real" world, one that is a product of our sensory systems and distinct from meaning and significance, is not really explained by science, it is explained by the observer. This implies that objectivity is more belief than fact.

One assumption commonly made is that RMM science is value-free and operates on the world in a way that is completely detached from human ideals and morals. But scientific detachment by itself leads to positions that are evaluative in their own right (Shapin 1996, 164). A natural philosophy based on reductionism, mechanism, and materialism (which implies determinism) that is presumed to be objective has this

---

11. Interestingly, if you examine the astrological literature, this duality is not only clearly recognized, it is actually thought to be measurable in terms of phase within a cycle. One example of this is the lunation cycle where, in a given process, subjectivity is thought to peak at the conjunction of Sun and Moon and objectivity at the opposition, with the phase angles between these extremes marking stages in the overall wave form that describes the full cycle. Regardless of whether or not this correlation can be demonstrated in reductionist terms, it is at least interesting that astrology attempts to quantify in certain ways the constant, but also regular, phase shifting of oppositional states of consciousness in a system such as a person or a collective entity. Astrology claims to precisely time these phase changes, but the process itself is only describable in a qualitative way, at least so far.

position as a subjective selection bias, and therefore it does establish values, in this case values that are not holistic nor participatory. Einstein noted that it is theory that decides what is observed, what instruments to use, and how to interpret data, all of which eliminate pure objectivity. The modern scientific dismissal and labeling of participatory practices, alternate realities, and traditional worldviews, and even consciousness itself as delusions, are examples of this hidden value system in science, values that came to the forefront and were expressed boldly in the philosophical movement called logical positivism. But positivism, which argues that only the observable is worthy of being a candidate for science, is contradicted by the fact that many scientific discoveries were made by scientists who were not strictly logical or who limited their investigations to observable matter and forces. Many discoveries came to be through speculative leaps and focus on the unobservable and areas lacking in evidence.

It is even possible to consider modern science as a species of religion. Astrologer Robert Hand observed that the clear and specific doctrines of modern science in regard to creation (Big Bang), time (linear), nature (objective), and reality (only one) mimic the frameworks of the major Western religions (Hand 2004). All of the above makes sense in terms of Pepper's epistemological analysis discussed earlier in which he argues that organicism and mechanism are two mutually exclusive ways of perceiving the world. When subjects like astrology, or even Lovelock's views of Earth functioning like a self-organizing system called Gaia, are judged by modern allegedly objectivist and mechanistic terms, they will inevitably appear to be impossible.

Systems science, which studies phenomena (systems) that are hard to quantify, is nearly a century old but not well understood by both scientists and the public. While efforts have been made to impose reductionist (bottom up, from specific to general) methodologies onto self-organizing systems, results have been mixed. More often, ideas of how systems work are adapted to reductionism. In her book *The Death of Nature,* Carolyn Merchant argues that ecological computer modeling and environmental studies that are useful to long-term sustainability are examples of what she calls "managerial ecology," an accommodation of organicism to reductionism and an application of the Biblical doctrine of human dominion over nature (Merchant 1980).[12] In this application of co-opted organicism, nature is at least seen as alive, but it has been reduced to a vegetative and pliable state. The dominant ideological force in modern culture is still reductionist, mechanistic materialism, forged during the scientific revolution of the seventeenth century, which regards the world as more or less a vast machine subject to predictable laws. Nature has been made a slave and progress has been left in the hands of scientists, technicians, and capitalists who operate in the context of an anthropocentric worldview. Science as human-centered progress (which gets grants) is hardly objective.

Unlike most subject matter studied by physics, systems contain innumerable variables. Reductionism applied to systems then presents many problems. Analyzing circular patterns and feedback loops in a system is possible, but explaining exactly how emergent properties emerge and what their "rules" are remains under investigation. Emergence is where something, apparently spontaneous, comes into being in such a way that prediction from the parts is difficult, if not impossible in many cases (i.e., the behavior of an ant colony, the origin of life, self-consciousness). It has also been a serious challenge to the average, conservative RMM scientist who thinks only in terms of physical laws expressed in formulas. RMM science randomly or insensitively applied to a system also tends to fragment its totality, making it difficult, if not impossible, to grasp the larger processes that play integrative roles. Doing this amounts to using a hammer to fix a watch. In a paper that addressed the limitations of reductionism in the investigation of a system (in this case a human social system), geographer Richard Wilkie distinguished between what he referred to as process science and hypothesis science. He described an inclusive methodology for making knowledge that is inductive (but not based on isolated parts) and called for an informed use of reductionism built on a framework established by data-driven pattern perception.

---

12. See also Chapter 10.

Instead of pursuing a "sacred ritual" where the entire analysis revolves around the central hypothesis, the Process Method lets hypotheses evolve out of the data that have been collected, as the nature of the relationships begin to take shape during the analysis. Thus, the Process Method does not exclude the testing of hypotheses, but they are not central to the analysis (Wilkie 1974).

The advancement of the systems science view (top down, from general to specific) is one of the major developments in science of our time, but more often it is sidelined while RMM science, its immediate practical effectiveness unquestioned, maintains a grip on scientific authority. This strong position guarantees grant money, which then drives scientific agendas and sustains a feedback loop that discourages change.[13] Urgent contemporary issues such as the importance of the microbiome to body and mind, the degradation of ecological systems, and the existential crisis of anthropogenic climate change, however, are topics now forcing a broader view of the interconnectedness of nature because these are problems that are not easily reducible to parts. Issues such as these, which fall into the domain of systems science and require multidisciplinary cooperation, haven't always received adequate support from the larger scientific institutions, though this is changing. The above problems, which have strong links to social issues such as socioeconomic disparities, gender equality, and overall quality of life, challenge how humans relate to nature and expose the shortcomings and assumptions of purist RMM science. I am generally hopeful about the future, however, as I see science, in the broadest sense, as an evolving social activity that experiences occasional paradigm shifts. With the possible exception of a global pandemic of authoritarianism and idiocracy, mainstream science will eventually be forced to shift, not completely away

---

13. RMM science, in its pure form as physics, is based on rigid boundaries established by its own rules in regard to verified data, quantification, and mathematical modeling. This gives it a watertight structure and rules of exclusion that translate into social hierarchies within the sciences and within the culture, where it maintains authority in matters of knowledge. RMM science as absolute law and the sole criterion of truth is the dogmatic position of positivism.

from reductionism, but to refine it and use it where appropriate. When studying complex systems, the scientific establishment will, hopefully, eventually come to allow systems science methodologies to first establish perspectives and then guide further experimentation using standard or adapted reductionism. RMM science is not wrong, it just has its limits and can easily make a mess of things when practitioners are myopic and overconfident.

9

THE CHURCH, SOCIETY, AND ASTROLOGY

"In regard to practicing the art of astrology diaries and other divine things: The Holy Spirit laughs [when astrologers] declare the things that will happen in the future. . . . [I]n such future contingencies inquisitive and random events prevent knowledge of the devil's operation to become involved in deception [taking] men from the way of salvation and into the snare of damnation. [Astrologers] adore and worship the sun and moon and all the stars of heaven by mistake. . . . But are the stars mini-slavery? It has been decreed that the bishops of the Council of Trent have been prepared to act diligently in order to censor books of judicial astrology . . . but judgments and natural practices which work for navigation, agriculture, or medical assistance are permitted. . . . We have received authority and a mandate against knowingly persons reading or retaining books and writings [on astrology and divination], or in books which contain such things, and the investigators will proceed freely and lawfully to punish and coerce them with worthy penalties."

PAPAL BULL 1586, SIXTUS V.

R elations between astrology and Christianity, each possessing great explanatory power regarding the nature and experiences of human life, were never comfortable. As the Roman Empire fell, Augustine attacked the "lying divinations and impious absurdities" of astrologers. While he accepted the powers of the Sun and Moon on nature in general, he didn't see their influences as necessarily leading to a rule of the stars over humans and thought that belief in astrology was a distraction from God (Augustine 1963, book IV, 3; book VII, 6). Since Augustine's modified Neoplatonic ideas dominated the thinking of Christian intellectuals during the first part of the Middle Ages, judicial astrology was rejected. Classical judicial astrology was, however, preserved by the Arabs. More than a few of the greatest Arab philosophers including Masha Alla, Al-Kindi, and Abu Ma'shar copied, studied, commented on, and added to the subject. Recall from Chapter 7 that their writings on astrology, Aristotle, and other related subjects made it into Europe during the eleventh century and trigged the Renaissance of the twelfth century. The intrusion of Aristotle and astrology into the Western Medieval world resulted in several condemnations by the bishops and theologians of Paris, the intellectual center of the time. They saw Aristotelian science as pantheistic, and astrology, thoroughly propped up by Aristotle's cosmology, as extreme determinism.

During the thirteenth century astrology was re-established, with some restrictions, as a mostly legitimate subject by the Christian philosophers Albertus, Aquinas, and Roger Bacon. At the same time the influx of Arab astrological texts enriched the knowledge of practitioners, including one of the greatest astrologers of the Western tradition: Guido Bonatti (c. 1210–c. 1290). Bonatti is known for applying the subject to military activities, that is analyzing the prospects for each side in a conflict and timing the initiation of a planned battle (Hand 2014). His massive *Liber Astronomiae* ("Book of Astronomy") covered the entire subject of astrology but was not just a compilation of previous works, he also added to it, based on his own experiences as a practitioner. In the first section of the book Bonatti gives a long list of attacks on astrology by the clergy, responding to each in Scholastic dialectical fashion and not showing respect to his critics. He stated blatantly:

Therefore it stands that the astrologers know more concerning astrology than the theologians do concerning theology, and are therefore much more able to judge than the theologians are to preach (Bonatti 1994, 10).

This pre-emptive strike against religious critics was not unprecedented (it could also be found in the writings of Abu Ma'shar a few centuries earlier), but it is an indication that practicing astrologers did have some standing in the culture at the time, and this grew over the next centuries as consulting work for the rich and powerful in Western Europe increased. The Church, however, issued condemnations of astrology from time to time. Some philosophers, like Nicole Oresme (1320/1325–1382), criticized the subject thoroughly, particularly in regard to judicial astrology and astrometeorology, which Oresme said had many failed predictions. Even though Albertus, Aquinas, and Roger Bacon were accepting of astrological effects on nature, including the weather, and were critical, though not completely dismissive, of judicial astrology, their conclusions amounted to a compromise and delicate balance between Christian theology and astrology, one that held for the next few centuries.[1]

The thirteenth-century synthesis of some parts of astrology and theology included an updated version of the fall-back position of Christian theologians since Augustine—that precise and accurate astrological predictions made by astrologers concerning people had to be due either to chance, the assistance of demons, or the workings of the Devil. On the other hand, natural astrology, and also medical astrology, which concerns the physical body, could be accepted, but any influence on the inner life that wasn't coming directly from God was a problem for theologians and clerical authorities. It was judicial astrology, which encompasses exterior/interior and body/mind distinctions, that flew in the face of their theology and challenged church control over interior or psychic space. In this regard, a distinction developed by Plotinus and then later used by Christians was a critical point: if planets were merely signs in the heavens, they could

---

1. This was especially true of Albertus, who wrote on astrology. See Thorndike (1923, vol. 2, 585).

be understood as having been placed there by God, and this is clearly stated in the Bible (Genesis 1:14 ASV-American Standard Version):

> And God said, Let there be lights in the firmament of the heaven to divide the day from the night; and let them be for signs, and for seasons, and for days, and years.

If, on the other hand, the planets were causes, then there was less room for God in a person's life and religion itself could be seen as just one type of social process modulated by the ongoing cycling of planets. Yet even if the planets were regarded only as signs, predictions made by astrologers could be seen as interference with God's plans, opening them to condemnation and accusations by religious authorities of practicing demonic magic. So, the Church could have it both ways and astrology was constantly on the defensive. Competent astrologers did respond to this dilemma in nuanced ways, but the details were generally ignored by the accusers. At the same time the words and actions of the less competent, and those that used astrology in natural magic, were a liability to the entire field. Patrice Guinard, in *Astrology: The Manifesto,* argues that it was the confusion between astrology proper (pure astrology) and astromancy (astral magic) that has served to sustain a negative reputation among the religious, and that this slur has been propagated by the enemies of astrology for centuries.[2]

Tensions between religious authorities and practicing astrologers were particularly high in the late sixteenth century, precisely the same time that witches were being persecuted in large numbers. For those of the clerical world, astrology was at best suspect, and the Church eventually came to officially reject it; in 1586 Sixtus V issued a papal Bull against it (see quote at top of chapter). In spite of this authoritative pronouncement, astrology was still so generally accepted by those in the upper echelons of society that some clergy, including Pope Paul III, personally used it. Another example is Pope Urban VIII who, in an attempt to evade what he believed were dangerous occult influences,

2. See Allen (1973) for the debate on astrology in Italy and England and Guinard (2002) for a discussion on the moral aspects of the conflict over astrology from ancient times through the seventeenth century.

was not above requesting the astrological aid of Tommaso Campanella, himself both a practicing Catholic and a practitioner of natural magic and astrology. This same pope also condemned judicial astrology again in another papal bull in 1631 (Walker 1958, 205; Yates 1964, 375). The founders of the Protestant Reformation were not at all sympathetic to astrology. Although the many Protestant sects battled among themselves regarding fine points of doctrine, reason was generally regarded as faulty, while faith, exercised through the use of the will, was seen as the best and most direct route to God. Martin Luther was an anti-intellectual and dismissed Aristotle as a thinker of any real value to a Christian, this distinction marking a clear separation of theology from philosophy in his thought (Kusukawa 1995, 35). If philosophy was deemed worthless in his mind, astrology couldn't be any better.

In spite of a few non-judgmental references in the Bible, such as God creating lights in the firmament for signs, and the three astrologers following a star (or a Jupiter-Saturn conjunction) to find the infant Jesus, the Church has mostly treated astrology as a dangerous competitor. In a world subject to the terrors of war and plagues, astrologers (practicing natural, not judicial, astrology) offered prognostications of such events and explained them as the effects of aspects between malefic planets that could be clearly seen in the sky. For example, the triple conjunction of Jupiter, Saturn, and Mars in 1345, accompanied by a lunar eclipse, was thought to be a sign of troubles to come. English astrologer John of Eschenden used those celestial events to forecast future misfortune: bad weather, crop scarcities, wars, and corrupted air that could spread disease. Most astrologers were in agreement, after the fact, on that conjunction being the astrological announcement of the Black Death, which reached Crimea by 1346 and Italy the next year (Thorndike 1923, vol. 3, 326).[3] Officially, the medical faculty at

---

3. A deteriorating climate in the early fourteenth century is now thought to have set the stage for the introduction of the bubonic plague (*Yersinia pestis*) into Europe from Asia. Extreme cold weather in 1346 and 1347, forcing people inside and close to each other, may have been a factor in the rapid spread of the disease. It is even possible that the conjunction modulated an already unstable climate system (in ways I've suggested in Chapter 5) and actually did play a role in the event (Schmid et al. 2015).

the University of Paris also concurred that this planetary event was the universal and distant cause of the plague, though earthquakes were also considered. A conjunction of planets was something that people could see with their own eyes and was an impersonal target for blame. On the other hand, how could God allow such a catastrophe? The Church did not predict the plague, many priests died, and the clergy's explanations of God's role in all of it were not so convincing to the laity, all of which substantially weakened their power over the populace for the next century, at least.

Judicial astrology was a major threat to the Church on several fronts. Natal astrology offered explanations for an individual's fate, fortune, and the possible meaning of a life, basically a completely secular perspective on the fate versus free will issue. Electional astrology was a way to control or modify the flow of personal time and the path of destiny, weakening the need for prayer. Another issue was that an astrologically informed personal philosophy and practice was a program (like Hermeticism) that eliminated the need for both religious authority and Sunday services. In response to this perceived usurping of their position, religious thinkers and ecclesiastical authorities have for centuries attacked astrology using mostly versions of the same two arguments. One is that it is not man's place to pry into God's secrets, and the other is that the planets cannot override human free will. Both arguments have a very long history. In regard to the first, St. Paul, in Romans 11:20, makes a statement about humility: *Non altum sapere* ("Don't look up to the sky for knowledge"). Although the passages in Romans that follow this admonition continue the theme of warnings against the "intellectual curiosity of heretics," the phrase was taken out of context and quoted often as a warning against the illicit knowledge of higher things (Ginzberg 1986, 61; Thomas 1971, 358–359). This is a stricture against knowing divine secrets, a rule that preserves existing social and political hierarchies and condemns intrusive and subversive thinkers who might consult the stars in the heavens, not the priests.

PICO DELLA MIRANDOLA AND THE
ANTI-ASTROLOGY ARGUMENTS

In regard to the free will issue, attacks on astrology from religious convictions were both intense and sustained, and they have to be seen as a major factor in the decline of all forms of astrology, even astrometeorology as a case of guilt by association. Already mentioned is that judicial astrology was attacked in Roman times by the Stoic Carneades, by Sextus Empiricus, and most importantly by Augustine.[4] But the greatest attack, probably in the entire history of astrology, came from the humanist Pico della Mirandola. A Renaissance scholar, he was originally favorable toward Hermetic thought, but near the end of his life, and apparently driven by religious motives and his position on human dignity, he changed his mind about these things and wrote a scathing criticism of astrology (Garin 1976, 83–99; Walker 1958, 54–59; Allen 1964, 19–34).[5] Later writers ascribed this reversal to the influence of the fanatical religious reformer Girolamo Savonarola (who also attacked the subject just a few years later). They also linked it to a prediction of his early death by at least one, and possibly three, astrologers, a prediction that did come to pass (Schumacher 1972, 17; Allen 1961, 18–35).[6]

Pico della Mirandola is best known for his Renaissance classic *Oration on the Dignity of Man,* which focused on human

---

4. Carneades was unusual in that the majority of Stoics were either supportive of astrology or silent on it. Sextus Empiricus (1949, 323–371) argued mostly from the standpoint that measuring the motions of the sky was impossible. Augustine used arguments of twins and, like Sextus Empiricus, the impossibility of exactly measuring the moving sky. Measuring the sky was a centuries-long project centered on the development of trigonometry, many contributors to it being astrologers.

5. Cornelius (2003) discusses Pico della Mirandola in Chapter 1 of his book *The Moment of Astrology* and summarizes the attack in an appendix.

6. Using standard astrological techniques of the time, Mars (directed by the standard 1 degree per year arc) moves to the square of his Sun at approximately thirty-one years. Predicting a life-threatening crisis at roughly that age would have been a no-brainer for any competent astrologer of the time. Pico della Mirandola died at thirty-one.

intellectual potential and the quest for knowledge in the context of Neoplatonism. Giovanni Pico della Mirandola's attack on astrology, published posthumously in 1497 (and edited by his nephew who was a follower of Savonarola), was comprehensive, twice the size of all his previous writings combined. Titled *Disputationes Adversus Astrologiam Divinatricem* ("Arguments Against Divinatory Astrology"), this work was a long list of arguments mostly against judicial astrology, and for the next century and a half, at least, opponents of astrology borrowed from it as needed. In the first section, Pico della Mirandola pointed to a select group of ancient authorities that condemned it (see below). He then argued that astrology is uncertain and not useful, and that many of the components of the astrological system, the zodiac and houses, for example, are not of substance. The first ten sections of the *Disputationes* describe traditional judicial astrological doctrines and criticize astrology as a science; the last two describe the ways of astrologers and criticize it as an art—all followed by rejection. Pico della Mirandola argued that man has the free will to change from brute to angel and that he is not controlled by outside forces like planets. He stated, consistent with the Hermetic notion of "as above, so below," that man contains the whole universe within—but that man has a separate mind (soul), and because of this nothing can control him. He argued that events ordained by God are not affected via the planets, God directs men rather through angels, and further, that prying into God's ways is wrong (*noli altum sapere,* "do not be arrogant"). Astrology thus interferes with both free will and God's plan. Here are the same two arguments again, based on unprovable metaphysical and religious concepts, and not on developed philosophical reasons or demonstrable facts. The unstated implication in all of it is that astrology may possibly work on some levels—but we know it is bad for you, so stay away from it.

CLASSIC ANTI-ASTROLOGY (JUDICIAL) ARGUMENTS FROM THEIR ANCIENT SOURCES

CARNEADES: ANTI-ASTROLOGICAL ARGUMENTS

Carneades: Anti-Astrological Arguments	Response of Astrologers
Precise observations of the heavens are impossible	Observations must be done carefully. Better technology improves observations.
People born at the same time have different destinies	People born at the same time are born under different life circumstances, but they will experience similar events at the same time
People born at different times may die at the same time (as in disaster)	The astrology of the greater (collective) subsumes that of the lesser (individual; Ptolemy)
Animals born at the same time as a person do not have the same fates	The influence of the stars is constant but received differently by men and animals (Ptolemy)
People are different due to race, religion and customs	The stars express themselves relative to these things

AUGUSTINE: ANTI-ASTROLOGY ARGUMENTS

Augustine: Anti-Astrology Arguments	Response of Astrologers
The stars cause evil things to happen	The stars incline but do not compel
Twins are different from each other	Twins are similar as they were conceived at the same time, but they are not born at exactly the same time
People born at the same time are different	People born at the same time are born under different circumstances, but they will experience parallel lives
The hour of conception must be considered	There is a method for finding this time, but it is not as descriptive of a person as the birth time
Electing times for action contradicts the fate of the stars	True, but only if you assume fate does not include the possibility of modifying it

Augustine: Anti-Astrology Arguments	Response of Astrologers
Every animal and plant must have its own horoscope, some the same as a person	The stars express themselves relative to these things
If men alone are subject to the stars, then the free will given by God cannot exist	Men are not the only beings subject to the stars, which only incline, not compel
If astrologers make true predictions, it must be due to the aid of demons	Is this true of anyone who makes a true prediction?

PICO DELLA MIRANDOLA'S ATTACK ON ASTROLOGY: *DISPUTATIONES ADVERSUS ASTROLOGIAM*

Pico Anti-Astrology Arguments	Comment
Astrologers are fakes that corrupt society	Some are
Ancient authorities have condemned it	Some did
Ancient authors like Ptolemy and Abu Mashar were fools	Pants on fire
Astrologers are aware of the uncertainty of their work	True
Astrologers disagree among themselves	True
Astrologers have always been ridiculed, none are honored	False
Astrologers make faulty predictions	Sometimes
Belief that astrology can see the rise and fall of religions is heresy	To the religious
The influences of planets are really those of the Sun	Maybe
Astrology is not founded on reason	False
God does not direct humans through the planets, he uses angels	What evidence?
If the stars are signs, then they cannot be causes	Why not both?
The star of Bethlehem was not astrology, it was made by God	Who said?
The timing/beginning of events astrologers choose are wrong	Sometimes
If the zodiac is false, then so is astrology	Pants on fire
Planetary correlations with the body are absurd	Experts disagree
Astrology arose among idolaters in the Near East	True

For the most part, Pico della Mirandola's anti-astrology arguments, which are mostly an attempt to separate the mystical from rational, are not entirely original. Thorndike, who we met earlier in the book, commented that Pico della Mirandola's presentation is rambling and he is not orderly in his arguments, making them less effective (Thorndike 1923, vol. IV, 529). Other historians suggest that Pico drew much of his material from an unpublished manuscript written by Marsilio Ficino, a scholar, priest, and translator of ancient manuscripts for the academy of Cosmo de Medici (Walker 1958, 57; Allen 1961, 4–18). Like Pico della Mirandola, Ficino was strongly influenced by Hermeticism, one of the most urgently translated subjects of this period, and his writings reflect this influence and the conflicts between it and Christian thought. In Ficino's book *De Vita Coelitus Comparanada* ("How Life Should Be Arranged According to the Heavens"), he wrote about astrology, describing the virtues of the planets, relations with the organs of the body, herbs, amulets, and so on, all from the perspective of a knowledgeable physician. However, Ficino was full of what appear on the surface to be contradictions, and while he argued against judicial astrology, he also read horoscopes and made predictions. His dilemma, and it is the same problem raised by astrology for millennia, was in regard to human free will. Ficino, like many other intellectuals of his time, believed the planets may be signs, but not causes, and at best they only inform matter, such as the human body, which makes perfect sense in the context of Neoplatonism. But as a priest he also believed the human mind, or rational soul, is free from planetary powers and only God can influence it, and that free choice is God's great gift to man (Allen 1961, 22). The problem here lies in designating exactly where the line between planetary influence and personal freedom is located, and that's not quantifiable if you take into account that some attributes of the body, like feelings and emotions, also operate on the mind. But regardless of where he got his ideas, it does appear that Pico's attack on astrology was essentially a defense of free will and therefore consistent with his humanist hyper-anthropocentric views on the inherent freedom of humans (Allen 1961).

The debate over astrology and free will may have occupied the minds of many Renaissance philosophers, but it is not well known to students of philosophy. As an undergraduate I was puzzled at how a survey course I took on the history of philosophy gave some attention to Thomas Aquinas but then skipped over the next three centuries to Descartes. I now realize that because astrology was so deeply embedded in philosophical thought after Aquinas, and that liberation from the clutches of the Medieval synthesis of Aristotelian philosophy and Augustinian theology was a major goal during the Renaissance, that something had to give if human freedom was to be salvaged from doctrines of determinism, astrological or divine. Because a working knowledge of astrology, in addition to knowledge of history, philosophy, and theology, would be required to properly discuss and teach the issues raised at that time, few, if any, philosophy professors are qualified to offer courses on this morass.

The discussion on astrology, theology, and free will got serious during the early Renaissance when Lorenzo Valla (1407–1457) outlined it by focusing on the Church doctrine of predestination, which, if it is absolute, means humans are not really free to choose. Pietro Ponponazzi (1462–1525), using Scholastic dialectic, took up the topic in this book *On Fate,* but he brought astrology into the mix as a deterministic factor ultimately orchestrated by God. Stoic thought on the subject figured into his thinking, but he didn't attack theology directly and talked of divine self-limitation as a way to keep free will as part of the human condition, that is God chooses to give humans free choice. At about the same time, Ficino was thinking that human will and knowledge can engage with the natural world (i.e., natural magic), but are transcendent and not of it. In his Neoplatonic view there are distinct levels of being, with material nature at bottom, which is where astrology, in some limited ways, can affect humans (Cassirer 1963). Pico, of course, denied any real astrological effect on the human soul and made sure it was taken off the table by damning the entire enterprise, except for some parts of natural astrology. Settling this metaphysical problem raised by the church and astrology, one involving intangibles like consciousness and the soul, was an impossibility, but taking astrology out of the picture to simplify the problem, as Pico did, was a major step toward a solution. During the Renaissance, relations between God, astrology, and human free will were compounded by the Medieval entanglement of Aristotelian philosophy and Christian theology. Only after these convoluted problems were more or less settled by drawing some boundaries (without astrology) could a new natural philosophy (science) make real progress.

While Pico stated that the stars are signs and causes of nothing, he admitted the Sun influences inferior things, but only through light, heat, and motion. He thought that the Moon has a similar force as the Sun, but less so, which is moisturizing, and that the Moon has nothing to do with the tides, as Galileo argued a century later. He attacked Galen's critical days, but thought that sailors and farmers should pay attention to the phases of the Moon.[7] All of this is really standard classical natural philosophy, the norm among Greek and Roman writings on natural history, Pliny being one example. It was judicial astrology that was the target of his defense of human freedom and exceptionalism, the branch of astrology that had associations with the Hermetic corpus from Egypt and the astrologers of Rome. It may be that Pico was actually taking a purist (racist?) stand by rejecting the infusion of astrology from the Near East. Because Plato and Aristotle lived before the influx of Babylonian astrology into the Greek world, and neither made any direct references to astrology, Pico possibly saw astrology as invasive (Cassirer 1963). But this just makes Pico seem even more conflicted in his thinking, as his previous writings were filled with numerology from Cabala and ideas from Hermeticism.

Pico della Mirandola's humanism and concern for free will, in the context of Christianity, was spoiled by the existence of astrology, so like Augustine, he argued against it. Still, many in Europe read *Disputationes,* and it can be argued that his attack on astrology, in which he distinguished rational physical causality from what he argued were illogical claims of astrological causality, was influenced by Greek skepticism and possibly even Sextus Empiricus, author of *Against the Astrologers* (who also wrote books against the grammarians, rhetoricians, geometricians,

---

7. The critical days were based on the quarters of the Moon and were used by doctors well into the seventeenth century. It was thought that every seven days, from the time a person took to bed due to illness, the condition they had would reach some kind of critical point.

arithmeticians, logicians, physicists, ethicists, and musicians). Pico's critical skepticism, regardless of how flawed it was, did play a role in the revolution of knowledge that became the Scientific Revolution of the seventeenth century. He wasn't forgotten; a century later Kepler was reading *Disputationes* and took ideas from it into consideration in his astrological reforms (Kepler 2008). Pico's acceptance of only tangible forces from the Sun and Moon therefore marks a step in the direction of a modern reductionist and materialist science of nature.

Astrologers, a diverse group that included some of the most enlightened and respected writers, mathematicians, astronomers, and philosophers of the Renaissance, responded immediately to Pico's broadsides against astrology offering both apologies and proposals toward the reform of the subject. In an age where scientific facts in the modern sense didn't exist, intelligent responses to Pico amounted to a series of analogies, or resemblances, that led "logically" from one proposition to another. One response was to appeal to authority, as many great astronomer-astrologers could be cited as favoring the subject, and another was to argue that experience (considered to be "facts") proved it. Lucio Bellanti, an astrologer who it was said forecasted Pico's early death, argued that astrology was a science. His abstract conceptual arguments were presented logically in Aristotelian Scholastic mode complete with references to authorities who supported astrological notions (Vanden Broecke 2019, 24–27; Schumaker 1972, 27–31). Giovanni Pontano, one of the greatest poets of his time, a scholar with political interests and insights into social psychology, and also a friend of Pico, stated that the latter's attack was full of errors. Pontano saw astrological influences as factors built into a person's corporeal body that required conscious attention, and the way around these embedded behaviors was to use astrology to raise awareness of them. Only in this way could one effectively express free will (Allen 1961, 36–43; Schumaker 1972, 31–34). Pontano also drew attention to the astrological charts of people who were especially fortunate, another important side of the debate over astrology. On the one hand differences in fortune were a kind of anecdotal proof that astrology worked, but these also touched on a much bigger question that engaged the philosophers and theologians: Why were some people lucky in life and others not? The theologians attributed this fact of life to both God and personal behavior; astrologers thought it was more complicated. In general, the astrologers saw astrology as a source of crucial information, based on natural laws, that was necessary for the construction of an accurate worldview and also knowledge of self. They accused their religious critics of fooling themselves about free choice, a very complicated matter they simplified by throwing astrology out of the equation. Astrologers could only resort to using dumbed-down responses to these uninformed dismissals of astrology from theologians, phrases like "the stars incline but do not compel" and "the wise man will dominate the stars," which they had been repeating ad nauseam to critics since Hellenistic times.

Pico della Mirandola's arguments against astrology also need to be understood in the context of the social world he moved in, and the social context of the next century. Propelled by the advent of printing, astrological practitioners during the last decades of the fifteenth century in Italy and France produced a steady stream of prognostications. Conjunctions, such as that of Jupiter and Saturn in 1484, were featured in published predictions that drew public attention. The distinction between prophecy, which could be described as information produced by imagination, and prognostication, a careful analysis of astrological data, was blurred. Much like the effects of information overload and fake news in early twenty-first-century internet-driven media, this lack of distinction leveled the playing field between serious mathematicians who worked out forecasts based on the geometry of the solar system at a given time (and how these had played out previously), and anyone who came up with a compelling vision of the future. Pico's arguments may be, as Vanden Brocke suggests, a response to a social situation (the bad behavior of some astrologers) and should be read as a call to reform—he was motivated by the negative social impacts of judicial astrology. This view is supported by the fact that Kepler was, to a large extent, in agreement with certain points along these lines made by Pico, as were many other leading astrologers (Vanden Broecke 2003; Rabin 1997, 64). A complete cancelation of judicial astrology wouldn't be a reform, however.

Pico's massive diatribe was not the only reaction against judicial astrology at this time; another is the near contemporaneous case of

Simon de Phares. In 1490, this physician and astrologer was consulted at Lyon by the king of France, Charles VIII, who testified publicly as to his expertise. De Phares, who was apparently too successful an astrologer, was quickly condemned and imprisoned by the archiepiscopal court, forbidden to practice, and moved to Paris. While the king was away in Italy and elsewhere, the Church court confiscated his books, a collection of about two hundred, and submitted them to the theologians of the university for inspection. It took until 1494 before they reported on his library and officially condemned eleven books, most of which were standard works on the subject. Their condemnation was mostly of judicial astrology and the astrology of precise predictions; a more general astrology they granted was allowable. De Phares was handed over to the bishop and the inquisition, but exactly what happened to him afterward is unknown, except that he published a historical defense of astrology in 1498. In it he attacked the "ignorant detractors" of the subject and pointed out that it was the advice of astrologers that made kings and built empires. Thorndike's appraisal of this situation was that the theologians were inconsistent and the matter was more suggestive of politics than any real doctrinal concern (Thorndike 1923, vol. IV, 529–540). De Phares's case was not a serious deterrent to the practice of astrology, but it was an example of the power of religious censorship and perhaps also an example of power politics between church and state. The arrest of Simon de Phares and the broadside of Pico della Mirandola focused the deep conflicts between astrology and religion as the Renaissance got into full gear and stimulated a reevaluation of the subject not only on the larger social level, but also among astrologers.

The Protestant leaders in European countries who attacked astrology drew from Pico della Mirandola's encyclopedic work. Jean Calvin wrote an anti-astrology book, translated into English in 1563, in which he condemned astrology for tempting men to know more than they should. He did not attack natural astrology, which he accepted as a study of how God works through nature, but he thought that humans should regard natural calamities indicated by the stars as examples of God's punishment. His contemporary, the Protestant humanist Thomas Erastus who, using the Bible literally as an authoritative scientific text, attacked all of natural magic and in the process argued against any astrological influence at all. The first English attack on astrology came from the Puritan William Fulke, author of a work titled *Antiprognosticon,* published in 1560. Fulke's attack was only against judicial astrology and, like Calvin's work, consisted of basically all the standard arguments propagated by Pico. However, he introduced these into England for the first time and they were, of course, consistent with the fundamentalist beliefs of Puritanism. Other attacks on astrology in post-Reformation England, directed toward judicial astrology and focused on the free will issue, came from anti–Roman Catholic Puritans. Very few came from those without connections to religion— and there were really no religious opponents of natural astrology by itself (Allen 1961, 148; Thomas 1971, 367).

The conflict between the perceived fatalism of astrology and human freewill could be reduced to the problem of where humans stand in regard to nature. Pico, the champion of the freedom and dignity of Man, had to separate Man from nature to be consistent. If humans are a part of nature, then they are subjected to the forces of nature, including planetary action at a distance. Destroying astrology, which had a general conception of reality and nature backed by Stoic physics and Hermetic doctrines, was one way to underscore the liberty of Man (and certainly simplify the debate). Astrology, with its implied fatalism on the one hand and the complications it brought to issues of human freewill on the other, struck at the heart of Ficino's and Pico della Mirandola's extreme anthropocentrism, better known as humanism. Renaissance historian Eugenio Garin places this intellectual conflict, with a focus on astrology, as one of three major debates during the Renaissance, the others being the problem of the soul and immortality (Ficino and Italian philosopher Pietro Pomponazzi), and the relations of the state and human society (Italian diplomat Niccolò Machiavelli) (Garin 1976, 93).

The revival of Hermeticism, Neoplatonism, Stoicism, and other ancient philosophies, most having associations with astrology, and also the advent of printing technology that allowed individual astrologers to speak to the public, moved the subject more directly into competition with organized religion. In response, the clergy and their allies attacked astrology in various ways throughout the sixteenth and seventeenth centuries, which impacted the legitimacy of the field as evidenced in the introductions to astrological publications of the time. These include abundant references (usually found on the first page) to God Almighty followed by assurances to the reader that the author is a Christian. For example, there is William Lilly's *Christian Astrology,* which is a standard textbook on the subject and has absolutely nothing to do with religion. Its title alone indicates the existence of religion-minded censors. The astrologers of mid-seventeenth-century England did consider themselves God-fearing interpreters of cosmic messages, and they saw astrology itself a divine kind of knowledge, a special way of reading the book of nature. Of a dozen or so books authored by astrologers from seventeenth-century England that I examined, none were even remotely atheistic. A 1642 publication titled *Astrology Theologized* makes the case that astrology, while basically the most powerful philosophical knowledge available to man, is not enough. Theology must be its companion (Weigelius 1642). This was the power of religion on the thought of the time, and it made the separation of theology and metaphysics from an understanding of nature on its own terms difficult—until Bacon, Descartes, Galileo, and others produced an entirely new way of making knowledge.

THE STUDY AND PRACTICE OF "PURE" ASTROLOGY IN THE FIFTEENTH AND SIXTEENTH CENTURIES

Astrology, both natural and judicial, had been discussed thoroughly by Albertus, Aquinas, and Roger Bacon in the thirteenth century, and each of them found ways to reconcile most of its doctrines with Christianity. By the fifteenth century astrology had become reasonably acceptable and was being taught at the universities. The university at Louvain (Belgium) was founded in 1425 and five years later hired Joannes Vesalius (great-grandfather of the anatomist), who was a doctor and astrologer, these professions overlapping widely.[8] Vesalius produced annual

---

8. See *The Limits of Influence: Pico, Louvain, and the Crisis of Renaissance Astrology* by Steven Vanden Broecke (2003, 29–53) for a history of Louvain astrology.

almanacs, which consisted of astronomical data, tables of the best times for medical procedures, and astrological predictions for the year ahead. This academic astrological publication format evolved into more specialized annual prognostications that enabled individual astrologers, like Johannes Laet, who worked in the region but had no professional connection to the university, to distinguish themselves and acquire patrons in high stations. With printing technology came the problems of publicity and information management. Previously, prognostications about notable persons, like the king or pope, might be handwritten private documents; but with printing, astrologers could reach many more readers, and this promotional bonus led to competition with each other (Grafton 1999).[9] One can imagine that a flurry of exaggerated predictions made by rival astrologers about an upcoming conjunction, widely distributed by this new information technology, would upset the public and invite the wrath of the authorities, civil and religious. Indeed, this sort of thing occurred frequently during the Renaissance, as it does in the internet-driven twenty-first century with its biased news, sloppy scholarship, and generally too much information. The situation wasn't totally out of control, however: in the sixteenth century there were annual prognostications written by the more restrained astrologers, mostly associated with universities, who were inclined toward a refinement of the subject. These practitioners used more detailed data, often restricted themselves to making only weather predictions, and paid more attention to theory (Vanden Broecke 2003, 188). Other non-academic astrologers weren't so constrained by institutions or a sense of social responsibility.

The understanding of astrology's place in natural philosophy evolved during the fifteenth and sixteenth centuries. Previously closely associated with mathematics, astronomy, and medicine, and broadly taken to be an effect of a transmission from God downward through the spheres, astrology came to be seen more as a specific type of natural phenomena operating in a closed system. Vanden Broecke argues that, among

---

9. The effects of printing on almanac production might be compared to the rise of the internet, which has brought with it lower standards and uncontrolled amateur publishing.

Lutherans at least, astrology was regarded as useful information seen in the celestial signposts that could be a guide for self-governance and information for minimizing the negative effects of the planetary signals. Astrological predictions were considered to only be possible in a loose sense as other causal factors had to be considered, which was Ptolemy's view also (Vanden Broecke 2003, 49–53; 2019). This constrained perspective was one held by Melanchthon (Martin Luther's right-hand man), who devised a college curriculum that included astrology. At the University of Louvain, astrology was first taught in the context of lessons in astronomy or mathematics, but this was a problem for the university theologians who were concerned about prognostication, which they considered to be divination. Potential conflict was kept in check by referring to astrological predictions as "future contingents" that did not negate the possibility of God's actions extending past the planets into the sublunary realm. Astrology was also taught at the university in Bologna where annual astrological prognostications were made by professors who taught in the arts and medicine (Thorndike 1923, vol. V, 234–251). Among this group of distinguished mathematicians and professors of astrology was Domenico Maria Novara (1454–1504), who had Copernicus as a student. The astrology taught there was conservative. In general, there was a trend among these university astrologers to favor Ptolemaic theory and practice, which would be the purist position, to that of the Arab astrologers who were known for their emphasis on revolutions (the return of a planet to a specific point) and the sequence of Jupiter-Saturn conjunctions. This cultural success and attention to theory didn't last long, and during the latter part of the sixteenth century astrology declined in parts of Europe; university astrology at Bologna ended after 1572 when the chair of astrology became permanently vacant.

A strong supporter of astrology after the Reformation was Philip Schwarzerde (1497–1560), Martin Luther's advocate, collaborator, and organizer. Better known as Melanchthon, he was a competent, intelligent, and rational person who promoted a moderate and rational astrology as part of a revitalized German natural philosophy (Kusukawa 1995; Schoener 1994; Allen 1961, 63). Luther generally distanced himself from astrology, but he did write an introduction to a book of astrological predictions, including the coming of a great religious reformer that could be construed as referring to him (Hoppman 1997, 49–59).[10] He was further caught up in the subject because his uncertain birth data led to speculation among the astrologers of the day as to whether he was the Messiah or Antichrist, this information being in high demand. Luther did tolerate Melanchthon's interest in astrology, saying that, while God made the stars, they were signs, not causes, and the interpretive art was fallible. Melanchthon, however, took the subject very seriously and regarded natural astrology as an obvious phenomena and good evidence of God in nature. He made astrology an important part of his reform of knowledge, a project coincident with the reform of religion launched by Luther. In 1545, at Wittenberg, he established a natural philosophy program, astrology included, that influenced many students including Michael Maestlin who later taught Kepler.[11] Rheticus, the student and promoter of Copernicus, was hired by Melanchthon to teach astrology and astronomy there. Melanchthon was also a friend of the notable astrologer Johannes Schöner (1477–1547), a multi-discipline expert known today as a mathematician and innovative globe-maker, and also the person who urged Rheticus to visit Copernicus. All of this points to astrology being at or near the center of some of the most important intellectual events of the time.

One of the most learned Renaissance scholars, Gerolamo Cardano was an internationally known polymath whose books were widely distributed and influential. In addition to being a master astrologer, though not an innovator (he practiced and confidently promoted a traditional style of astrology), Cardano was one of the creators of modern algebra and probability theory and was also a medical doctor. Cardano annotated Ptolemy, wrote a defense of astrology, collected birth charts (genitures) of the rich and famous, and published them with commentary in an effort to both promote himself as an astrologer and put the subject on what he believed would be more secure foundations. This method of

---

10. The book was the *Lichtenberger Prophecy,* published in 1488.
11. Kepler referred to himself as a Lutheran astrologer, something explained by this connection and highlighted in the title of Field's (1984) monograph on Kepler. See also Kusukawa (1995, 170).

gathering case studies, an empirical approach, was one that many other astrologers continued over the next century and a half in Europe and England. Cardano was critical of certain parts of astrology and was well aware of its limitations, though he also cited the weakness of the human intellect as a factor in wrong astrological prognostications, which were common and often embarrassing (economists, pollsters, and weather forecasters in modern times often have these same problems). As a practitioner of astrology, one event in his career, in 1552, stands out: his failure to predict the early death of Edward the VI after spending what he said was one hundred hours working on his birth chart at the royal court. This was more than a minor embarrassment, and it generated much discussion about how accurate astrology really was. Cardano said he had purposely avoided investigating the parts of astrological analysis that could predict the possibility of Edward's death (which occurred just a few months after he had done the reading). Not doing those calculations (or pretending not to) was probably a smart move in the world of royal power politics. He faced a typical astrologer's dilemma: any prediction in that situation (i.e., the king would die or not die) could focus attention on him and possibly generate accusations or worse (Grafton 1999, 115–123).[12] Like Ficino, Cardano mostly lived his astrology, seeing it as useful to self-study, a guide to choice and action, and an interpretation of the passage of time. Working for royalty, however, paid the bills.[13]

Jean Bodin (1530–1596), born in France, was a lawyer, philosopher, and an early social scientist who initiated many modern ideas in matters of law, political economy, and government theory. He wrote major works on history, political theory, and economics. He was not an astrologer himself, but he held strong opinions on natural law and linked the cosmic order to the terrestrial political order (as was done

---

12. Of note is that the inclusion of Uranus and Pluto in the birth chart of Edward VI strongly suggests (in modern astrology) the possibility of an early death, something Cardano would not have known.

13. Anthony Grafton's book *Cardano's Cosmos* (1999) delves deep into Cardano's life in astrology, describing in detail his personal life in a time of intense competition with rivals and ever-present, powerful religious controls.

in ancient Mesopotamia) through a kind of astrological numerology he thought would lead to better government. In his method of cyclical historical analysis, he referred to planetary movements as the framework on which history moved. Bodin had a plan for reforming astrology that included the comparison of chronologies of events with planetary cycles and eclipses, which would then, presumably, allow for translation into specific patterns organized by numbers. He didn't actually do this, however; it was just a suggestion (Campion 1989, 89–136). Bodin is also known for his thoughts on demonology and sorcery, significant because he was living at the time when the witch trials were near their peak and one of his primary interests was legal matters.

Tycho Brahe, who was introduced earlier, studied astrology early in his life but became frustrated with the inaccurate tables that were available to him—so he did something about it. Using huge self-designed and sophisticated sighting devices, he mapped the stars and the planetary motions in great detail; his labors allowed Kepler to solve the problem of the planetary orbits, which put the heliocentric hypothesis on solid foundations. Brahe wrote astrological forecasts for his high-ranking patrons, and he regarded the subject as generally reliable and worth improving through greater astronomical accuracy. In delineating the astrological effects of the comet of 1577, he offered practical information and called for a rational exercise of free will informed by astrological interpretations of the comet's significance (Christianson 1986, 130). Brahe was an eccentric who lived an extravagant life surrounded by a cast of characters and has been the subject of a number of biographies, one even arguing that Kepler poisoned him (Gilder and Gilder 2005)!

EXPERIMENTAL SCIENCE AND CAPITALISM

By the cusp of the sixteenth and seventeenth centuries, the Renaissance natural magic that was infused, if not integrated, by astrology had become perceived by many as a serious threat to society. This mixing was done by a number of prominent natural philosophers, some being examples of the Renaissance magus. The physician, scholar, and occultist Heinrich

Cornelius Agrippa (1486–1535) wrote on natural magic, including the critical role played by astrology in that subject. Much of his thought was built on the writings of Albertus, Ficino, Pico, and others who had previously written on the subject, and like them he sought compatibility with Christianity. The Neoplatonic and Hermetic thought of these writers on natural magic was sophisticated and complex, but potentially heretical because distinctions between it and demonic magic were not always clear to church authorities. This and related issues later led to Giordano Bruno, another advocate of natural magic, being burned at the stake in 1600.[14] John Dee, the Queen's astrologer who was also involved in natural magic and the calling down of angels, lost credibility, had his house ransacked, and died in poverty in 1608/9. In 1542 and 1604 witchcraft laws were passed in England, a way of prosecuting and executing people who practiced what was believed to be demonic magic. The peak of the witch trials, which led to the death of thousands, mostly females, occurred from roughly 1580 to 1630, the same period that astrology slipped into serious decline. It was in this volatile and dangerous intellectual and social environment that Marin Mersenne (1588–1648), a mathematician, music theorist, and devout Christian of the Order of the Minimes in France, came to lead a major attack on Renaissance natural magic, Neoplatonism, and the other holistic and animistic philosophies of nature, portraying them as the chief enemies of both Christianity and true science (Yates 1964, 432).

In 1623 Mersenne published a work called *Quaestiones Celeberrimae in Genesim* ("Questions Celebrated on Genesis"). Using Bible text as the basis of his attack on Renaissance naturalism, which was already in decline, he completely dismissed astrology, sympathy, and microcosm/macrocosm and condemned the doctrine of the anima mundi. Mersenne was a major supporter of mechanical philosophy, and he worked tirelessly, networking with those sympathetic to his cause, including his influential friends Descartes and Gassendi, through correspondence and regular conferences (Yates 1964, 432; Berman 1981,

---

14. Yates calls this period "some of the worst mapped and most forbidding territory known to intellectual history" (Yates 1964, 130; Grafton 1991, 145).

109–111; Butterfield 1957, 83–88; Bono 1995, 256). At the same time the Rosicrucians, an underground brotherhood essentially Hermetic in outlook, was gathering steam as a movement, and Robert Fludd (1574–1637), the last of the great Renaissance magicians, was desperately attempting to prop up Renaissance natural philosophy and sustain the worldview of the Renaissance magus. Fludd, in a 1617 publication, essentially reconstructed the Renaissance outlook of the previous century, hoping to give it new life by combining Ficinian Hermeticism with Cabalism, and keeping it vaguely Christian by the addition of a hierarchy of angels.

Much of the prestige of Hermeticism came from its alleged antiquity, one that challenged the precedence of the Judeo-Christian tradition and revelation. This ancient pedigree was undermined by Isaac Casaubon who, in 1614, dated the Hermetic writings to post-Christian times. Using style, vocabulary, and mentions of events and authors as proof, Casaubon argued that the writings were forgeries and the contents were basically reinterpretations of Platonism. This revelation shattered the authority of Renaissance naturalism, lowering the social position of the Renaissance magus, and it gave Mersenne a major weapon in his attack. A debate between Mersenne and Fludd, well-publicized and closely followed by intellectuals in Europe, ensued, and it focused on the clash between these rival worldviews. Even Kepler joined the debate. On the one hand he defended astrology, while on the other he attacked Fludd's Pythagorean numerology in his major work *Harmonies of the World,* arguing for a quantitative use of mathematics. In the 1630s Gassendi also weighed in the debate with a Christianized version of a revived atomism and argued that only measurable things were real.[15] This period, from 1615 to 1630, may well be the focal point of the larger paradigm shift or episteme mutation that disenfranchised astrology. While Renaissance naturalism, and its holistic worldview, was defeated publicly, it was not completely dead—it went underground. Fludd lost the public debate, but

---

15. Gassendi, in his book *Vanity of Judiciary Astrology,* argued that Ptolemy's *Tetrabiblos* was a fake, though that association has not been held up by subsequent scholarship (Bowden 1975, 5; Rosen 1984, 256).

he was a Rosicrucian, the secret society originating in Germany about this time that was a discrete form of Hermeticism. Like Freemasonry, which may have been in part an offshoot, Rosicrucianism has preserved elements of Renaissance naturalism to the present day (Yates 1964, 404). Renaissance natural philosophy has also surfaced in the form of Theosophy in the nineteenth century and the "New Age" movement of the late twentieth century.

PURITANISM, CAPITALISM, AND EXPERIMENTAL SCIENCE

Puritanism was another factor in the rise of experimental science. In spite of the fact that there were as many as 180 Puritan sects, they all shared some central religious and ethical convictions, and the differences were in the details. Calvinism, perhaps the most classic type of Puritanism, promoted the view that the world was evil but humans could remake it through hard, practical work. Calvinism thus favored utilitarianism, a value that was shared with experimental science, making the latter congenial to Puritan tastes (Merton 1970, 80). The practical study of nature then became an exercise in the appreciation of the works of God and, further, success in the world by doing good works was evidence of being one of God's elects, not the result of well-placed planets in one's horoscope. Contrast this with Catholicism, in which the material world was generally perceived as evil and salvation involved retirement from the world. In Puritanism, strong emotions were to be controlled and avoided, while reason and education were promoted. This formula clearly favored experimental science with its reductionist thinking, disciplined hands-on approach to problem solving, and its practical goals. The most compelling evidence for the connection between Protestantism and experimental science is in regard to the composition of the Royal Society, founded in 1660, whose original members were strongly religious, and mostly Puritan. In 1663, forty-two of the sixty-eight members were Puritan, a ratio much higher than found in the English population. Protestantism continued to produce scientists. By 1869 Europe was populated by 140 million

Catholics as opposed to 44 million Protestants, but there were many more Protestant scientists of note than those practicing the Catholic faith (Merton 1970, 114, 178). Experimental science and the Puritan religion had no quarrel during the formative stages of the Scientific Revolution, and the early experimental scientists of the Royal Society were most anxious to show, through their new methodology, the existence and glory of God as revealed through the study of his creation, the book of nature.

Historians have drawn connections between the Scientific Revolution and the rise of capitalism in Italy. Although the origins of capitalism could be located as early as the tenth and eleventh centuries, settlements during Medieval times were, for the most part, self-sufficient and commerce was limited (Cipolla 1994). After recovery from the Black Death plague of the fourteenth century, which had wrecked the localized feudal system, there was an increase in commercial economic activities that included long voyages to the Far East and later the New World. This expansion, which led to the commercial revolution of the Renaissance, required improvements in technology. For the status-conscious bourgeoisie, the new experimental science and the practical technological offshoots it produced offered money, prestige, and upward mobility, and this generated a faith in progress. The rising population density of the times and climate change (the Little Ice Age, a factor that should not be underestimated in terms of political and cultural history) produced an energy crisis in the form of timber depletion. This created a need for solutions to problems that experimental science, not astrology, alchemy, or natural magic, could solve. The seventeenth-century push to industrialization in England involved coal mining, a topic addressed by the Royal Society, and pumps like those described by Robert Boyle and Robert Hooke became urgently needed. Since Puritanism's religious doctrines were compatible with science, and its middle-class congregations were very interested in opportunities to move up the socio-economic ladder that capitalism provided, a positive feedback loop between religion, the new science, and wealth-generating trade, further described in the next paragraph, was established.

Between the fifteenth and seventeenth centuries a general congruence between RMM science and capitalism developed. Technology, like that used by alchemists or cathedral builders, had long existed but lacked a theoretical basis—until Francis Bacon's proposed scientific methodologies made it relevant to the production of useful knowledge. At the same time the need for improvements in mining, navigation, and the like, and the erosion of the gap between scholar and craftsman epitomized by Galileo, accelerated technological progress. Then, in the seventeenth century, with a mechanical theory of natural science in place, expanding commercial activity driven by growth, a cooling climate, and resource depletion, the stage was set for intellectual activities that were compatible with capitalism. At this point the static beliefs of Medieval religion began to be replaced with belief in unlimited growth through free market economies. The new science community, dominated by ambitious middle-class Protestants, developed a social identity built on control of nature and material rewards. Alchemists or astrologers could not fill these needs. In alchemy one asks why a mining activity is conducted, and no sharp distinctions are made between mental and material events in the process of doing a transformation of substance. The alchemical attitude (based on a living world) is not driven by a profit motive. In the competitive capitalistic mindset, the need for capital accumulation drives a need to know practical things, such as how to extract metals from a mine or how to navigate around the world and trade goods. The demand for solutions to material problems replaced the alchemist or astrologer with someone more practical—the scientific engineer, the cartographer, and the maker of navigational instruments. The post-Renaissance worldview came down to a focus on reason, objectivity as a source of truth, and a reality that is located in matter. This is a view of the world, itself given to humans by God, that is dead and exists to be exploited. For the people of the time, still steeped in Christianity, the Renaissance notions of the central importance of the individual and the necessity of free will were triumphant, and this led to a sustained belief in progress. For the most part, we still live in this world.

SOCIAL FACTORS

During the sixteenth century some astrologers worked as advisors to royalty, what was then the highest level of politics.[16] Famous examples were Luca Gaurico in Italy, Michel Nostradamus in France, and John Dee in England. In Italy, in the papal court, astrology was routinely used, though the activity became increasingly constrained by the seventeenth century. Predictions of the success or failure of important personages made by these advisors, sometimes right and sometimes wrong, could be perceived to be a problem—depending on what side of the prediction one might have been. An astrological prediction of the early death of a royal heir or defeat in an impending battle was not welcome in all quarters, and such predictions could easily become politically destabilizing. In political turf wars of any kind, a clever advisor to a ruler may appear to be an asset, but only if that ruler remained a dominant force in their world. In such a situation the astrologer had two options: One was to make unbiased predictions, which, if these were negative to the ruler, could lead to the loss of a job or worse. The other was to predict success for the leader in a description of the future that could change, due to evolving circumstances, over time. Either way, the astrologer would eventually get into some seriously awkward situations, and given the introspective poverty generally found among political leaders (then and now), advice and forecasts that were anything more than superficial would fall on deaf ears. Kepler, who was an astrological advisor to General Wallenstein and Emperor Rudolph II, complained about exactly this in his writings. Decades of astrological advice to leaders during the Renaissance proved to have a neutral influence at best and were counterproductive at worst—and the reputation of astrology suffered. The complications of advisory politics as a social factor in the decline of astrology were significant (Hand 2014).

Dee, advisor to Elizabeth I, wrote that there were three enemies of astrology: those who gave it too much credit, those who granted it too

---

16. See Baigent, Campion, and Harvey (1984, 61–68) for a review of politics and astrology during the Renaissance.

little, and worst of all, the vulgar practitioner who gave it a bad name (Allen 1961, 182). In the debate on astrology, the one point both sides agreed on was the despicability of charlatans. Dee's sentiment was shared by serious astrologers like Cardano, Brahe, and Kepler, and the most vicious attacks on astrology from its enemies generally focused on those who made outrageous claims for judicial astrology. These people undermined the credibility of the subject itself and surely contributed greatly to its downfall. A major factor in all of this was the advent of printing. It opened the floodgates of opinion, which had previously been restricted and restrained—or, more accurately, controlled—by the patronage system. With printing technology, an individual astrologer's prognostications could be published and widely circulated in an annual almanac or as a separate prognostication. Through their publications, astrologers routinely criticized each other in regard to method as well as the accuracy of their specific predictions, each claiming to have the best product, and as I pointed out earlier, this ushered in waves of mostly low-brow information.

In the sixteenth century, competition among astrologers, on all levels of society, was a part of the profession, and often those making the most extreme forecasts sold the most almanacs. In this way astrologers became easy targets for ridicule and mistrust by the thinking classes and, to a lesser extent, the public. One example of this trend was the great conjunction of 1524. Thorndike devoted forty-five pages to this event, and his report demonstrates wide differences in opinion and a deep lack of cooperation among astrologers (Thorndike 1923, vol. II, 178). There was an unusual series of planetary conjunctions in that year, including that of the slow-moving planets Jupiter and Saturn in the water sign of the two fish, Pisces, that caught the attention of astrologers some years before it occurred. A large number of publications appeared in the years before the event, and while some of these writers predicted 1524 to be only a very wet year, which it turned out to be in some regions, many predicted much worse—a Noah-scale flood. Some people who read the more extreme forecasts were motivated to take drastic precautions, like uprooting and moving their family and personal property to higher ground. Both the public and the rulers perceived these forecasts in retrospect as not only wrong, but socially disruptive. Another similar example were predictions by some English astrologers of trouble for their country in 1588. Spain did attempt an invasion then, but England managed to beat the Spanish Armada, and that year turned out to be one of the best for English prestige and government. Massive failures at exact predictions like these, at least in regard to public perception, did much to undermine the credibility of astrology, yet this tabloid-like fear-mongering astrological literature was persistent and abundant. Kepler expressed his concern for the situation, which he saw as a kind of vicious circle where strong almanac sales encouraged astrologers to dramatize predictions more than they probably should, giving the public an unbalanced literary diet (Kepler 2008, T-116, 176). A more subdued version of this can be seen today on the internet where consulting astrologers compete for product sales and clients. For those doing unconventional self-employed work in a capitalistic society that rewards hustling, using the media for self-promotion is a necessary part of making a living, but it doesn't guarantee quality.

The decline and marginalization of astrology accelerated during the second half of the seventeenth century (the dawning of the Enlightenment) as the various holistic beliefs and practices of Renaissance natural philosophy, previously overlooked or tolerated to some extent by the Church and the intelligentsia, were attacked, abandoned, or driven underground. In France astrology was officially excluded from the Academy of Science in 1666, and court patronage of astrologers ended. However, elements of natural astrology were appropriated by some natural philosophers and almanacs with weather predictions continued to be published, but not without censorship and controversy. According to Patrice Guinard, what changed at this time, in France at least, was not any refinement of the standard anti-astrology arguments; it was an amplification of anti-astrology opinion due to the formation of a rare social consensus. He argues that it was the enforcement of this consensus among the dominant intellectuals of the time, not scientific proof or philosophical analysis, that removed astrology from educated conversation (Guinard 2002, 80).

This view of the decline of astrology in mid-seventeenth-century France as accelerated by a societal reconfiguration, and also its decline in England following the restoration of the Stuart monarchy (see page 315) is yet another strand among several histories. Astrology was not compatible with the new mechanical philosophies that fostered reductionism, and consequently suffered from a lack of interest by scientists. It was not socially progressive, and its methods, which require interpretation, were not easily learned. In addition, astrology was driven into exile by hostile clerical authorities, was discredited due to outrageous claims and failed predictions of astrologers, and was found useless to the needs of capitalism and modern urbanization. These reasons have already been discussed and, collectively, they do seem to have made conditions extremely difficult throughout Europe for astrology to survive as a serious subject in the seventeenth century. What I find compelling about Guinard's explanation, however, is its implied socio-biological view of the historical process. People will often make history when personal drives for dominance and social position are exercised fanatically. It only takes the right social conditions for a group, or even a single persistent and committed zealot like Marin Mersenne, Harry J. Anslinger, or Rupert Murdock, to change the course of history.[17] Attention given to the dynamics of territorial maneuvering within human primate pecking orders at crucial turning points in history (on a decadal scale) will often fill in details otherwise missed when reviewing history. Large-scale historical change takes time and many factors set the stage, but the great turning points in history, the kind of history that most people learn about, are often driven by dominant individual humans, and the key players are not always fully visible to the public eye. They forcefully promote their personal agendas, including their own definitions and descriptions of reality, and attempt to steer history when conditions are at a tipping point. Fox News, part of a right-wing info-entertainment empire built by Murdock, is a modern example of this sort of single-handed social manipulation.

In England, the seventeenth century was broken up by civil war. Discontent over increased taxes and duties, exercises of absolute power by the king (including dissolving Parliament), harsh treatment of Puritans, and Catholic-friendly attitudes led to a rebellion and civil war in 1642. The Puritan leader Oliver Cromwell organized an army,

---

17. Consider the war on marijuana waged by Anslinger who almost singlehandedly outlawed cannabis in the United States. His legacy in the "war on drugs" ruined many lives and halted scientific studies for decades.

defeated the royal forces of Charles I, and in 1649 had him executed. Cromwell ruled until 1658 but was hated for his Puritanical standards, and within just two years of his death in that year, the Stuart monarchy was re-established. The disruption of the English civil war had a powerful effect on the practice of astrology. Until 1641 astrological publications and their contents were regulated by the Company of Stationers and also certain religious and university censors. The breakdown in political and ecclesiastical restrictions that occurred when the civil war began, as well as cheap printing, led to an explosion of publications (another case of media lawlessness) peaking in the 1650s. During this time of little or no censorship, hundreds of almanacs and textbooks dealing with astrology appeared, and its secrets were open to inspection by the public. The most famous astrologer of the time was William Lilly (1602–1681), a moderate Parliamentarian who was known for his more or less accurate predictions of Royalist military defeats that he issued in almanacs or as separate publications. In the 1640s, sales of Lilly's almanacs were selling in the range of thirty thousand a year, and sales of almanacs in general are estimated to have reached four hundred thousand a year by the 1660s. This uncontrolled flood of astrological publications and the rising power of astrologers was generally perceived as connected to radical politics, and astrology became identified with Parliamentary revolution, regicide, and radical sects—associations that had historical consequences when order was restored.[18]

Astrology in England was at its peak of popularity and influence from the late sixteenth and through the first half, at least, of the seventeenth centuries, a kind of golden age for consulting astrologers. It was common for people to consult astrologers about practical things such as property matters and, at a time when wooden boats driven by the wind sailed great distances, the safety of sea voyages was of great concern as lives could be lost should things go wrong. (Another factor in the decline of astrology was the rise of the insurance industry in the late sixteenth century, which made such inquiries unnecessary.) Casebooks of a few

---

18. For accounts of Lilly and other Civil War astrologers see Thomas (1971, 288–89), Capp (1979, 39, 182), Curry (1989, 19), and Parker (1975).

successful practitioners like Simon Foreman (1552–1611), Richard Napier (1559–1634), John Booker (1603–1667), and others reveal that these consultants were a combination of problem-solver, advice-giver, doctor, and prophet. They answered questions concerning property, marriage, the fate of ships at sea, legal matters, political outcomes, the possibility or legitimacy of offspring, winners of contests, and so on, and they served all classes of society. Lilly was seen as a spokesman for the radical elements of popular culture, yet his friends included some advanced astronomers and some royalists, an indication of the fluid boundaries between the subject of astrology and society itself at the time. Elias Ashmole, one of the founding members of the Royal Society, and a royalist conservative, was his close friend and a fellow astrologer. (Ashmole also collected the casebooks of many astrologers of the time that are now preserved in the Bodleian Library at Oxford.) Still, astrology was far from uniformly socially acceptable, and the subject was considered by many, especially the fear-mongering clergy, to be diabolical, practiced by conjurers and wizards. The title of Lilly's *Christian Astrology,* in which he outlined in ordinary English how natal and horary astrology were practiced, says much about his wit and social survival skills.

The mid-seventeenth-century English almanacs, full of astrology and political predictions (prognostications), had become wildly successful and widely distributed, and the power of astrology to incite the public had become obvious to many. When order was reestablished, the government moved to censor almanac content again, and by the post-Restoration period a line had been drawn, but not between astrology and reason, nor one within the broad subject of astrology itself, that is between judicial and natural astrology. The distinction had become, by this time, a social one, a case of status-sorting by referral to a controversial subject. In England this split, which occurred around 1660, did not separate rich and poor or aristocracy and commoners, it was more a case of respectable astrology (nobility, gentry, rising middle class) as opposed to vulgar astrology (country folk and laborers). In an account of the social status of astrology in seventeenth-century England, Patrick Curry argued that to understand the fate of the subject, it is best divided into three categories: popular astrology, interpretive judicial astrology, and high cosmological astrology (Curry 1989, 95). Popular astrology included the astral beliefs and practices of the rural and semi-literate laborers and also the urban artisans and working class. This astrology, consisting of lunar wisdom and weather lore, offered to the masses both an explanation of natural phenomena and a means of making choices based on planetary positions. Some astrological medicine that included home remedies and healing methods was part of this mixture. As doctors were not always available, the local wise "cunning" man or woman would typically use this kind of astrology in their practice. Popular astrology was sustained by farmer's almanacs and other secular publications that included weather forecasts based on astrometeorology and provided current tables to the phases of the Moon and planet's places. These kinds of publications have never ceased to exist and continue to sell today, though most offer only a very simplified version of astrology.

Judicial astrology, a more complex interpretive and technical astrology that focused on the lives of individuals and their problems and decision-making pressures, appealed to the better educated. Its practitioners were informed, but no longer at the cutting edge of science, as it was basically applied math and interpretation learned from experience, not a theoretical study. Members of this community included almanac makers, antiquarians, and physicians that were mostly middle class and living near London. Several astrologers of this general category actually attempted to reform the subject more or less in line with experimental science. John Gadbury (1627–1704), a royalist and Lilly's competitor, published in 1658 his anecdotal evidence for the subject, a collection of 153 natal horoscopes, the largest collection published in England to that date, which he analyzed in great detail. It was about this time that astrology ceased to make any real improvements, became mixed with other occult subjects, and gradually began to lose some of its more complicated techniques. A century later, Sibly, the English physician and astrologer discussed earlier in the book, and author of *New and Complete Illustration of the Celestial Science of Astrology,* only continued the traditions of sixteenth- and seventeenth-century judicial astrology with virtually no advances or reforms. This level of astrology, some of it mixed with other strange and exotic subjects, persisted in England until the late nineteenth and early twentieth century when the subject resurfaced, mixed with ideas drawn from theosophy and psychology and built around a modified, and simplified in many respects, methodology. More on this later.

In the seventeenth century there was also a high philosophical and cosmological astrology, essentially the purer elements of natural astrology, that was speculative and overlapped with physics, at least in regard to the presumed influences of the planets and the mechanics of the solar system. Although there was a deep distrust of astrology by the elite, due in part to the uncontrolled explosion of astrological political predictions following the Civil War, the Royal Society did include a few who did know something about astrology. If astrological ideas were discussed, however, they were probably redescribed in terms of experimental science. One astrological reformer, Joshua Childrey (1623–1670) was, like John Goad, inspired by Bacon and worked closely with the Royal Society in regard to the improvement of meteorology. His work on tides was important as he kept records of observations and noted the role of wind and the effects of lunar perigee. In astrology he argued for the importance of aspects that were coincident in both geocentric and heliocentric perspectives, these being the conjunction and opposition (Childrey 1652; Bowden 1975, 164–75). About the same time as Goad and Childrey, the scientific polymath Robert Hooke was noting correlations between the phase of the moon and weather records, and Robert Boyle was considering an explanation for astrological influences that involved magnetism and air pressure.

To some extent certain components of astrology were considered legitimate topics in the official scientific community. The effect of the Moon on the tides, for example, and also Newton's redefinition of action at a distance, of course known as gravity, could be seen as natural astrology without the name. A reformed version of medical astrology combined with astrometeorology (which might be called environmental medicine) was developed by the medical doctor Richard Mead (1673–1754), who counted Newton as one of his patients. He wrote on lunar and solar influences on the human body and how their motions correlated with patient symptoms (what he called alterations in the body), but none of his ideas were classified as astrology. He did reference Goad's work, however. Cometography, the official discussion on comets and their effects on the Earth and its inhabitants that took place among members of the Royal Society, is an example of the appropriation of astrological methods without using the word astrology (Schaffer 1979, 219). Astrology also offered a system of historical explanation, and the concept of historical periods grew out of astrological interpretations of long planetary cycles (Abu Ma'shar and Bodin being example predecessors). Isaac Newton's *Chronology of Ancient Kingdoms,* in which he attempted to date Biblical kingdoms and events by the slow precession of the equinoxes, even looks a bit like astrology without any claim of causation. The idea of authorized prophets, rather than an uncontrolled stream of enthusiastic and questionable people all calling themselves astrologers, was far more appealing to this new group of natural philosophers who were concerned with maintaining a strong and respectable social position. Remember that experimental science was a way for mostly middle-class Protestants to move up the socio-economic ladder. So, astrology was not disproved, it was removed from high and respectable society and kept at a safe distance by those in control, or it was renamed and modified.

In general, there was a state of permanent crisis in European and English politics, society, and culture during the sixteenth and seventeenth centuries. During periods such as this one, when thought systems collide, paradigms shift, and institutions are attacked, skepticism flourishes and solutions to the problem of what exactly is proper knowledge, and how it could stabilize society, become urgent. Knowledge that a person holds to is, in nearly all cases, a form of faith based on a sense of collective truth. One knows something to be true because one trusts someone else who, presumably, has the proper knowledge, speaks or writes convincingly, and has a good reputation. People-knowledge precedes thing-knowledge. Knowledge is really quite tenuous; the claims of science in regard to objectivity, and the trust in personal experience, however appealing, are only claims registered at a distance from the source. The perception and establishment of borders and boundaries among people is a fundamental social process related to tribalism, ranking, and pecking order, and it can be seen to operate on many levels throughout history. Humans are domesticated primates, and social maneuvering, driven by deep instincts for rank and status shared with other apes, is the means by which most come to accept those things that are regarded as truth. (This is made particularly evident today by the legions of voters that have been influenced by warped versions of reality promoted by some media and some politicians.) The reputation of astrology having been diminished during this period meant its truth was compromised as there were no longer many astrology-friendly authorities to trust.

Indicators of reliable truth-tellers in seventeenth-century England were birth station, wealth, and behavior. According to Steven Shapin, a gentlemanly culture of honor was transferred to the new domain of experimental philosophy, and the financially unburdened Robert Boyle became the greatest model for the seventeenth-century experimental scientist, a new kind of respectable identity in English culture. Boyle was a master of scientific credibility, a Christian gentleman, and a scholar virtuoso. His persona was a careful and purposeful assembly of cultural elements that defined respectability and truth in his world. Yet Boyle was known to have had an interest in astrology, consulted judicial astrologers, and even proposed a mechanistic explanation for its effects that involved magnetism, air pressure, and other unknown forces he referred to as "determinate effluvia" that acted on substances. However, the model experimental scientist he was living or portraying (probably both) made sure that this interest was handled discreetly and kept within a scientific context. So astrology, for Boyle and a number of other early members of the Royal Society who favored the subject, was not a badge to wear under any circumstances because it would associate them with enthusiastic, uncontrolled political prophecy produced by charlatans (Shapin 1994).[19]

---

19. For Boyle on effluvia, see Bowden (1975, 202–209). It is also true that a number of members also spoke out against the subject (Shapin 1996).

This separation into insider and outsider groups was no doubt in part due to the failure of astrology's symbolic language, which is the doctrine of aspects and the many symbols for planets and signs, for example, to be translated into the evolving language and conceptual categories of natural philosophy. Ann Geneva argues that astrology's language was meant to conceal, and in an environment of demystification it could not survive (Geneva 1995, 271).[20] This problem had concerned Francis Bacon and others who advocated for a universal language that would reflect nature accurately and become a medium for democratic scientific discourse. A committee was even established in the Royal Society to work on this matter. Astrological language, being symbolic (semiotic) and requiring considerable study to fully grasp, found itself in a situation within which it could not attract the brightest minds of the time, and interest in it declined among elite intellectuals.[21] To be fair, the symbolic language of astrology, which is essentially a kind of taxonomy, built on pattern cognition and synthesis, was not easily translatable because its subject matter was not amenable to quantitative measurements, these being similar to some of the problems faced three hundred years later by an emerging subject that came to be called psychology.

A contrast between medicine and astrology during this period further illustrates how the differences between these two social activities led to one achieving acceptance, the other rejection in the context of the social conditions of the time. At the beginning of the seventeenth century there was much overlapping between astrology and medicine, and both were often practiced by the same person. Regulation was minimal, though it was more focused in the

---

20. From an evolutionary psychology perspective, all of human history is about group bonding and maneuvering for status. In this case, one group, the experimental scientists, sanctified their exclusive activities and thus were able to get past the dominant censoring group, the religious authorities, while the astrologers remained outsiders. Tribalism is quite possibly a kind of social immune system, which had a critical role in pre-historic times, but now operates as both a binding and excluding social force.

21. This argument makes better sense if applied to alchemy, but astrology's symbolic language was, to a large extent, part of the public dialogue as shown by its frequency in the writings of Chaucer and Shakespeare and the popularity of almanacs.

case of medicine by several regulatory organizations. At the end of the century the status of medical doctors (physicians) had gone up and that of astrologers had gone down, and not necessarily because one was more effective in getting cures than the other. The medicine of the time, which included blood letting and leeches, was quite primitive, and the knowledge of astrologers in regard to herbal cures or other non-invasive actions was considerable in some cases. What seems more to be the case is that medicine, a classifying and labeling activity using a technical language, but not a symbolic language, was intrinsically more compatible with the methods of experimental science, particularly in regard to anatomy and chemistry. It also had a regulatory organization, the Royal College of Physicians, that became dominant and was able to define and police the subject in ways that suited the trends of the times: mechanical science, technology, and capitalism.

There were social factors at work here as well. Doctors classify patient symptoms in the context of a model, and then they dispense or act on the situation in a more or less consistent way. Diagnosis accompanied by remedy gives doctors a certain authority that is not found in the case of the astrologer who responds to whatever questions (often demands) the client brings to them. The doctors also operated in a collegial way, the astrologers were not organized and individually responded to clients. Further, astrologers also did not use models that were scientific, at least relative to the times, and they had no regulating entity. There were territorial conflicts as well. Nicholas Culpeper (1616–1654), a younger contemporary of Lilly and a medical astrologer (his astrological herbal *The English Physician* has never gone out of print), published the first English translation of the London College of Physicians Dispensary, which was formerly in Latin. By doing this, Culpeper, in a personal campaign against the monopolization of medical practice by the Royal College of Physicians, opened their medical secrets to the public. This was not authorized by the doctors, who were furious at the herbalist-astrologer. In the end, the doctors were far more successful than the astrologers in defining their boundaries, and they protected themselves collectively. The downfall and marginalization of astrology from this perspective was not simply a case of effectiveness in actual practice, but more a case of better socially defined boundaries and the power of one organization to hold its ground (Wright 1975).

10

REFORM, DECLINE, AND SURVIVAL

. . . [H]ow suddenly the celestial knowledge would be advanced, if our ancestors' defect herin could be made up by some private re-search, or voluntary contribution.

JOHN GOAD, *ASTROMETEOROLOGICA*, 1686

. . . [I]t is striking to observe the way in which a sharply ideological awareness of astrology gradually subsided, and a less conscious, more automatic, class mentality towards it arose. . . . [I]n 1708, a change is already perceptible: from enthusiastic (dangerous, radical) into vulgar (common, crude). By the early eighteenth century, no one could have seriously represented astrologers as a real threat to the state or civil order; yet it had become a mental habit of the edu-cated elite to scorn them. . . . In other words, what had begun as ideological had become a class mentality.

PATRICK CURRY, *PROPHECY AND POWER: ASTROLOGY IN EARLY MODERN ENGLAND*

REFORM

Following the attack by Pico della Mirandola, the improvement of astrology became a serious concern for the field. Not all astrologers

were onboard, however, and some thought astrology did *r*
be overhauled in any way. Many thought that only better *i*
accuracy was needed; others argued there should be a retu.
damentals of Ptolemy and a rejection of the methods added by the A1a.
astrologers. Some proposed more critical analysis of astrological charts
in order to discern which practices worked and which didn't. The tools
and technologies that could have been used to study astrology scientifically during the Renaissance were either nonexistent (statistics), hard to
obtain (such as the thermometer and barometer that Goad needed), or
impractical (as in the mirrors of John Dee), and so quantitative data
was not generated. The most radical reformers advocated fundamental
changes in traditional practice including the abandonment of the zodiac
or the houses, a disassembling of the system that had been remarkably
stable over the previous fifteen hundred years or so.

One low-tech and low-budget approach toward improving astrology
was in correlating planetary patterns in natal charts with life events.
Collectors of nativities, like Cardano in the sixteenth and Gadbury in
the seventeenth centuries, engaged in this work and published their
results. Charts were calculated, though exact birth times were not
always certain, and planetary configurations were compared with biographical notes. Correlations such as these were only anecdotal evidence
of something, yet they were usually taken to be confirmation of existing practices. These attempts to "test" judicial astrology were impossible as there were no units to measure anything, except those that made
up the astrological chart itself. Not much came out of this approach,
though it has preserved interesting data for both modern historians and
practicing astrologers who study classical methodologies and historical
personalities.

How to test natal astrology rigorously using reductionist methods
would have challenged any of today's social sciences had they existed then.
It should be noted that psychology, in its form of behaviorism, did adopt
reductionist-mechanistic scientific methodologies in the twentieth century. By reducing the study of behavior (using mostly pigeons and rats) to
externals only, behaviorism avoided dealing with personality, mind, and
consciousness, subjects almost impossible to quantify. Psychoanalysis and

psychotherapies were, and are, a lot more like astrological practice, but because conflicts with religion are minimal, and both are aligned with the medical establishment, which itself is based on measurable biological causes, they are generally accepted.

Pressures to reform astrology were high during the period between Pico della Mirandola and Newton, especially among those who had connections with the scientific community, but only a few individuals made attempts to do so. Most astrological activity was conducted by practitioners with little or no interest in theory who simply followed the rules of interpretation and delineation that had been established over the centuries. No coordinated reform movement was ever launched because astrologers were barely organized; only a discussion and dinner party group were formed in London, eleven years before the formation of the Royal Society in 1660. This group, the Society of Astrologers of London, met periodically, though the society only lasted a decade and failed a second time some years later when an attempt was made to revitalize it. It appears that this organization, which included members of rival political parties, mathematicians, astronomers, herbalists, and psychics, did not lay out a research agenda but only discussed astrological technicalities (Curry 1989, 40–44). It was left to individual astrologers to plan and execute any serious, fact-based overhauls of the subject.

DEE AND BACON

Dee, who was the English mathematician and astrological advisor to Elizabeth I, was for a time part of the Louvain group of mathematicians and philosophers that included the cartographer and globe mapmaker Gerard Mercator (1512–1594). Dee's book, *Propadeumata Aphoristica,* first published in Latin in 1558 and dedicated to Mercator, is a collection of aphorisms on astrology that points the subject away from Hermetic symbolism and toward greater precision and conformity with mathematics and the laws of optics (Dee, 1568). His proposed astrological reforms not only called for more rigorous measurements but also the use of technology, specifically mirrors. For Dee, astrology was essentially a practical application of astronomical geometry with a goal toward capturing, or more accurately, harnessing (in his case through the elaborate use of mirrors) a planet's light, which he thought was the carrier of planetary effects to the sublunar realm. He appears to have studied Al-Kindi, who we have seen hypothesized astrological influences as rays of light propagated by the planets. Dee took that idea further by describing how mirrors could focus planetary rays, like a concave mirror can focus the Sun's ray to light a fire. This method, he thought, would allow for a quantified comparison of the rays from the planets—and it could also be used to charge a talisman with planetary power. While Dee, at least in his astrological aphorisms, limited the subject by keeping it mathematical and as scientific as possible, he also used it in his experiments with natural magic. He was a classic example of the Renaissance magus: steeped in Hermeticism and knowledgeable of many subjects, some philosophical and scientific and some wildly experimental, like alchemy and angel magic, all of this making for an abundance of fascinating biographical literature on his life and escapades.

In his astrology, Dee drew attention to astronomical facts. He considered the distances of the planets from the Earth at perigee and apogee, and also the precise length of time a planet spends above the horizon during its diurnal cycle (an effect of declination), as important factors by which astrological influences could be better known, and therefore candidates for quantification. He actually proposed a kind of computational method that involved assessing what he determined to be 25,335 variations of astrological effects based on the strength of the planetary rays and their commingling—and this figure applies to conjunctions alone! Clearly such an attempt to put astrology on a more mathematical foundation by quantifying astrological influences would require splitting hairs and years of training. It is most likely that Dee was being facetious and drawing attention to the fact that astrological interpretation required both generalizing and subtlety (Dee, 1568). His contemporary, Jofrancus Offusius, who Dee accused of plagarism, also tried to reform astrology by introducing a method of quantification (Bowden 1975, 78–85). In his 1556–1557 *De Divina Astrorum Facultate,* Offusius presented numerical values for qualitative factors used in astrological interpretation. Using the four qualities (hot, cold, wet, and dry) and figures for planetary distance, he arrived at specific formulae that one could use to calculate the strength of a planet in an astrological chart. His approach did not catch on, though he wasn't the last to propose a more quantitative method of doing astrology by assigning numbers to planetary placements and aspects.[1]

While Francis Bacon, the father of scientific reductionism, was not a student or practitioner of the subject, he was open to an astrology that studied general conditions. Bacon was critical of the astrological doctrines concerned with nativities, elections, and inquiries, nearly all of which, he noted, were plainly refuted by obvious physical conditions. Essentially, he wanted to remove anything occult or mysterious from the subject and make it a "sane astrology," as reasonable as any of the other sciences. Bacon actually laid out a plan for the reform of astrology with a focus on aspects (phase) and also on elevation or altitude, which draws attention to a planet's distance from the meridian in its diurnal cycle.

> First, Sane Astrology should include the doctrine of the mixtures of the rays, namely, the conjunctions and oppositions and the other alignments of the planets amongst themselves.

> Second, there should be consideration of the planets nearer to the perpendicular, or recedings from it, according to the latitude of the region—for each of the planets are like the sun in having their summers and winters that modify the strength of their rays (Bacon, 1623).[2]

Bacon also called for attention to the apogees and perigees of the planets, their changes in motion from direct to retrograde, their

---

1. A method like this, called astrodynes, developed by Elbert Benjamin, has been used in modern times and is one of vast arrays of astrological calculations for users to experiment with that can be found in the more advanced commercial astrological software packages.
2. The vertical he refers to is the meridian, the north to south great circle that crosses the zenith. In the same section of the text (not shown) Bacon appears to advocate for the sidereal zodiac by suggesting that it is the particular stars located in the signs of the zodiac that account for the differences between signs, this only being the case with a zodiac fixed to the stars and not the tropic zodiac that is fixed to the vernal point.

distance from the Sun, any increases and decreases of light, and so on. All of this is strictly astronomical, quantifiable, and measurable, but Bacon was practical and had broad vision. He stated that the ideas of the historic astrologers ought not be completely rejected and that the particular natures of the planets as handed down by tradition should be considered carefully; further, he thought that the subject needed to be approached experimentally and include a search for physical causes. Although the usual subject matter of judicial astrology, such as people or questions, was included in Bacon's proposed reform, he criticized it. He completely rejected celestial magic, such as that practiced by Ficino, put limits on the use of elections, and called for establishing reliable rules of prediction. For the most part, however, Bacon thought astrological methodologies, specifically in terms of making predictions, could be most confidently applied to the domains of natural astrology.[3] These include:

> . . . [A]ll kinds of meteors, of floods, droughts, heats, ices, earth movements, water-bursts, eruptions of fire, and major winds and rains year round by various storms, pestilences, diseases that prevail, by the abundance and charity of the crops, wars, seditions, sects, and deportations of the peoples (Bacon 1970, 465).

Bacon's plan of reform did not enter into the great debate over astrology in his time (freewill and the precise location of God relative to the planets), and most astrologers probably weren't even aware it existed. His plan was just an idea that couldn't be easily implemented because astrology doesn't lend itself to quantification and the work involved to study it in this way didn't pay. Most astrologers of the time were consultants and healers, not scientists, and they preferred to stay with the traditions that worked for them, Simon Forman and William Lilly being examples. Two men with interests in astrology and the revolution

---

3. The term *mundane astrology* is used more frequently today among astrologers and, in addition to astrometeorology, includes historical astrology (correlating historical changes with planet cycles), political astrology (analyzing popular trends and predicting the outcome of elections), and financial and economic astrology (stock market predictions).

in natural philosophy were Brahe and Kepler, both scientific in their thinking and concerned with precision and clear, rational thought. Brahe's astrology incorporated precision, but it was traditional. Kepler, however, was a particularly enthusiastic astrological reformer.

Kepler completely rejected some standard components of Ptolemaic astrology, including houses and signs, regarding these as artificial divisions that were useful for positioning planets on a grid, but not much else. Kepler published two books completely on astrology, *More Certain Fundamentals of Astrology* in 1602 and *Third Party Intervening* in 1610, and two that included astrology, *De Stella nova in pede Serpentarii* ("The New Star of 1604") in 1606 and *Harmonics of the World* in 1618 (Field, 1984).[4] In *Harmonics of the World,* Kepler presented his cosmic theory, a detailed and all-embracing synthesis of geometry, music, astrology, astronomy, and epistemology that he considered to be in the tradition of Pythagoras and Plato. It was his magnum opus, and in it are sections in which he presented his resonance theory of astrology.[5]

Kepler thought that the aspects, analogous to the ratios produced by musical tones, made astrology work through resonance. When aspects were exact they produced a non-material tone that could be "heard" by a responder, these tones being the music of the spheres. His unique reform proposal was both practical (the addition of new aspects based on the fifth and eighth harmonics[6]) and theoretical; astrological effects

---

4. *De Fundamentis. (De Fundamentis Astrologiae Certerioribus)* Field, trans., (1984) is a lengthy introduction to the almanac for that year, and it includes his weather forecasts. *Tertius Interveniens* (Negus, trans., (2008) is a passionate argument in favor of astrology in which Kepler responds to criticism of the subject. At present, only *The New Star of 1604,* which concerns a nova that appeared and its possible astrological meaning has not been translated into English.

5. The usual translation, *Harmony of the World,* or *World Harmony,* completely misses the point of the work, which presents a Neo-Pythagorean theory of resonance as the key to understanding the universe. Much of the astrology in the *Harmonics* is found in Book IV.

6. The fifth harmonic series, based on division of 360 by 5, produces aspects of 72 and 144 degrees. The eighth harmonic series, 360 divided by 8, produces aspects of 45 and 135 degrees. These aspects were adopted by astrologers and are commonly used by modern practitioners.

occurred because of the existence of an animal-like soul faculty by which humans can sense occult changes in the sky produced by the resonances of the aspects between the planets, a kind of Platonic sensory system.

Kepler approached astrology as a careful scientist with data collection (weather records) and correlation testing with planetary aspects, and he never doubted that there was a real, perceptible astrological effect. But in his rejection of other components of traditional astrology he was almost alone among his astrological contemporaries, and he also had doubts about the subject ever being an exact science. Kepler complained that judicial astrology was based on personal experience and expertise; it was complicated, abused by malpractice, and misunderstood by the public (Field 1984, 260). He compared astrology to medicine, which he says is also based on experience and therefore not a true science, and it too had its own share of questionable practitioners. In *Third Party Intervening,* Kepler's account of an imagined debate over astrology, he defended almanacs and weather predictions and spoke of the wide gap between the complexity of the astrologer's work and the common language he had to use to transmit his interpretative reports to the readers.

And as the common man usually knows nothing of the abstractions of generalities, he perceives only concrete things, and often praises an almanac when it hits the mark, and he scolds the almanac when the weather does not come about as he had anticipated, even though the almanac really had hit the mark in its generality covering many possibilities (Kepler 2007 T-133, 194).

Kepler was here expressing frustration with astrology as a practice (almanac writer and consultant to the wealthy and powerful) because it was hard to synthesize so much qualitative data, impossible to be precise, and likely to be misunderstood and criticized. He was also frustrated with the methodologies of the subject and was ready to abandon many traditional astrological notions and forge ahead into a limited kind of astrology based on a theory of resonance. This never happened because he had other things to do in the way of data processing and mathematical modeling. But he did defend the subject (he actually says not to "throw the baby out with the bath water"), and he proposed a deep reform of the astrological toolkit. Ultimately, Kepler had very few followers in astrology, and his impact on the subject at the time was minimal (Beer and Beer 1975, 439–448).[7]

During much of the seventeenth century the social boundaries between the scientific and astrological communities overlapped to an extent. Some astrologers were associated with the leading natural philosophers of the time, and some members of the Royal Society, the world's oldest scientific institution, were sympathetic to the subject. Robert Boyle was known to consult with the astrologer John Bishop, and members, such as Robert Hooke, included astrological data in some of their own experiments (Capp 1979, 189). Astrological studies and early experimental science had a few things in common, and it has been argued that during the seventeenth and eighteenth centuries experimental science adopted some elements of the natural astrology tradition, particularly in regard to the measurements and discussion of newly sighted astronomical phenomena like comets (Capp 1979, 138). The physician Richard Mead mentioned earlier, along with Franz Anton Mesmer, should both be considered as contributors to the reform of astrology, though they didn't classify their ideas as such. They incorporated notions of cosmic influences, taken from astrometeorology and medical astrology, into theories of solar system influences on human health (Campion 2009, 186–191). By the eighteenth century the pressure to provide a mechanism for proposed effects such as these was very high, and both men pointed to gravity as a cause.

There were other factors that prevented scientific reforms in astrology. One was the strange taxonomic symbolism that made it a challenge

---

7. Morius, a contemporary of Goad and an astrological writer who had a great influence in the field of astrology, rejected Kepler. In the twentieth century a reform of astrology resulting in a substantially different methodology (The Hamburg School, or in the United States, Uranian Astrology) was based partly on Kepler's emphasis of the planets and aspects. So, it could be said that his reforms took four hundred years to be taken seriously by astrologers.

to translate findings into the vernacular.[8] Social problems existed, being a lack of institutional support and funding. Another was a brain drain. By the end of the seventeenth century the best minds of the socially advantaged were working in experimental science. Astronomy, as a subject, was transformed by the telescope and the discoveries made with it. This development alone greatly expanded the split between the activity of simply mapping nature and the socially complicated work of consultations. As astronomy moved on, the ranks of astrologers came to include many of lower social ranks, a factor that limited opportunities for research due to lack of connections and also education. Further, the practice of natural astrology, which was in theory more amenable to the scientific method and may have been improved, was mostly conducted by the same people who practiced judicial astrology, which was itself in conflict with religious conservatives and with Renaissance humanist notions of free will.[9] In addition to having a tainted reputation, these judicial astrology practitioners didn't understand experiment and testing. It was the case that during the course of the seventeenth century, astrology moved from being generally accepted knowledge with many practitioners and advocates in high social positions to a small collection of middle-class amateurs and almanac writers that wrote for the lower classes. And throughout there were also the quacks busy soiling the reputation of the field. More of this later.

Reforming astrology was simply too hard to do without institutional support. Kepler was deeply discouraged and complained that the subject was misused by believers and critics alike. He did eventually stop writing meteorological almanacs, probably because they were no longer his job and he could afford to do so (Kepler 2007, T–130, 131, 193–194). When John Goad attempted his monumental Baconian reform

---

8. Translation from astrological symbolism to the vernacular presents real problems. Those versed in astrological symbolism will understand a given situation from within the context of the specific planetary configurations involved. Such an understanding involves a multivalent taxonomic system utilizing symbols, not ordinary language, and as such, limits understanding to those who know the code. The study of symbols today is called semiotics.
9. This third point will be considered in the discussion of the social factors in the decline of astrology. The other two points will be discussed later in this section.

gy, he only found some correlations that he regarded
ultimately all he had was a list, not data that could
produced. Without instrumentation for recording
deal with the complexities of a system, he was left
any anecdotal evidence for astrological weather rules.[10] Other
is had been suggested by various astrologers, some of whom were
respected natural philosophers, but there was no collective and coordinated effort. Historian of science Mary Ellen Bowden conlcuded that it wasn't bad methods that sunk astrology, it was that they were attacking a problem that simply could not be solved (Bowden 1975, 223).

The actual scientific reform of astrology, both natural and judicial, obviously didn't happen. In my view, the major obstacle was that no single astrological influence could be precisely quantified, and so there were no units to be analyzed. You just can't do reductionist science without units of measurement. Descriptions of alleged astrological effects, including possible explanatory mechanisms, were by necessity qualitative and of infinite variety because planetary patterns do not repeat, and this makes specific predictions extremely difficult and exact replication impossible. In the final analysis, attempts to reform astrology scientifically in the seventeenth century, that is to place it more or less in the context of RMM science, failed and the project was abandoned.

DECLINE

Over the previous three chapters I have tried to show that the decline of astrology is only describable with multiple strands of history, each important in establishing a reasonable understanding of what happened. These histories show that astrology, both judicial and natural, which were normally practiced by the same person, steadily slipped in reputation from the latter part of the sixteenth through the

---

10. Experimental reproducibility has proven difficult with astrology—as it is with any system that can't be broken into parts. One exception has been the Gauquelin studies (discussed later) that correlate planetary diurnal cycles with profession and other personal characteristics, which, though challenged many times, appear to be holding up (Ertel and Irvng 1996).

seventeenth century and became marginalized in respected society by the eighteenth. From the progressive perspective of modern science, that is the RMM worldview that replaced Renaissance naturalism, astrology had lost its previously authoritative "scientific" foundations with the collapse of Ptolemy and Aristotle. Astrology therefore found itself without a rationale and was abandoned by the astronomers and other elite thinkers of the period and became a study and practice unrelated to the other modern acceptable fields of inquiry. Because most astrologers were practitioners and not concerned about theory, this view doesn't fully explain astrology's fall. Other reasons, including the complex effects of the rise and fall of holistic perspectives and organicist philosophies, and the regulatory power of the Church on thought, need to be considered as well.

Astrology always had unusual relations to other fields of knowledge. Since ancient times it depended on astronomy, itself considered to be a branch of mathematics, and until the seventeenth century most astrologers were also mathematicians and astronomers. Astrometeorology could be considered a topic in natural philosophy, but its close associations with judicial astrology left it in the cracks between disciplines. Astrology, although built on empirical observations over the centuries, was not natural history, neither was it medicine, although it informed that subject, and many people practiced both. It had applications to agriculture and geography, and also to history in that it mapped historical cycles. It could be said, then, that astrology has consistently been its own subject, a complex interpretive and predictive study of the natural environment and the behaviors of groups and individuals. Most important, it is a study that includes a practice. I consider it a subject that maps out and measures the trajectories of self-organizing systems, and it does so using a mixture of symbolic language and geometry. Finding a mechanism that can satisfy the demands of RMM science has been elusive because self-organizing systems just aren't that simple.

Assuming astrology is a legitimate (though academically banished) subject in and of itself, which is the argument of this book, it can be said to be in the same league as philosophy, religion, sociology, and evolutionary theory—all of them being subjects that have extensive explanatory power. It could even be said that astrology was the earliest attempt at understanding universal laws that were not specifically religious (Thomas 1971, 324). Like religion, it can be used as a worldview and offer answers to big questions, but without all the ritual and priesthood. Historically, and in the present, it is a comprehensive guide to self-knowledge, social understanding, practical information about personal choices, and a broad perspective on life that includes insights into one's deepest personal inclinations and, consequently, destiny. Recognizing this, the clergy, for the most part, have viewed it as an unacceptable alternative to their program, and for centuries they either rejected it or discouraged its use. There was also the fact that astrology offered a worldview that contained within itself the rise and fall of religions, and also the possibility of making an astrological chart for Christ's birth. This by itself was enough to put it on the firing line from Rome and from Protestant leaders.

The major philosophical argument over astrology that engaged intellectuals during the entire sixteenth century and into the seventeenth was basically the major theme of the Renaissance: the dignity and freedom of humans. This was championed by humanists like Erasmus and Pico della Mirandola who had no place for astrology's nuanced notions of planetary influence on human destiny in their worldview. Proponents of astrology admitted the importance of free will, and they argued that foreknowledge gained from astrology actually permitted a better use of it. This rebuttal, however, required psychological caveats that were too subtle and complex for most to understand, especially those not at all knowledgeable of astrology's symbolic language and repulsed by even a whiff of fatalism.

By the seventeenth century, however, astrologers were in trouble with more than a few parts of society. Their social prominence and power, driven by political predictions fueled by printing and almanac sales, was not well-received by the secular establishments. Intellectuals were rapidly replacing their holistic and Hermetic notions with atomism and moving away from metaphysics and on to projects with more certainty and bigger payoffs. Astrologers' services to the public had been cutting into those of the physicians, and also the counsel of the church. To slip around the church blockade they published books with titles that made it seem like astrology was a Christian activity and another way of reading the book of nature, but it wasn't enough. By the eighteenth century astrology had become irrelevant to those who were driving the intellectual and social trends of the time, and the subject nosedived in status.

In evaluating the "truth" of astrology, historians, while acknowledging its ubiquity in certain periods of the past, generally dismiss it. Keith Thomas said repeated failures in almanacs showed that astrology was utterly incapable of providing accurate predictions—except, one supposes, for the many that were accurate, like those of Kepler, Lilly, and countless others before, during and after the Renaissance.[11] But the question remains: If the forecasts in almanacs, which were mostly weather and politics, were just guesses, why were astrological almanacs in demand? One answer, a variation of the Barnum effect, is that the public is basically gullible and will believe just about anything. Psychological tests have demonstrated this over and over, and it does appear to be validated by the existence of those segments of contemporary American culture who listen only to distorted opinions that pose as news and believe what they hear. But for how long can the public be fooled? Does the public eventually sour on nonsense, or can the public be fooled with the same things for centuries? Exactly how stupid can people be? Thomas sidestepped this problem by first declaring that almanac astrology didn't work, but he ignores astrometeorology, which always was the primary astrological content of an almanac and the subject on which most of the pre-industrial economy depended. Instead, he attacked judicial astrology, which amounted to very little almanac content. He said that, to account for failures, astrologers would typically fall back on exceptions like divine intervention, complain about censorship, point to mistakes they made in calculations, or lodge complaints about

---

11. Most predictions made by astrologers were for patrons and have only survived as second- or third-hand reports. Having published his own almanac, Lilly's prognostications have been preserved, including his political predictions and those of the London fire and plague. Many of these are (wisely) couched in symbolism and astrological jargon (which people understood then)—camouflage that will only convince skeptics that the entire business was either a scam or self-deception.

other bad astrologers (Thomas 1971, 334).[12] So the assumption here is that any prognostications that were accurate happened by chance.

What people or a culture believe to be true does shape what is possible within it. Today a number of worldviews dominate Western society and shape perception. Christian nations process information and set priorities very differently from Islamic nations. There are the competing socio-economic views of capitalism and socialism, which in turn shape politics. In economics, globalism competes with nationalism. The dominant worldview of modern times, at least in regard to nature, is RMM science. The modern cosmology of science (until the 1960s it was more or less a pseudoscience) is centered on Big Bang Theory, which is now supported by evidence and offers a realistic, but not necessarily comforting, explanation for why we are here. This cosmology is an outcome of the highly successful approximately four-hundred-year development of RMM science that is now championed through the STEM (science, technology, engineering, and mathematics) programs taught in schools and supported by governments. In addition to its cosmology, science has produced the theory of evolution by natural selection, both of these amounting to a major intrusion into the domain formerly held by religion. Catholicism has gradually been accepting of these theories, but some fundamentalist Protestant sects have retreated into a religious worldview that is far less sophisticated than that held by theologians of the Middle Ages, though is maybe closer to the views of the common people back then. It's apparent then that, in the world today, a wide range of thought and belief systems, religions, and philosophical narratives based on science all compete

---

12. Thomas may have produced a good social history, and his chapters on astrology are full of interesting details, but he is out of his league in judging the subject. The evaluation of the predictive accuracy of consulting astrologers of centuries past is not so easy, as the primary sources are limited. Many astrologers simply gave advice and did not limit themselves to predictions. Their casebooks may contain astrological charts but usually don't reveal what they told clients in private, or what happened afterward. (Thousands of astrological charts have survived, however, which nearly all academic historians are incapable of reading.) Thomas simply doesn't know that much about the nuts and bolts of astrology. It is only fair to expect a writer of the history of geology, for example, to actually know something about the subject they are reporting on. See Campion (2009) for a more balanced social history of astrology.

in the postmodern world and define "truth" in their own ways. None have any place for astrology, which is labeled either a pseudoscience, dangerous divinatory activity, or entertainment.

The decline and marginalization of astrology in the sixteenth and seventeenth centuries was also driven by changes in the availability of information (printing), urbanization, and trade. The problems of theology, politics, and economics, the challenges of assimilating the New World discoveries and opportunities, and the social conditions of the age demanded the attention of the best minds. There was certainly a brain drain during this period away from practices such as astrology and toward work that was more tangible and had the support of institutions, or conferred social status in some way. Fewer and fewer talented individuals entered the field. Consider Kepler, a master astrologer who would rather solve the complex astronomical problems of his time than cater to powerful, wealthy, and self-indulgent patrons who were incapable of understanding the subtleties of their own astrological charts and not at all interested in raising their level of self-knowledge. Who could blame him for turning away from this? The job of doing astrology for powerful patrons, similar in some ways to that of a consulting practical psychotherapist, was not what a person of Kepler's caliber was seeking in a career. It is extremely frustrating to communicate nuanced information and suggest life strategies to people who are set on getting what they want and see the astrologer as only a tool for them to use.

The creation of a new, respectable social identity for the natural philosopher exemplified by Boyle was surely a factor in the decline of astrology. To be associated publicly with astrology—which, by the late seventeenth century, implied political recklessness, enthusiasm, and links to the vulgar classes—was to be stigmatized. Although Lilly had exceptional social skills, astrologers as a group did not conduct themselves well during the English civil war and consequently set themselves up for marginalization by royalist factions that eventually returned to power.

Astrologers, the intellectual elite, and the religious authorities were not that far apart in regard to the charlatans who were offensive to respected society and an embarrassment to the astrological community. Once protected by royal or wealthy patrons, astrologers were set loose on the public by printing technology in an increasingly trade-driven culture, and they failed to agree on standards of conduct. The new freedom of individual astrologers to promote themselves with printed prognostications and almanacs led to abuses like the rash of wild predictions for a great flood in 1524 and the drama of politics during the English civil war. By the end of the seventeenth century the astrologer as a public figure had become a figure to be mocked, which Jonathan Swift did in his satire of an almanac writer (Campion 2009, vol. 2, 170–173). Unlike the natural philosophers of the seventeenth century who formed study groups like the Royal Society, astrologers did nothing of the sort. Their failure to organize (which the physicians did successfully) left each practitioner free to do whatever it took to make a living; there was no institutional support or regulation of the subject. The post-patronage astrologer was essentially a sole-proprietor who ran a consulting and publishing business and was in competition with his peers. Like the television advertisements of lawyers and dentists today, standards were lowered in order to reach a wider range of potential clients. We cannot underestimate the effects of rogue individuals on the public image and the long-lasting damage that results. Or maybe we can, as shameless self-promotion has become very much a part of modern life in print, on television, and in politics, even at the highest levels.

SURVIVAL

The events and trends of the sixteenth and seventeenth centuries cited above were devastating for judicial astrology, and it was stripped of respectability in the academic world and in educated society in general. Astrometeorology, however, was rarely attacked even during Newton's time. Goad's book, published in 1686, was favorably reviewed and considered respectable by some in the Royal Society. There was not an abrupt, clean break between natural astrology and the interpretation of cosmic influences on the Earth by those within the Royal Society. It was more the case of the appropriation of method and some content by one group from another. From cometography to William Herschel's (1738–1822) observations on the correlation between grain prices and sunspot numbers, an interest in connections between celestial and terrestrial phenomena continued. Some scientific meteorology utilized astrological indicators, including the Moon and the signs of the zodiac, up to the late eighteenth century, and the practice of keeping weather records like Dee, Kepler, or Goad, was appropriated. Traditional astrometeorology survived in the farmer's almanacs, which were successfully transplanted to the colonies. The first press in America opened in 1638 in Cambridge, and the second document printed there was an "Almanac Calculated for New England." The press was set up by Henry Dunster and it became the Harvard University Press. During the later part of the seventeenth century, New England printers and publishers came up with new formats for almanacs, and one from Saybrook, Connecticut, broke tradition and began the almanac with January, not March. This one also included weather forecasts based on planets and presented the most ubiquitous feature of the popular almanac, the zodiac man (Man of the Signs; Sagendorph 1970, 46–47).

Many other annual almanacs were published during the colonial period, including Benjamin Franklin's *Poor Richard's Almanac,* which continued the tradition of astrological weather forecasting. Franklin apparently employed others to do the astronomical calculations that made up much of his almanac's content, which included the forecasts based on planetary aspects. There is no question that astrology was behind the forecasts. For example, in his 1753 almanac, it is noted that the Sun was in opposition to Saturn on June 24 and the forecast for the twenty-third to twenty-sixth was "thunder, then cooler." On December 29 of that year the Sun was conjunct Saturn and the forecast was "cold and cloudy." Both forecasts are consistent with astrological tradition tested in Chapter 5. In the 1790s Benjamin Banneker published annual almanacs, with astronomical calculations done by himself, that contained weather forecasts. These also predicted cold weather when aspects formed between the Sun and Saturn and also Moon and Saturn. To a large extent the almanacs preserved astrometeorology, and other parts of astrology as well, which then experienced a major resurgence during the twentieth century in both England and America.

The case of Franz Anton Mesmer (1734–1815) needs to be mentioned in any discussion of eighteenth- and early-nineteenth-century astrology, in this case specifically natural astrology. Mesmer, a controversial doctor known for his theories on healing through magnetism and as a pioneer in hypnosis, used astrological notions to explain his work, which he did not identify as astrology probably for reputation protection purposes. His doctoral dissertation, however, was titled *Physico-Medical Treatise on the Influence of the Planets,* in which he argued that planetary tides affected the body and their cycles account for rhythmic changes in health. There was no zodiac in his theory, only planetary forces transmitted to Earth through a cosmic fluid he called animal magnetism. It appears that many of Mesmer's ideas about the planets, including gravity as the mechanism behind cosmic effects on human health, were taken from an earlier work by the English physician Richard Mead, who was also Newton's doctor (Campion 2009, 186–190; Simon 2010). Both Mesmer and Mead utilized material taken from both astrometeorology and medical astrology, but not judicial astrology.

Traditional natural astrology was preserved in other ways. A few English and American astrologers in the nineteenth century wrote on natural astrology, but more as a review of established methodology and less as an ongoing investigation.[13] An astrometeorological society was formed in London in 1860, but it lasted only two years. Within the small middle-class community of English and American astrologers in the late nineteenth and twentieth centuries, natural astrology came to be commonly referred to as mundane astrology: the analysis of weather, earthquakes, storms, plagues, crop harvests, migrations, historical cycles, politics, market fluctuations, and the like. The use of the time slice (the astrological chart), most often calculated at equinoxes, solstices, and lunations, became the preferred technique for this branch of the subject. (Given the precision of modern astronomy, Kepler and Brahe would have had no complaints about the use of this technique.) Actual research in natural astrology, aside from case studies, was, for the most part, abandoned as few understood proper scientific methodologies, it required a lot of work that didn't pay, and institutional support was lacking.

---

13. Pierce (1890) is an example. He summarizes, more or less, Renaissance astrometeorology with occasional references to John Goad's observations.

During the nineteenth and twentieth centuries, modern subject areas replaced the public and personal needs that astrology had filled for centuries. Astrometeorology was replaced by modern meteorology, which is based on data from scientific instruments. It was during the seventeenth century, just as astrology was in steep decline, that the thermometer and barometer were invented and rain and wind gages began to be used. Weather diaries became more common then, as precise quantitative data could be recorded for the first time, and this stimulated the establishment of weather stations. This trend grew, and the usefulness of weather data led the activity to become a part of regional and national governing bodies, the first ones being established in the mid-nineteenth century in Austria and England. With government and institutional backing, meteorology progressed, and today it is considered a rigorous physical science that studies the atmosphere, with forecasting being an application of that information. Prediction of weather is what the public wants, and short-term forecasting has steadily improved, as has the media delivery of this information.

Natural astrology also provided information on the behaviors of nations (i.e., war, peace, prosperity, depression), the value of currencies, and also harvests produced by agriculture. Brahe and Kepler made these kinds of forecasts for their wealthy and highly placed patrons in the form of written reports. Almanac writers also issued prognostications for the public that concerned these topics. Much of this information has been replaced by modern economics, which shifted the subject matter from being controlled by planetary cycles to changes explainable with economic theories, many based on cycles of one length or another. By the late eighteenth century and into the twentieth, the ideas of a few philosophers (i.e., Adam Smith, Thomas Robert Malthus, David Ricardo) became the basis of the modern social science of economics, with its emphasis on resources, capital, labor, and production. Today, investing in global markets is of great interest to the public, and information on current events and future trends in economics is offered in print and other media programming.[14]

---

14. Economists (most of them), it should be noted, generally failed to predict the market crises in 1987, 2008, and 2020, though most of the leading financial and mundane astrologers saw these coming.

For much of its history judicial astrology answered questions about one's character, though this was seen more in terms of one's fate and fortunes. A person born under Saturn, with the Sun conjunct or opposite Saturn, or having Saturn prominently placed in the birth chart in some other way would know that they were naturally inclined to be serious, prone to melancholy, and pressed by circumstances to work hard, among other constraints, and to live what most would consider to be a difficult life. Ficino, who was born under Saturn, struggled to understand his fate and in the process created a kind of astrological psychology. His ideas were steeped in Christianity and natural magic, but there were others like Pontano and Kepler who also used astrology in a secular way to better understand themselves. The study of the mind, traditionally in the domain of philosophy, was revolutionized in the eighteenth century by Immanuel Kant, who argued that the mind conditions sensory perceptions and not the other way around. This led to nineteenth-century attempts to scientifically test the mind (e.g., Wilhelm Wundt). Introspective studies like those of William James, and the psychoanalysis of Freud, Jung, and others were additional foundation blocks in the construction of twentieth-century personal psychotherapies and social psychology. The application of psychological findings in the form of individual therapy became a kind of personal consultation, which was what astrologers did previously. So, in modern times, meteorology, economics, and psychology came to replace most of the social and cultural needs that astrology had formerly filled—and without objections from religion.

The twentieth century saw a significant revival of astrology, much of it in England where it had never died out among those in the middle and lower classes. Comprehensive textbooks on the subject were published, and attempts were made to reconcile astrology with psychology, especially that of Carl Jung, who was interested in the subject. For much of his life Jung studied astrology, and he drew from it when developing his theory of personality types. The four functions (intuition, sensation, thinking, and feeling) he proposed correspond neatly with the astrological elements fire, earth, air, and water. These form the original basis of the Meyers-Briggs personality assessment, which has become widely utilized. Jung also proposed a non-mechanism to explain the workings of astrology and certain other phenomena that he called synchronicity (more on Jung later). He conducted several studies on astrology, but abandoned this line of inquiry and theory when hearing of the work done by people like John Nelson who were proposing physical mechanisms for observed correlations between planets and solar phenomena.

Astrology produced a few intellectuals in the twentieth century, notably Dane Rudhyar, who attempted to place the subject on the foundations of holistic thinking and later integrate it with humanistic psychology. On the other hand, popular astrology in newspapers and magazines was developed and has now become ubiquitous, though considered lower class. Professional astrology, benefiting from progress in psychology and computer software, is now quite sophisticated and used by people on all levels of society, even those at the top, though in those cases very discreetly. Innovations in technique, some based partly on Kepler's ideas, have extended or replaced some traditional methodologies. Professional organizations are now found throughout the developed world, and older non-Western traditions, such as what is now called Vedic astrology from India, have made inroads into Western astrological culture. Progress in natural astrology, still referred to as mundane astrology and focused more on interpreting and forecasting political and economic trends, has occurred as well, though the public hears little about it except in regard to presidential elections and stock market forecasting. More recently astrology has been riding a wave of self-interest generated by the internet and social media. An entire generation has now been imprinted with basic astrological notions, mostly Sun signs, but also a few notions about general conditions involving Mercury and the Moon. One could say that astrology is making something of a comeback, but it is not taken seriously by the media and is still under attack from academics, skeptics, and religious fundamentalists.

PART 3

The viability of an astrological restoration

II

EVIDENCE OF
ASTROLOGICAL EFFECTS

Hence despite much progress there remain few concepts in astrology that are not disputed among astrologers. To date the most significant research results have sometimes supported tradition but have more usually contradicted it. The picture emerging suggests that astrology works, but seldom in the way or to the extent that it is said to work. Obviously genuine research must start with as few assumptions as possible.

DEAN AND MATHER, *RECENT ADVANCES
IN NATAL ASTROLOGY*

We do not now have to worry about the astrological hypothesis, because it is probably true. The Gauquelin results, limited as they may seem to one familiar with the numerous items of lore in astrology, are already enough to confirm the hypothesis.

ROBERT HAND, "THE EMERGENCE OF
AN ASTROLOGICAL DISCIPLINE"

If astrology was never actually disproven but went down with Renaissance naturalism in a perfect storm of radical ideological change and social catastrophes, then we are left with the question, *Is it real in a modern sense?* Any answer has to take into consideration the numerous scientific studies that have demonstrated that both the Sun and Moon have ancient, deep connections to life, information that indicates life has evolved in a temporal, as well as spatial, environment. These two astronomical bodies are also known to drive certain weather, climate, and geological processes. The first two chapters of this book are an inventory of these correlations between the cosmic and terrestrial environments, a review that makes the case that life doesn't evolve in a closed bubble and that weather and climate are not entirely random processes. Some of these correlations—such as temperament and season of birth, positions of the Sun or Moon in their diurnal cycle, and the tracking of phase angle between Sun and Moon—are very suggestive of what astrology has been getting at for millennia. The gap between these scientific findings and many traditional astrological methodologies is really not that wide. In addition, Chapters 1 and 2 also serve to inform the reader how linkages between astronomical factors and the subject matter of chronobiology and climatology are often extremely difficult to map or quantify. Consider that research into the effects of Earth's magnetic field on life, or cosmic rays on climate phenomena (both living things and climate are systems), can be complex, costly, and time-consuming, and not much would be learned without funding and institutional support.

In the largest sense, the activity of astrology, or doing astrology, comes down to the observation of correlations between the solar system bodies and people, society, market behaviors, weather, or other self-organizing systems, and then generalizing these findings, a process, passed down over the centuries, that has established guidelines for interpretation. Astrologers, knowing that sky patterns never repeat themselves in exactly the same way, have settled on general rules that may then be modified by other rules, a process akin to mixing paint colors. If it gets cold in a particular region most of the time when the Sun is opposite Saturn, then there would be a high probability of cold at the next opposition of these two bodies. But suppose the Sun happens to be conjunct Mars at the same time, which the tradition has settled on as an indication of heat. Then, two rules will need to be blended; in this case a canceling effect may be indicated. This is a combination of quantitative and qualitative thinking, a blending and synthesis of numerically constrained information in the form of astronomical positioning. The general rules of astrology are not made up, they are based on empirical observations and pattern recognition, refined and extended over the centuries from the experiences of practitioners. Developing astrological methodologies is a kind of scientific activity similar in ways to the practices of animal biologists who work in the field. Nature is studied, behaviors are generalized, and predictions are made based on a synthesis of observations that are then tested multiple times by others. Astrology is then a group project using an early form of natural science that prescribes close observation, description, and pattern recognition as a basis for conclusions that are built on logical deductions.

Aside from rules accumulated over the centuries, the earliest attempts to prove astrology using another early component of natural science, critical comparison, were in the form of anecdotal evidence. During Roman times some astrologers (e.g., Antigonus of Nicca and Vettius Valens) assembled collections of nativities that they used to illustrate how astrological charts correlated with personality and destiny. The same thing was done by Cardano and Gadbury in the Renaissance, but collections don't really prove anything in a quantitative sense. During the seventeenth century the tradition of science in the West was revolutionized. Critical thinking, observation, description, and comparisons were enhanced with a brilliant method built on hypothesis construction, reductionism, experiment, and mathematical modeling. As recounted in previous chapters, astrologers did make an effort to bring astrology into this new model for doing science; Goad's massive weather research project in the mid-seventeenth century is probably the best example. It turned out that fitting astrology into the new scientific methods proved too difficult at the time, and it wasn't until the early twentieth century that a new round of astrological studies was attempted. One of the earliest of these was done by a woman associated with the teachings of Rudolf Steiner (Waldorf schools, biodynamic agriculture, etc.). She was not part of the larger astrological community, and her studies, and the replications of them, remain an anomaly in the history of astrological research.

In the 1920s Lilly Kolisko (1889–1976) observed and noted the crystallization patterns of certain metallic salts (silver nitrate, lead nitrate, iron sulfate, gold chloride, tin chloride). These compounds were dissolved in distilled water and placed in contact with filter paper. When the silver solution was mixed with the iron solution in equal parts, she found the patterns produced on the filter paper differed if the experiment was done during day or night, but also in light or dark. Further, the patterns differed in terms of reaction rate and general form during precise astronomical events, such as during an eclipse and during a triple Sun, Saturn, and Moon conjunction. Her experiments with the metallic salts and filter paper, which she called capillary dynamolysis, were published by Steiner's Anthroposophical Society and they have been replicated, along with variations based on them, several times since then (Kolisko 1928).[1] These findings, and they are unquantified observations, suggest that light and solar system bodies may be modulating matter on the molecular level, but these studies are known only to those who participate in and read the publications of the alternative holistic and scientific methodologies built around the thought of Steiner. In principle, however, Kolisko's studies are similar to those of Georgio Piccardi, an Italian chemist who, beginning in the 1950s, studied the precipitation of certain substances (i.e., bismuth oxychloride) in water. Variations in the trends of precipitation were found, which he linked to magnetic field variations and the solar cycle. Piccardi's work remains controversial.

The central problem in testing astrology in a reductionist way, using statistics to draw boundaries and establish probabilities, has to be the difficulty of assigning units to the subject material which is not easily quantified. This point cannot be overemphasized. Paul Choisnard (1867–1930) may have been the first to apply simple statistical techniques to natal astrology using sun signs as data points.

---

1. A replication testing a Mars-Saturn conjunction was done in 1949 by Theodor Schwenk (described in Pelikan (1973). See also Kollerstrom (1976), Rohde (2003), and Dean and Mather (1977, 228–234). Kolisko also experimented with planting during the lunar cycle, finding more rapid growth to occur when plants were started prior to the full moon in comparison to those started at the new moon.

In one of his studies, he took a sample of noteworthy intellectuals (n = 123) and found a majority of them were born with the air signs (Gemini, Libra, Aquarius) rising (on the Ascendant). But he didn't account for the astronomical fact that the twelve signs of the zodiac do not rise at the same rate, and he had no control group, so his study wasn't done properly. Swiss astrologer Karl Krafft (1900–1945), who worked for the Nazi SS and propaganda ministry, was another early researcher who applied statistics to astrology (Howe 1967). While he collected large samples for his many studies, he also failed to take into account certain astronomical factors and claimed significance when his data didn't show true statistical significance. In spite of their naive methods and flawed results, Choisnard and Krafft initiated the immense task of testing astrology in the twentieth century, a project that the astrological community found to be difficult, complicated, and unfunded. (This would be expected given that astrologers were few in number and profoundly marginalized during the previous approximately 250 years). Of the many studies published since their time, the best known, longest, and most sophisticated were those done by the French statisticians and psychologists Michel Gauquelin (1928–1991) and his wife Francoise Schneider-Gauquelin (1929–2007). The work of the Gauquelins has been described in detail in many other publications, and my account will therefore be relatively brief.[2]

THE GAUQUELIN STUDIES

Michel Gauquelin, who had an early interest in astrology, recognized the serious flaws in the studies of Choisnard and Krafft and became intensely interested in testing the subject in a proper reductionist manner. He began what turned out to be his life's work in the late 1940s when, after having studied psychology and statistics (he received a doctorate in the latter), he traveled through France to collect birth data

---

2. For more on this topic, I suggest Michel Gauquelin's many books, especially *Written in the Stars, Birthtimes,* and *Cosmic Influences on Human Behavior.* In addition to these, I also suggest *The Tenacious Mars Effect* (Suitbert and Irving 1996) and *The Case for Astrology* (West 1991).

for a study of astrology. This was followed by the laborious calculation of astrological charts, the assembly of data points, and the slow refinement of his methodology. By the mid-1950s, now with technical and financial assistance from his wife Francoise, he published *The Influence of the Stars,* the first of many books that described his research. This was just the beginning of a multi-decadal research project on astrology with results and conclusions that were met with stonewalling or furious resistance from skeptics every step of the way. At the same time there was a significant measure of ambivalence from the astrological community as the results were not particularly supportive of some traditional doctrines. After the death of Michel in 1991, Francoise continued the work until her own passing in 2007.

The Gauquelins' research focused on planetary positions in the diurnal cycle, the daily rising and setting of planetary bodies. Their studies led to the discovery of statistically significant correlations between planetary positions in the diurnal cycle and three measured factors: profession, personality traits, and heredity. Since ancient times, astrologers have claimed that the location of a planet relative to the horizon at the time of a person's birth has descriptive value in assessing personality and significant life experiences (these being related by the astrological axiom "character is destiny"). The diurnal cycle of rising, upper culmination, setting, and lower culmination, was traditionally mapped by dividing this cycle into twelve spatial sections; each planet would then be located in one of these houses or domiciles. This spatial quantification (house position) in celestial longitude of the diurnal cycle served to standardize interpretation and was one of the key sources of information to be derived from an astrological chart. For example, a planet near its upper culmination, located within a zone called the tenth house, would offer information in regard to a person's profession, or general social status, the two overlapping in most cases. A planet about to rise, which is, according to tradition, the most potent place for a planet to be located, would be in the first house and be descriptive of the person's basic nature (personal identity) and physical body. If a planet was near setting, on the western horizon, it would be in the seventh house and describe significant others, including partners and relationships. A planet approaching lower culmination would be in the fourth house and signify home, family, and ancestry. A more general distinction (sect) made in regard to a planet's position relative to the horizon was the dirurnal arc (moving from rising to setting) and the nocturnal arc (from setting to rising), the former descriptive of externalities, the latter subjective factors.

The experiments that were undertaken by the Gauquelins were immensely time-consuming, self-financed, required calculations done by hand (personal computers didn't exist then), and took place well before the internet and the consolidation of records in national departments of vital statistics. Work began by first accumulating birth data on thousands of persons, something that required travel to towns and cities, locating the appropriate government agencies, and requesting vital information that had to include the time of birth. On the plus side, since the time of Napoleon, French birth records were required to include the time of birth, something not done in many other countries. But this request also usually included a fee, which, along with everything else, the unfunded Gauquelins paid themselves. When the data had been collected, astrological charts were then calculated by hand (multiple calculations are needed for a single astrological chart) to determine exactly where each planet was located relative to the horizon at the time and place of birth. The traditional twelve-house astrological methodology was replaced by their own spatial sectors (twelve, eighteen, or thirty-six) that were calculated to quantify planet positions in the diurnal cycle.

In one study, done in 1955, Michel plotted the position of Mars relative to the horizon for a sample of 570 French sports champions. According to traditional astrology, Mars is associated with conflict and competition. If there were no planetary effect and all astronomical variations had been accounted for, one would expect Mars to be randomly distributed in the sample and the probability of Mars being located in any sector should be equal to that of the others. Instead, the study showed that Mars was found more often to be just past rising and upper culmination and, to a lesser extent, just past setting and lower culmination (see figure below). A control group of people who were not sports champions showed an even distribution of Mars in its diurnal and nocturnal arcs. This correlation, which has survived numerous challenges from skeptics, has become known as the Mars Effect.

A description of how the experiments were set up will clarify how rigorous they were. First there was the selection of data. In studying sports champions, the Gauquelins only considered those who had achieved eminence in their sport, and this choice was based on published criteria and sports records. For example, only athletes who won medals in the Olympics, or at world-class competitions, might be

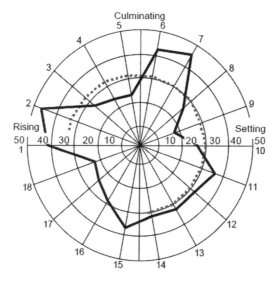

Figure 26. Typical Gauquelin 360-degree graph showing the results of a Mars study (sports champions). The diurnal cycle is designated by lines that mark rise, set, and culmination (lower culmination is shown but not labeled). In this study the diurnal cycle is divided into 18 sectors running clockwise from the rising point. The dotted line is the calculated probability for the position of Mars in the diurnal cycle, the heavy line is the actual position of Mars in the sample. Numbers on the horizontal line connecting rise and set denote the number of instances Mars was found in each sector (e.g., Sector 2 = 47, sector 3 = 29, sector 4 = 15, etc.). (McRitchie 2011 via Wikipedia, CC BY-SA 3.0)

included in a data sample, not just anyone who played their sport on the weekends. Different sports were included, but were also examined as separate samples and differences were found. Once the sample was selected, mathematical calculations, using time and coordinate system data, were worked to produce an astrological chart. This exercise also required adjusting for summer time or war time and the entire procedure was done twice to catch errors. Second, after hundreds of charts were calculated, and the positions of the planets in their diurnal cycle were known, the statistical analysis began. The expected frequency of a planet in a diurnal cycle sector needed to be calculated to compensate for annual length of day variations. Then both the mean annual and daily birth rates, and variations in planetary motion (planets can be retrograde) had to be taken into consideration. In this way a baseline was established against which results were analyzed. Finally, statistical calculations were performed and the results written up. This is a phenomenal amount of work to do with no funding or institutional support, and it says much about Michel's determination (he was born with the Sun in Scorpio, a sign known for relentless persistence in the face of adversity).

The Gauquelins conducted hundreds of similar studies, many of which found statistically significant correlations between factual commonalities in a sample of people and the diurnal cycle position of planets at the time of birth. Findings included correlations with Saturn in the charts of scientists, Jupiter with actors, and the Moon with writers and politicians. Another finding was that the Mars effect distribution (see figure) just past rise and culmination was not found in samples of artists, painters, and musicians—they had Mars prominent in the zones where the athletes had it less frequently. In later studies by the Gauquelins, personality traits showed an even clearer correlation with the diurnal cycles of the planets than profession did. These studies suggested to them that the real link with the planets is with personality, not profession, the latter being a choice dictated by the former. In general, they found substantial statistical evidence that different professions and personalities are linked to different planets, evidence that becomes stronger with samples of only those who have achieved eminence and weaker or non-existent for the non-eminent. In addition, they also found what appears to be evidence of a kind of planetary heredity: a planet located in one of the emphasized sectors of the sky tended to be duplicated in the astrology charts of offspring.

The Gauquelin studies hardly went unchallenged. It is the case, however, that after intense scrutiny, their findings were successfully replicated many times, both by themselves and by independent statisticians. But an enormous resistance to accepting the findings persists to this day—in fact, *The Tenacious Mars Effect,* a book that describes the multi-decadal battle between the Gauquelins and their skeptics, is dedicated to uraveling the situation (Ertel and Irving 1996; West 1991). Consequently, these studies have, as far as I can tell, made no noticeable impact on the scientific community. On the other hand, the reception of the Gauquelins' findings by the wider astrological community hasn't been particularly warm either as they tend to challenge some traditional notions about the astrological houses.[3] In spite of all their meticulous work, the Gauquelins found themselves between a rock and a hard place, particularly in regard to the resistance coming from the skeptics. Perhaps this persistent rejection, in combination with their divorce, took a toll. At age sixty, Michel, suffering from depression, had his files destroyed and committed suicide.

The following is a short version of how the Gauquelins' studies were received by a certain segment of the scientific establishment. It is a story of dishonesty and frustration, not unlike hardball politics, and also an example of how irrational some scientists can be and how science is not supposed to work. In 1956 the initial results of the Gauquelins' studies were sent to the Belgian Para Committee (*Comité Para*), a panel set up in 1949 to evaluate paranormal claims. After a multi-year delay a member of the committee responded (in 1961) saying that because the sample was limited to France, it meant nothing. The Gauquelins then collected data for a sample of over twenty-five thousand from five

---

3. While it is the case that few consulting astrologers deviate from the traditional rules of chart interpretation, and the actual findings of scientific research are of interest to only a small minority in the field, the Gauquelins did have very good relations, on an international level, with some of the most knowledgeable astrologers.

countries and got basically the same results. After another year's delay they got a response from a member of the Para Committee who stated there was nothing wrong with the methodology, but it was their opinion that the Gauquelins had selected only supportive data. Four years later a new study was done with new data, the collection of which was overseen by the committee, and the results were the same. More tests were done transparently with the committee, and still the results verified the Gauquelins' findings. The committee did not publish these results, saying only that, while the methodology was acceptable, they could not accept the conclusions. But in 1976, twenty years after the process of independent verification began, a report by the committee was published that stated the results were caused by certain, though unspecified, demographic errors.

In 1975, *The Humanist* magazine dedicated the one-hundredth anniversary of the American Ethical Union to the discrediting of astrology, which they apparently believed to be a thought virus infecting the minds of the public. On the very first pages of the September/October issue was a list of 186 scientists, including 18 Nobel Prize winners, who had signed a statement saying that astrology was essentially superstition and that there was no verifiable scientific basis for

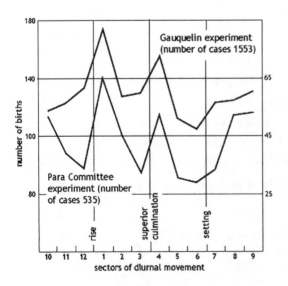

Figure 27. Here, in a different graphic format, the results of a Gauquelin experiment are compared to the results of the same experiment done with the participation of the Para Committee with different data. The pattern, peaks just after rise and upper culmination, are consistent in both studies. © 1988 by Michel Gauquelin.

a belief in it.[4] The issue contained a few anti-astrology articles including one that attacked the methodology of the Gauquelin's' astrological experiments. Michel responded by challenging the skeptics to replicate the experiments. The Committee for the Scientific Investigation of the Claims of the Paranormal (CSICOP), founded immediately after the success of the attack on astrology in *The Humanist,* agreed to do this. Marvin Zelen, a statistician, led the investigation of the Mars studies that also involved publisher Paul Kurtz and astronomer George O. Abell, hence this "test" of the Gauquelin Mars Effect is often referred to as the Zelen test. Their analysis began by first taking a random selection of the Gauquelins' sports data, which was then compared with a control group; the results were the same. This apparently unacceptable result led to them manipulating the data in various ways, including reducing the original sample of 2,088 to 303, and then reducing this to just over 100 by taking out female athletes. This being accomplished, the Mars Effect was reduced to chance. The same group also tested a sample of American athletes and found nothing statistically significant, though Michel complained their sample included a preference for mediocre athletes and was loaded with basketball players, which made it a less representative sample. But the Zelen group publicized their conclusions, loudly broadcasting their findings through the popular media. The Gauquelins' claims were proclaimed as false and astrology was again pronounced dead.

For several years, numerous articles and stories appeared in magazines and newspapers reporting that the scientists were absolutely correct about the Gauquelins' statistical studies being faulty. But also during this period, one of the original CSICOP skeptics, astronomer and historian Dennis Rawlins, resigned from the committee and published an article in *Fate* magazine, October 1981, stating that the CSICOP replications of the Gauquelins' experiments actually agreed with their conclusions, and that this fact had been covered up (Rawlins 2001). A couple years after this embarrassment, a tiny notice appeared

---

4. The statement and signers are reproduced in Grim (1982, 14–18), followed by scathing criticism of its presumptuous and authoritative attitude by Paul Feyerabend.

on an obscure page in the Spring 1983 issue of the *Skeptical Inquirer* reporting that the Gauquelins' studies had actually withstood their challenge. Most of the general public, and the media, which had heard all the negative attacks on astrology for months, never became aware of the eventual outcome of this "debate."

In the mid 1980s, German psychologist Suitbert Ertel became interested in the matter, specifically in regard to the problem of data selection. He examined the Gauquelins' data looking for hidden biases and, in the process, introduced more objective ways of determining the eminence of athletes (Ertel 1988). After rating eminence in five categories and compensating for the Gauquelins' personal judgments in certain cases, the resulting statistics were the same as before. His finding was published in scientific papers and in a book, but they don't appear in the Wikipedia article on the Mars Effect—at least at the time of this writing—which appears to be monitored on an hourly basis by militant skeptics. There have been a few other tests by skeptical groups or individuals since Ertel's reviews, and these have been accused of being flawed. The debate on the Gauquelin studies continues to this day. The level of discussion, kept alive by a handful of people, is complex and rife with disagreements over statistics, and boring to the rest of the world. Ertel and Irving, in a paper that unravels a paper by skeptics, summarizes the position of those who have been most honest about the Gauquelins' work, which anyone capable of unbiased critical thinking will conclude are obviously the astrologers and their allies. In their view the Gauquelins' evidence for an astrological effect continues to gain in strength while the skeptics' dogmatic assertions that deny the evidence mysteriously persist.

A major criticism of the Gauquelins' results has been in regard to data selection. In fact, frustrated skeptics today have slipped so far from reason as to argue that parents (who know nothing about astrology) fudged their children's birth times to make them look better in the study (a scientific paper they probably would have no interest in reading). Unlike my own studies with undisputed daily temperatures, the astrology (or psychology) of people is not so easily quantified. The Mars Effect does appear in samples of achievers in their sport or profession, but exactly how eminence is to be quantified is the issue. And getting a clean sample based on eminence is not easy; that is how the CSICOP committee's American study failed to find significance and is what Ertel looked into and improved upon. In selecting data about who to include in a sample, Michel Gauquelin did make subjective judgments that were based on his own understanding of what constitutes a true successful sports professional. He no doubt knew far more about this than the average researcher as he was an athlete himself, at one point ranked among the fifty best tennis players in France. Because it is true that some outstanding athletes suffer injuries, die early, or are the victims of circumstance at crucial times in their careers and never achieve the medals or honors that they actually deserved, should they be included in a sample? This is a good question and, if so, each case should probably be considered on its own terms. Still, as Ertel found, when the Gauquelins' personal selections were taken out of a sample leaving it based only on a strict ranking of eminence by published facts, the effect was still visible and statistically significant.

Another factor discovered by the Gauquelins, and one that complicates an evaluation of their work, was that the planetary effects they discovered diminish to some extent for births after about 1950 and diminished considerably for those born by intervention, such as in drug-induced or Caesarian births. Michel Gauquelin was very concerned about this, not just because his studies lose significance, but because he saw it as a case of humans tampering with a natural process they really don't understand. The science behind birth is complex and evolving, it being a case of multiple chemical feedback loops, but studies suggest that, under normal conditions, the fetus initiates labor and in a sense chooses the time of its natural birth. One theory is that the lungs, the last organ to develop, produce a protein that signals readiness for labor. Another is chemical signals from the fetus (specifically cortisol) stimulate chemical production and even epigenetic changes in the placenta that regulate the length of pregnancy. Also, the adrenal gland of the fetus may send a signal of readiness, or the mother's pituitary gland secretes oxytocin, which stimulates labor. Regardless of chemical pathways and exactly what is initiating them, today medical intervention is becoming the norm and a high percentage of births are now occurring only during doctor's hours and not on weekends when the doctor may be playing golf or on vacation.

Most of the Gauquelin findings are usually displayed in a circular grid (see figure above), which clearly shows peaks just past rise and culmination, and lesser peaks just past set and lower culmination. These were interpreted as a fourth harmonic pattern where there are roughly four equidistant peaks (or waves) spread over the 360 degrees of the diurnal circle. This is apparent in many of their figures, but some are more distorted and even triangular. English astrologer John Addey, a contemporary of Michel Gauquelin, saw a combination of the third and fourth harmonic in these circular grids. Addey used some of the Gauquelins' data for sports champions (n = 1485), noting in which of the diurnal sectors (thirty-six in this case) Mars was most often found. He then ran a harmonic analysis (Fourier transform) on this data to obtain numbers on amplitude and phase. The results confirmed that the amplitude of the third and fourth harmonics were highest, that is these waves rise well above the mean. In addition, the location in the diurnal circle of the peak phase was also found, that is where these peaks were anchored relative to the horizon or meridian. Addey thought these results explained quite well the pattern shown in the Gauqeulins' studies, which seemed to be a combination of these harmonics, and he did many other studies that found correlations between other harmonics and professions. More on Addey later.

The Gauquelins' studies, which remain the longest and most thorough astrological research program to date, show that certain, though limited, astrological principles can be demonstrated through highly focused, complex, and time-consuming statistical studies— basic reductionism. There are two major findings that do support traditional astrology in their work, the first being the meanings of the planets, specifically Mars, Jupiter, Saturn, and the Moon (Gauquelin 1982). Strong correlations were found between the planets and personality, and, by extension, between planets and profession. It is highly significant that what these planets signify has remained essentially constant since Babylonian times. The second finding has to do with the importance of positioning in the diurnal cycle, specifically near rise and culmination. The axis created by the horizon (rising and setting) and that of the meridian (upper and lower culmination) together establish the primary framework for both field observations of planets in astronomy and the layout of the astrological chart. Ptolemy, and all the other ancient, Medieval, and modern astrologers, accepted that planets located near these points, both after and before, have more power in shaping personality or events than planets in other positions. (These four points are called the angles by astrologers and the positioning of planets near any of them is called angularity.) The Gauquelins found that this was true, but the focus was shifted a bit clockwise, which some astrologers rejected. In the end, the Gauquelins called their findings "neo-astrology" and argued that astrology must be built on the factual foundations that only rigorous reductionist scientific studies can establish, not anecdotal information.

An odd twist on the Gauquelins' research is found in a study done by Suzel Fuzeau-Braesch (1928–2008), an insect biologist associated with the French National Center for Scientific Research (CNRS), who researched and wrote papers on biochemistry, insect behaviors, and circadian rhythms. In 1970 she purchased computer-generated astrology reports for each of her children as a novelty, but found them so accurate she decided to take a closer look at the subject. When she realized that astrology was much more than simple horoscopes in the newspapers, she decided to study it scientifically and eventually published a number of papers that reported her astrological findings on, among other topics, twins, SIDS, human sociability, and dog personalities. In her study of the latter, she collected data on five hundred dogs from twelve breeders who, as part of their job, meticulously recorded the birth time and date for each dog and, as they grew, made notes as to their personalities, this information being of great value to buyers (Fuzeau-Braesch 2007). Using categories of extraversion and neuroticism (based on the Eysenck Personality Inventory and their variants, which produce six types of behavior), along with the breeders' descriptions, she determined a personality type for each dog. Then, using the birth data of each dog, she calculated astrological charts and focused on planets located within 10 degrees of the four angles: rise, upper culmination, set, and lower culmination. These four 20-degree segments roughly approximate the dominant sectors in many of the Gauquelins' studies. She found strong and statistically significant correlations between dogs with extraversion as the dominant trait and the position of either Jupiter or the Sun at rise or upper culmination. She also found that nervous and introverted dogs were born when Saturn was in these positions. This latter finding replicates what the Gauquelins found in their study of scientists. Perhaps this is an indication of what lies below the surface in many a scientific personality.

BLIND TRIALS AND THE CARLSON TEST

Testing astrology is difficult, in part, because of the constantly changing cosmic environment and, consequently, the individual nature of each astrological chart. If every chart is unique, where is the standard against which differences might be assessed? Only with large sample sizes, like those of the Gauquelins, can these problems be overcome. Another problem has to do with the complex nature and variations of human personality. Many critics assume there are but twelve standard astrological types, but that is hardly the case. The Gauquelins were able to tease out a few very general personality types and found correlations with profession and heredity, but these don't capture the nuances that are found by experienced astrologers in natal charts. Astrologers claim to describe a wide range of general types, each composed of a collection of behaviors and attitudes, that are subject to infinite variations that may or may not be internally integrated. While it could be said that natal astrology provides a general multi-level model of personality, it is also used for synthesizing and organizing a given set of psychological traits in ways that offer insight into the self (a phenomenon that resists reduction). These astrological measurements and interpretations differ significantly from the far more simplistic models used in psychology, and they make for difficulties in the design of experiments.

At present, a wide range of astrological methodologies remain to be tested, largely due to lack of institutional support and funding. One alternative approach to this immense problem is the testing of astrologers in a blind trial, conducted as either a test of the abilities of astrologers to match astrological charts with a set of personality profiles or for them to distinguish between charts of very different people. Several studies of this type have been conducted. Some of the first studies in this category were conducted by Vernon Clark, a psychologist who studied astrology, and his results were published in 1960 and 1961. In three studies, titled "An Investigation of the Validity and Reliability of the Astrological Technique," Clark tested the abilities of individual astrologers to essentially fit an astrological chart to a described person. The trials were well designed, thoroughly analyzed, and met high standards for this sort of research. Each test included an experimental group (the astrologers) and a control group (which consisted of psychologists and social workers). The first test required twenty astrologers to match five astrological charts with five case histories (focused on professional history) of males, and then do the same with a second group consisting of females. In the second test, twenty astrologers were given ten case histories and two charts for each of these. They were asked to choose which of the two charts (one of which was based on a random date of the same year) best matched a detailed case history. The third test asked thirty astrologers to distinguish between the charts of persons having an IQ of over 140 and charts of persons with incurable brain damage (cerebral palsy). The results of these tests were quite spectacular, statistically speaking, in favor of the astrologers, and the research was published in an obscure astrology journal.

The results of Clark's blind trials did get attention in the astrological community, and a number of replications were attempted (Dean and Mather 1977, 544–54; Cornelius 2003, Ch. 4). Most of these came in with results closer to chance, though still slightly above it (better than chance). What happened? The details, statistics, and criticisms are too complex to get into here, but my observation is that Clark's original set of astrologers was a particularly distinguished and highly competent group who were able to excel in this test. Subsequent samples of astrologers used in replications included individuals with a wide range of expertise between them. This was probably a result of the fact that very few astrologers who are asked to participate in studies like this bother to actually do so, and among those that do, their experience and the methodologies they employ vary. In a replication of the third test (which considered high intelligence versus brain damage), I was one of twenty-three astrologers challenged to make this decision. The results of this trial came in at barely above chance (50 percent). I remember when taking the test, I analyzed only the positions of the Moon and Mercury, both of which are traditionally associated with the brain and mind, and I employed a wide variety of methodologies (I was told later that my performance in it was very good). It has been my experience that most astrologers are not so inclined to experiment with techniques and will tend to stay with the system they learned from their teachers. Like doctors and therapists, the abilities and skillsets of astrologers vary widely; some are excellent at counseling but not so good at making the sharp distinctions needed in a blind test—and some are good at both.

In 1985 the prestigious journal *Nature* published the results of a study that tested the ability of astrologers, given a set of psychological profiles, to match astrological charts with their owners (Carlson 1985). Shawn Carlson, an undergraduate in physics at the time of the study, stated he was testing the fundamental astrological thesis that the planetary positions at the time of birth can be used to determine the personality traits of a subject. The participating astrologers were selected by a leading American astrological organization, the National Council for Geocosmic Research (NCGR). A total of twenty-eight astrologers from the United States and some from Europe were selected and asked to calculate and prepare a natal chart interpretation for a number of volunteer subjects. Next, the subjects were given the natal chart interpretation for their own birth data, plus those of two other persons, and asked to select the one that they found to best match their own personality. In the second part of the study, the astrologers were given an astrological chart of one of the subjects along with three reports generated by the California Personality Inventory (CPI), which offered them eighteen personality trait scales generated from 480 questions given to each subject. They were then asked to select the natal chart that best matched the CPI. In both cases two selections were made, a first and second, but no ties were allowed. The study was double-blind, and all tests were coded and known only to Carlson's graduate advisor, physicist Richard A. Muller.[5]

Of the subjects recruited for the study, 70 percent were college students. Subjects were asked questions about astrology and those who were strong disbelievers were rejected as were those who had previously had astrological chart readings. After these and other factors were accounted for, a total of 177 subjects were assembled, comprising 83 in the test group and 94 in the control group. The results of the first part of the study, in which subjects select the natal chart interpretation they thought best fit them, came in at the level of chance. The control group, having been asked to choose the CPI that fit them best, came in at chance also. In the second part of the study, where CPI reports and natal charts were matched, the astrologers came in at below the level of chance. Carlson concluded that his study clearly refuted the astrological hypothesis (that astrology is valid), and the study went on to become a first-ranked, frequently-cited scientific paper and a solid resource for skeptics. It has been called a devastating verdict for astrologers.

The Carlson study has been contested, however (Vidmar 2008; Currey 2011; McRitchie 2011). Hans Eysenck, who has a controversial legacy but was a leading personality theorist and creator of his own psychological personality inventory, objected that the CPI was a bad choice for the study and that a psychologist, not a physicist, should have been involved with the experiment. The format, a choice of three rather than two selections, has also been argued to be an unnecessary bias. Astrologers who participated claimed the CPI reports did not distinguish between male or female and the reports were more

---

5. Muller is known for his involvement in climate change issues and also his hypothesis that the Sun has a companion star (Nemesis) that periodically perturbs the Oort cloud every twenty-six million years, causing extinctions.

similar to each other than not, making confident choices impossible. Astrologers also complained that Carlson did not listen to their suggestions in regard to what they were actually capable of doing and what they needed to do their work properly. In addition, the astrologers were required to do an immense amount of uncompensated work, as a written natal chart interpretation from an experienced professional astrologer was worth upwards of a hundred dollars in the market at that time. Carlson's failure to cite similar previous studies, even if flawed, or even mention the Gauquelins' findings, is inconsistent with introductory references to previous studies that are typically found in scientific papers.[6] A reappraisal of the study done by Ertel found significant errors in Carlson's use of statistics, and he judged the study to be very weak due to small sample size, far below what was expected of the Gauquelins, for comparison. And, when the study was analyzed properly, Ertel found that the astrologers actually performed slightly better than chance (Ertel 2009).

What is to be made of this situation? As one of the twenty-eight astrologers who participated in the Carlson study, I did find it a time-consuming, uncompensated, and extremely frustrating task, the major problem being the differences between how personality is assessed and organized by astrology and by the CPI, or any psychological inventory for that matter. The study assumed that these two methods of personality description (personality being a nebulous thing to begin with) would be interchangeable—but they aren't. An analogy would be to test the abilities of two groups of surveyors to measure a complex land formation, one using the metric system and the other the U.S. customary system, and pretend that one is legitimate and the other isn't. This situation raises another significant problem mentioned earlier. Studies like Carlson's assume that astrologers, even ones selected by a reputable organization, will perform similarly, and this is far from the truth. Consider that, aside from the fact that differing specialties

---

6. Studies similar to Carlson's, in which astrologers were asked to distinguish between charts of persons with mental disabilities and those of superior intelligence, were conducted by Vernon E. Clark. The astrologers performed well above chance ($p = 0.01$; Clark 1961). $P=0.01$ means one chance in a hundred.

require different knowledge, doctors, psychotherapists, and other consultants are known to arrive at different diagnoses or assessments, and some are clearly much better or worse than others. At the time of this writing, for example, I am on a fourth diagnosis for a chronic physical problem, after previously seeing three other doctors whose assessments proved false.

My experience has been that the interpretive abilities of astrologers, like those of teachers, doctors, psychotherapists, and artists, are all over the charts, so to speak. Talent is a major factor and, unlike major league baseball players or jazz musicians whose abilities are in plain sight or sound, there hasn't been a sorting out process in astrology, with the exception of personal business success, which depends largely on a specialized set of social skills. This is partly because many in the astrological community regard certification tests as objectionable and follow whatever path their teachers or personal interests have set them on. Certifications do exist in the astrology subculture but vary widely, and most fall a bit short, in my opinion, of what might be expected of a person in a normal academic environment. Most other fields rely on quantifiable achievements, such as degrees and certifications, but due to its peculiar historical circumstances, the field of astrology still lacks strong institutions and ways of ranking expertise in different specialties, an important issue and too easily glossed over when studies like Clark's or Carlson's are criticized. While this situation has been improving in recent years, it is still the case that anybody with a business card, Facebook page, and enough hutzpah can be an astrologer. Getting astrologers to agree on a set of standards is like herding cats, and the result is that people in the non-astrology world have no idea who is who in the field. For example, one study similar to Carlson's, conducted by two members of the psychology department at the University of Indiana, used six "expert" and "cooperative" astrologers drawn from a relatively unknown local astrology group with no indication of their qualifications except membership in a very small organization and a recommendation by a numerologist (McGrew and McFall 1990). Despite these points, this pathetic study, which concluded that astrologers cannot match natal charts to a personality profile, continues to be cited. That being said, there are some very intelligent and highly competent astrologers out there, it just takes some time to sort them out from the noise.

THE ZODIAC PROBLEM

For most people today, the zodiac is all there is to astrology and your "sign" (Sun-sign) is your "horoscope." Of course, a delineation for your Sun sign is not a horoscope and any delineation of this type, based only on the Sun's position within a range of 30 degrees along the ecliptic, is exactly the same for anyone born under that sign. This kind of popular astrology is to the entire field of astrology as the "Dear Abby" column is to the entire field of psychology. It just doesn't amount to much of anything except that it has kept serious astrology under the radar. The high visibility of the zodiac has made it a target for attacks on astrology in general, these occurring on what seems to be a regular cycle. Typically, an astronomer from a second- or third-rate college makes an announcement to the press that astrologers have it all wrong; there are actually thirteen constellations of the zodiac, the extra one being Ophiuchus, a mythological man wrestling with a snake whose foot dips into the zodiac between Scorpio and Sagittarius. Right up front, though, the astronomer's ignorance is on display, because astronomers use a different zodiac than astrologers—but the media doesn't know this. The next news flash from the astronomer is that the zodiac has shifted over time and astrologers are found guilty of confusing Pisces with Aries. The actual situation is a bit more complex and is therefore of no interest to the media, which assumes a viewer attention span of about three seconds, so they generally defer to the authority of the professor who almost always gets the most air time or print space. Sometimes astrologers do get to respond, but most are not qualified to explain the astronomy behind precession clearly and with authority. The key issue here, the difference between the tropical and the sidereal zodiac, is a huge one, and it is subject to distortion by people who are uninformed. Even the astronomy magazine *Sky and Telescope* gets it wrong (Beatty, 2011). You may recall this topic was previously raised in Chapter 3, and the discussion will be continued here.

Conventions for mapping the sky have evolved through the centuries. In late Babylonian times the zodiac, which demarcates the Sun's path (ecliptic), was sidereal, that is fixed to the positions of stars and constellations. It was essentially a grid composed of roughly equal named spatial sections of sky designated by star patterns through which the Sun moved over the course of a year. But during the three hundred years between Hipparchus and Ptolemy (~150 BCE–150 CE), the idea of the zodiac being permanently attached to the celestial framework established by the equinoxes (intersection of Earth's celestial equator and the ecliptic), began to be accepted. This newer framework defines the *tropical zodiac,* a zodiac of twelve signs, each of exactly 30 degrees, that shift over time against the stars and constellations at a rate of about 1.4 degrees per century. In other words, it is the Sun and the Earth that makes the zodiac, not the constellations. Modern astronomers, however, use a thirteen-constellation (of varying sizes) *sidereal zodiac* as a positioning framework for objects they study, this being a zodiac of constellations. The boundaries of these constellations were set in 1930 by the International Astronomical Union. Confusion arises because the same names are used in both zodiacs.

For nearly two millennia nearly all Western astrologers have used the tropical zodiac and have been very aware of the difference between it and the constellations of the same names. However, a twelve-sign (of equal segments) sidereal zodiac is used by astrologers in India and also a few Western astrologers known as siderealists, though in both cases it is used more as a framework for the positioning of planets and less as an interpretive factor. Due to the precession of the equinoxes, the sidereal zodiac is today offset from the tropical zodiac by about 24 degrees. The exact figure of this displacement (called the ayanamsha in Hindu astrology) is a contested point among practitioners of sidereal astrology and usually depends on the choice of star that anchors (marks a key position in) the sidereal zodiac. While this explanation should make it clear that there are two very different zodiacs, the distinction is apparently too complicated for television, newspapers, internet, and other popular media to manage, and the public remains confused. This is unfortunate because this problem, as noted above, has become a launching point for modern attacks on astrology, generally coming from astronomers who should know better.

Given that the zodiac gets so much attention, how has it tested out in real scientific studies? There are a few things that need to be made clear in understanding the studies that have been done. The tropical zodiac, which is the one usually tested because it is what nearly all Western astrologers use, is a spatial division of the Earth's orbit anchored by the equinoxes, the points where the Sun, as seen from Earth, crosses the celestial equator. There is no physical reality to these points in space that are the vernal and autumnal equinoxes (created by an intersection of two imaginary planes) except in terms of geometry, but they have consequences in terms of seasonality, which is driven by the varying levels of incoming solar radiation. The Sun moves along the ecliptic annually, and this cycle has four key points where solar motion (as perceived geocentrically) and incoming solar radiation (relative to terrestrial latitude) either crosses a threshold or shifts—these points being the equinoxes and also the solstices, which are the boundary markers of the seasons. In astrology, the quadrants formed by the equinoxes and solstices are divided into three signs of 30 degrees, these being the components of the tropical zodiac. Because it is the steady, annual apparent motion of the Sun that makes the zodiac, knowledge of a person's Sun-sign only requires a birth day, not even a year. Knowledge of one's Ascendant, Midheaven, Moon, or planet signs is rare among the general population in part because they must be calculated and, because birth times are often unknown, most scientific tests of the tropical zodiac are limited to just Sun signs.

A little more information about the zodiac will be useful in understanding how scientific studies of it often fail to take into account its organizational scheme. The signs of the tropical zodiac are 30-degree sections of the ecliptic anchored to the equinoxes and solstices. The sequence of signs begins at the vernal equinox with the sign (not constellation) Aries and ends with the sign Pisces. Given that the tropical zodiac anchors the seasons, it follows that it also is a framework for photoperiod, that is the cycle of the day-to-night ratio as shown in the first figure below. The twelve signs are traditionally subdivided into polarities, qualities, and elements. These divisions could be seen as being produced by the wave forms of the second, third, fourth, and sixth harmonics as shown in figure 29 (below). In regard to interpretation, the tropical zodiac is thought to be a symbolic sequence of developmental-like phases that modifies, or filters, the influence of the Sun over its annual cycle. From this perspective the zodiac could be considered a time-template of photoperiod, the changing day/night ratio, which is witnessed in time and measured spatially in celestial longitude and also in declination, the distance of the Sun's path north or south of the celestial equator.[7]

The signs of the tropical zodiac are traditionally said to alternate between active-masculine (fire/air) and receptive-feminine (earth/water) beginning with Aries, this division being called polarity. The cycling variation is much like the yin/yang principle in Taoism. Since Hellenistic times the signs have been further divided into different elements. Aries is the first fire sign and is followed by Taurus/earth, Gemini/air, Cancer/water. Leo/fire begins the second series of elements and Sagittarius/fire the third. In astrological delineations the elements are thought to modulate the basic properties of the planets; fire enlivens, earth solidifies, air communicates, and water reacts. The sequence of four elements repeats three times in the zodiac (of course, three multiplied by four equals twelve). Three qualities or modes further focus the principles of signs that are spaced 90 degrees apart, these being produced by the fourth harmonic wave. Cardinal signs are thought to initiate and the four of them begin at the equinoxes and solstices. Fixed signs, which designate stability, follow each cardinal sign, and mutable signs, which oscillate, follow the fixed. There are also subdivisions of

---

7. The fact that the Moon and planets are also thought to be modified by their positions in the tropical zodiac suggests that declination, that is distance north or south of the celestial equator, must be a significant factor in assessing a planet's effects. Although the planets are only reflecting sunlight, planetary declinations in any zodiac sign and degree are always close to those of the Sun when it is in that sign and degree. In addition, the zodiacal sign and degree placement of a planet accounts for rising and setting positions along the horizon and therefore the altitude and duration of its diurnal and nocturnal arcs. This idea can be found in classic astrological texts over the past two millennia.

Annual day to night ratio for NYC

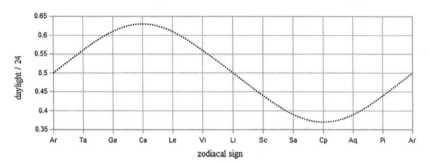

Figure 28. The changing day-night ratio (y axis) is depicted here over the course of a year (x-axis) at the latitude of New York City (40 N 43, 73 W 56; these are coordinates with degrees and minutes).

Harmonics 2,3,4, and 6

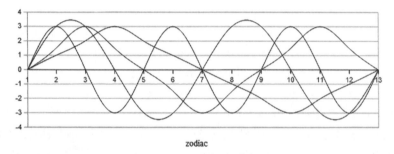

Figure 29. Any given length (here the length of the year as a circle of 360 degrees) divided by a number will break it into equal parts. Division by twelve (not shown) will produce twelve 30-degree parts, eight of which will be also produced in division by two, three, four, and six. This demarcation based on number has been suggested to be the basis of the twelve-sign zodiac.

each sign that are regularly employed in making delineations by astrologers that potentially could be employed in a test as well.[8] Finally, each sign has traditional associations to a specific planet, which is thought to resonate with that sign and is traditionally called its ruler. Starting with

---

8. Other divisions of the zodiac exist, including the 36 decans of 10 degrees each and the 144 dwads of Hindu astrology, which divide each sign into a micro zodiac.

ZODIACAL SIGN CORRESPONDENCES

Sign	Polarity	Element	Quality	Planet
Aries	Male	Fire	Cardinal	Mars
Taurus	Female	Earth	Fixed	Venus
Gemini	Male	Air	Mutable	Mercury
Cancer	Female	Water	Cardinal	Moon
Leo	Male	Fire	Fixed	Sun
Virgo	Female	Earth	Mutable	Mercury
Libra	Male	Air	Cardinal	Venus
Scorpio	Female	Water	Fixed	Mars
Sagittarius	Male	Fire	Mutable	Jupiter
Capricorn	Female	Earth	Cardinal	Saturn
Aquarius	Male	Air	Fixed	Saturn
Pisces	Female	Water	Mutable	Jupiter

the Sun ruling Leo and the Moon ruling Cancer, the rest of the classical planets fan out on either side, each ruling two signs. Regardless of whether it works or not, the tropical zodiac is not a simple measure, it is an ancient symmetrical, structured schema of photoperiod with each increment described in terms of fundamental qualitative principles.

Dean and Mather critically reviewed dozens of zodiac studies in their book *Recent Advances in Astrology,* which is limited to 1900–1976, and find that nearly all of them failed to find significant correlations, were not reproducible, or were flawed in some way (Dean and Mather 1977, 75–164).[9] Most of these and more recent studies attempt to link a psychological personality profile or a description of personal traits to the Sun signs, while others looked for correlations with profession or eminence of some kind. Setting up a proper study of zodiac astrology

---

9. Geoffrey Dean is a rare critic of astrology who actually knows something about it. He has been dedicated to forcing the subject into a reductionist straightjacket that it (almost) consistently fails to fit (Dean and Mather 1977, 7).

turns out to be difficult. Astronomical corrections for Earth's elliptical orbit need to be considered (the Sun spends more time in signs near the summer solstice than the winter solstice when Earth is at perigee and moving faster) and the distribution of births in the year must be accounted for, something that varies by season and country. Another is the problem of self-assessment; a very high percentage of subjects will agree that a uniform description of what they are told is their unique personality is, in their opinion, accurate, this being known as the Barnum effect. (Self-knowledge is not taught or encouraged in modern culture possibly because it is not valued in Western religion and it is of no use to capitalism.) Other problems include data gathering and sample size, precise trait definitions, and what statistical methodologies to employ (the chi square test is most often used). It is true that few significant correlations have so far been found between the predicted personalities of the signs of the zodiac and the lives (i.e., professions, personalities, etc.) of people.

In contrast, consider that psychologists routinely employ personality models like the nine-type Enneagram or the Myers-Briggs Type Indicator. Models such as these are built on personality assessments (self-reporting), a kind of inductive approach to the slippery topic of personality, and not on linkages to a common parameter (a fixed unit of measurement) such as date of birth. How accurate are these self-reported personality inventories? In regard to the Enneagram as a psychological test, one website says you have to determine for yourself if your results are accurate. The Myers-Briggs test is routinely criticized for multiple reasons and is often called pseudoscience (a thorough account of problems can be found on the Wikipedia page). Given that self-knowledge has low value in the American culture, what these personality assessments find is probably useful information and I believe many people benefit from them. But I would also argue these categories of personality are more or less on the same level as Sun sign astrology when interpreted by competent astrologers.

Michel Gauquelin ran a few experiments on the zodiac. In one he began with a sample of 2,088 sports champions and selected those whose personalities were described as being intensely competitive, aggressive, and so on, which are qualities that would be associated with the Sun in the signs Aries or Scorpio, both planets ruled by Mars. When the sign positions of this group were tested, the resulting totals in these two signs was actually less than chance, but Gemini and Aquarius, which are not considered fighting signs, scored much higher. In addition to looking for Sun sign totals, he also included other planets, the Ascendant, and the Midheaven and still found nothing of significance. Another study, this one by Robert Pellegrini, used an even sample of twelve women and twelve men born in each sign who had taken the CPI (Pellegrini 1973). The eighteen traits produced by this test were correlated with their Sun signs, but nothing significant was found for seventeen of these. However, the trait of feminity was later found to be strongly correlated with all the signs from Aquarius to Cancer, roughly births that occurred between January and July. This may be an indication of seasonality, but it doesn't support Sun sign divisions of the seasons.

One of the best-known studies of the zodiacal signs, and one with replications, was done by Jeff Mayo, an astrologer, along with psychologists Hans Eysenck and O. White (Mayo, White, and Eysenck 1978). Mayo was an English astrologer known for his correspondence school of astrology and for publishing a set of concise and useful texts on astrology that are factual and devoid of metaphysical notions. Eysenck (who I noted previously had criticized Carlson's test) was a respected, though controversial, English psychologist who studied intelligence and personality. (His work and models in regard to extraversion and neuroticism are precursors to the domains that make up the Five Factor personality model that is popular with psychologists today.) Recall that the zodiac is subdivided into alternating polarities where the fire and air signs (odd numbered signs starting from Aries) are considered masculine, active, and outgoing while earth and water signs (even numbered) are thought to be feminine, passive, and internalizing. These descriptions, similar to the two major dimensions of personality, extroversion and introversion, led to a test of 917 men and 1,407 women who supplied their birth dates and answered some questions on the Eysenck Personality Inventory (EPI). The results were striking, a close correlation being shown between the fire and air signs and extraversion as established by the personality test, and likewise for the earth and water signs and introversion. A second part of the study tested for emotionality versus stability and found a strong correlation between emotionality and the water signs (Cancer, Scorpio, Pisces) and stability and the earth signs (Taurus, Virgo, Capricorn). At least two similar studies were conducted in New Zealand: one produced similar results; the other showed no correlation (Veno and Pamment 1979; Dean and Mathers 1977, 125).

One major issue with the Mayo study was in regard to subjects that had prior knowledge of astrology, a suggestion that self-attribution may have played a role in the results. Mayo and his team then divided the original sample into two parts, those with knowledge of astrology and those without, and the results were the same. In order to completely rule out self-attribution, 1,160 children, who had previously been tested and had scores for extraversion and introversion, were grouped by Sun sign. No correlation was found. A German study divided a sample of 799 adults into three parts in regard to their knowledge of astrology: non-believers, believers, and strong believers (Pawlik and Buse 1979). The personality scores of the believers and nonbelievers matched their astrological signs, but that was not the case for the non-believers, which makes a strong case for self-attribution. A 1988 study, controlled for self-attribution by using data from a previous study that had nothing to do with astrology, found a correlation with Sun signs, but when the zodiac signs of other planets were added the trend disappeared (Van Rooij and Jacques 1988). In yet another study, 190 first-year students took the EPI and also supplied their birth data. No mention of astrology at the time of the testing presumably eliminated the possibility of self-attribution (Clarke, Gabriels, and Barnes 1996). The results were correlated with Sun, Moon, and Ascendant signs in various combinations. Only the group with both Sun and Moon in the positive fire and air signs (correlated with extraversion) differed significantly from those with Sun and Moon in the negative earth and water signs, but other combinations were within normal expectations.

The controversy over the Mayo team's findings has been long-lasting. In 1986 Geoffrey Dean published the results of a study that used a sample of 1,198 subjects, the majority born in the southern hemisphere, a serious issue if the zodiac is considered to be an effect of seasonality. The subjects first took the EPI, measuring extraversion and neuroticism and their respective opposites introversion and emotional stability (Dean 1985; 1986). From this large sample Dean selected only those who scored very high or very low in extraversion and likewise in neuroticism, that is the top and bottom of the sample, creating a new sample of 288. This sample was then sorted according to the four possible extreme scores where E+ and N+ are the top scores in extraversion and neuroticism and E– and N– correlate with introversion and emotional stability. These four factors produce four combinations: E+N–, E+N+, E–N–, E–N+. Each of these four combinations, in this case psychological types defined by words selected by the psychologist from a test he designed, were then tested against the signs of the zodiac as defined by thousands of astrologers over the course of two millennia. The test failed to find a significant correlation. He then had forty-five astrologers attempt to match these test results with astrological charts, but the results were not statistically significant.

Dean's study was reworked thirty years later by Robert Currey, who was able to obtain the original test scores for the 288 subjects from Dean (Currey 2017). Currey criticized Dean's study for its weaknesses in organizing Eysenck's psychological definitions of extraversion and neuroticism (and their opposites, introversion and emotional stability) into a form that would be consistent with the concepts and words astrologers routinely use. By failing to make such adjustments, some parts of the test were like comparing apples to oranges, exactly the same problem that astrologers had to deal with in Carlson's test. Using textbooks written by contemporary astrologers with substantial backgrounds in psychology, Currey reorganized, but did not change the substance of, Eysenck's trait words for the four categories so as to better match the four categories commonly used by astrologers in descriptions of the signs as elements: fire, earth, air, and water. Then he examined the data, including the three fire signs versus extraversion. The results showed statistically significant correlations between the two factors, a completely different result from Dean. Both papers were published in the very small, little-known peer-reviewed astrology journal *Correlation*, so the debate on this complicated study continues to occur far from the mainstream of scientific research, which is not particularly interested in this sort of thing anyway. And until astrologers come up with their own comprehensive personality inventory to use in studies like this one, which is no easy matter due to the fact that astrology measures a lot more than psychology does and astrologers disagree among themselves on how this information should be organized, I'm sure this will not be the end of it.

John Addey (1920–1982), known for his harmonic wave theory of astrology mentioned earlier, approached the zodiac differently. Unlike the majority of astrologers who focused on signs and houses as discreet segments or containers for planets, he saw the distribution of planets along the ecliptic, regardless of sign position, as points in waves of varying lengths and amplitudes. Addey conceptualized the zodiac not as twelve discreet signs of 30 degrees, but as six sine-like waves with the fire and air signs being the positive phase of a wave that is followed by the negative phase, these being the water and earth signs. In many of his studies Addey took the Sun (or planet) positions, distributed along the ecliptic on a single-degree basis, and subjected the sample data to a mathematical process called a Fourier transform in order to determine the amplitude and phase of the various harmonics, or waves. In a sample of 4,465 clergymen the dominant wave was one of about 51.5 degrees or one-seventh of the full 360 degrees of the zodiac. In another study of 7,302 doctors Addey found the fifth harmonic to be strongest; after that the twenty-fifth and then the one hundred and twenty-fifth; the latter two figures being five to the second and third powers. Again, another way of visualizing this is that the fifth harmonic wave emerges from the data with the zodiac being divided into five sections of 72 degrees that are then stacked and averaged. For Addey, who studied philosophy and was a committed Pythagorean-Platonist, his findings were evidence of the power of numbers operating behind astrological phenomena.

Sun sign studies have looked at a wide variety of human behaviors and interests. One study took into account the Sun and Moon signs of World Cup soccer players (n = 704) drawn from the teams that competed for the 1998 World Cup. The author, recognizing that the Sun sign distributions in sports may be distorted by age eligibility, focused solely on lunar distance from the Sun and Moon. The study found the Moon was more often found in one of the two signs adjacent to the Sun, which happens near the new Moon (Verhulst 2003).[10] A study of suicides and Sun signs (n = 7,508) found a correlation between thoughts about suicide, not the act itself, relative to four possible situations (incurable disease, bankruptcy, dishonor, and extreme malaise) and the sign Pisces (Stack and Lester 1988). Many examples of bad science on the part of both astrologers and scientists have been published, some even comical. One study took one hundred celebrities and, using what seems like an excess of statistical measurements, found the sign Aquarius (the authors call individual signs "zodiacs") to be significant. No information was given about how the celebrities were actually chosen, though the study was then repeated with two and three hundred of them, with the same results. This publication was quickly attacked for its methodology, yet one author apparently found the study, by itself, still to be all that was needed to pronounce astrology invalid. Not long after, another more credible researcher, one who seemed to actually know something about astrology, discredited those criticisms and pointed to a seasonal pattern that is found in the data (Adel, Hossain, and Johnson 2014; Genovese 2014; Hamilton 2015). Meanwhile, amateur astrologers go about compiling long lists of people linked to one category or another and then organize them by Sun sign, many not aware of seasonal variations (more people are born in Virgo and fewer in Sagittarius) or exactly how to do a statistical study. But some do know what they are doing, some are building large databases, and some are now tapping into the data produced by the internet where it is available. Perhaps some interesting correlations with Sun signs will be found in the future.

At present, the testing of astrology amounts to essentially a list of conflicting statistical studies that are no longer of interest to

---

10. The issue of birth earlier in the year giving an advantage is thought by skeptics to be the reason doe = for a large majority of Olympic medal winners (gold, silver, and bronze) being born in Capricorn.

mainstream science or even to practicing astrologers. Only a few studies, the ones that invalidated astrology and appeared in major journals, are regularly cited, protected by zealots on Wikipedia, and are therefore assumed by many to be definitive. Most of these are flawed at least in part due to a lack of understanding of astrology on the part of the researchers and the peer reviewers. Some studies done by skeptics actually butcher the data to get the results they want. The studies that purport to show support for astrology are found in very small and virtually unknown journals mostly published by astrology organizations. Not all of these journals are peer-reviewed, but some are, and *Correlation: Journal of Research in Astrology* is probably the best of them. Many of these studies have tested traditional notions, have been replicated and checked, and should probably be done with much larger samples, but because there are no institutions in astrology, this amounts to a lot of work for no pay. One example is a correlation found between people having red hair (n = 500) and Mars located within 30 degrees of their Ascendant, this being more or less one of the so-called Gauquelin zones in the diurnal cycle. The low probabilities of this genetic-planetary correlation were criticized, but after a number of replications, it appears to be valid (Hill and Thompson 1988–89).

The problems inherent in testing astrology are obviously complex, and a very good case can be made that just a few self-appointed defenders of the status quo exert a disproportionate effect on the perception of such studies, this being another example of how a single person, like Mersenne, Anslinger, or Murdock, can almost single-handedly steer the course of events in their time, with enormous repercussions for generations that followed (McRitchie 2016).[11] Since, so far, the only strong evidence for natal astrology is statistical, and statistics are suspect at best and befuddling at worst, it has come down to one side claiming

---

11. Geoffrey Dean has been a prolific writer and scathing critic of astrology. A wealth of information on his application of RMM science to astrology can be found on the website "Astrology and Science." To develop a more balanced and nuanced appreciation of skepticism in astrology visit Robert Currey's pages on the website "Astrology for Critical Thinkers: Evidence Based Astrology."

victory and the other crying foul. It may be more productive to consider possible mechanisms and then devise models that can be tested, rather than mix psychology's apples with astrology's pomegranates and give astrologers baffling mix-and-match quizzes.

EXPERIMENTS TRIED AT HOME

Without the funding and institutional support needed for rigorous research, what's an astrologer to do? Mostly assemble data and correlate planets with behaviors and events, it seems. What passes for research in the larger astrological community are anecdotal studies, and there are plenty of them. These compilations of correlations (or pattern observations) are cheap and can get a practitioner published, but they aren't much of an improvement over what Valens, Cardano, Gadbury, and others were doing centuries ago. Another variant of low-budget research are reports on personal field observations. For much of the 1970s I played rock music in a bar-band several nights a week, usually from nine o'clock at night to one o'clock in the morning. While the band played more or less the same material each night, it became apparent to me that audience interest and reactions varied considerably from night to night. I began to keep records of the times during the gig that the band-audience connection was particularly strong and enthusiasm high; the next day I would calculate planetary positions and look for correlations. What I found was that audience interest had a correlation to the Moon (its sign and aspects) and that moments of intense interactions correlated to the Moon or a planet either rising/setting or culminating (upper and lower), more or less in the Gauquelin zones. Taking the study further, I adjusted a pocket watch so that it ran four minutes fast per day and set twelve o'clock to zero hour Greenwich Mean Time. This allowed me to read local sidereal time, track the rotation of Earth, and also track the diurnal cycles of the planets. With this device, and a prepared list of the sidereal times that planets would occupy these four positions (the angles) for each night, I could then observe in real time variations in crowd behavior. (Today there are apps that do all of this instantly on a phone.) One observation was that the size of the audience mattered: with fewer people, group response was inconsistent, but a kind of human quorum sensing kicked in with large groups, and these followed the angularities of the Moon and planets in behavior quite closely. My observations eventually produced knowledge that informed choices (an exercise of free will) regarding the best time to play a certain song and also when the band should take a break. Useful information, but lacking units and a figure for probability.

Over the next decades my field observations became more sophisticated and were extended to many other behaviors, individual and group. I obtained an astronomer's sidereal watch, which made things a bit easier, and memorized the sidereal times that the positions of each of my natal planets, Sun, and Moon would cross the angles in their diurnal cycles (these crossing points shift four minutes earlier each day and cycle through the year). I would then make observations of what might be called micro events and trends. While attending astrology conferences I was able to obtain the birth data of many friends and acquaintances, and I stored this information in notes. Then, while sitting in the back of large lecture rooms during long lectures, I would observe when any of these "subjects" got up and left for a snack or visit to the rest room. It became apparent to me that people were having (presumably) involuntary responses when the angles coincided with significant positions in their natal chart. In more than a few cases I was able to predict exactly when one of the subjects would get up for a break, a feat (some might say an invasion of privacy) for which I earned a reputation.

Computers became available in the late 1970s and soon software made the calculations easier. My friend Barry Orr wrote a program that displayed these angle crossings with a special emphasis on the simultaneous angularity (rising, setting, and upper and lower culmination) of two or more planetary bodies, these being called parans, short for the Greek word *paranatellona*. I used this information to do personal experiments on event initiation. For example, one ongoing study involves the effects of leaving my house to run errands at specific times. Over several decades I have left to run errands, take multi-day backpacking trips, and begin air travel vacations either without thinking about it (the controls), or by making a conscious choice based on the angularity of planets (the experiments), recording my observations in either case. It became apparent early on that leaving when Saturn was angular nearly always correlated with delays or restrictions, but not so with Jupiter. I extended my observations to other people, asking them to only give me the time they left on a trip, and noted similar patterns. In 1980, about a decade into the study, I wrote a book (*The Timing of Events*) reporting on what I had found and how to apply it, but the project has continued into the present. These field studies are both a subjective and objective test of electional astrology—you learn something about the temporal environment and then you use this knowledge, so its also an application of free will.

Another set of field observations has relevance to a biological connection to the diurnal cycle of the planets and zodiac. Following a bad accident, two surgeries, and a series of iatrogenic problems, I was left in a condition where some parts of my body system were damaged and displayed symptoms that appeared in unpredictable and completely involuntary ways. Tracking the causal chain of events from a behavior (moving a certain way, eating a particular food, etc.) to a symptom proved complicated, so I decided to keep detailed records.[12] (This exercise might be called autoastrometry.) What I found was that, when my physical condition was chaotic, my symptoms correlated strongly with astrological aspects. But when my body was stable and able to tolerate interruptions (trigger behaviors) and symptoms were absent, predictable circadian rhythms were dominant. What this suggests to me is that when my system was chaotic, it was not only more sensitive to identifiable triggers such as food or stress, but it was also sensitive to the positions of the current planets. The take-home point here is that my

---

12. Over time I found that symptom flareups did correlate with certain behaviors— but these were always timed and apparently amplified by transiting planets to my natal chart. Specific events or points in the diurnal cycle, most often lunar angularity or the angularity of the zodiacal degrees of my natal Moon and Ascendant, were observed to correlate with sudden flareups. While behavioral triggers could be linked to symptoms (i.e., physio-chemical cause and effect), just as often planetary signals alone would correlate with symptom flareups, which makes the mapping of causality in the body very difficult.

body, specifically the nervous system and the microbiome, appears to respond to both behavior triggers and planetary signals in ways that are not consciously controlled. This finding complicates medical diagnosis in regard to cause and effect, but it is not really new knowledge. These observations fall into the category of medical astrology but are also relevant to the subdiscipline chronopharmacology that Franz Halberg pioneered.

12

Mechanism or Magic

. . . [T]he sun always functions as an attuned gravitational resonance aggregate when the planets are clustered around [it] in configurations corresponding to the oscillations or harmonic vibrations of the gravitational waves.

THEODOR LANDSCHEIDT, *COSMIC CYBERNETICS*

The birthchart is simply the whole universe focused at one point in time and in space. If we feel that some "scientific" explanation is necessary to explain why it defines the role of a person in the universe, and thus the pattern of his individual selfhood, it is because we have a fragmented and mechanistic view of the universe; that is, we fail to realize that the universe is and operates as an organized Whole of structured activities.

DANE RUDHYAR, *ASTROLOGY FOR NEW MINDS*

An astrological theory that offers an explanation of how the subject works is barely on the radar screen of the vast majority of astrologers today. While part of this is because the intellectual terrain that must be traversed can be difficult and requires specialized knowledge, another part is that it just doesn't matter to them. As I see it, most people drawn to astrology today are motivated by a need to better understand themselves, their social world, and the world at large. What they find is that natal chart interpretation can fill these needs at a level that is at minimum comparable to, but more often greatly surpasses, what psychology or other social sciences have to offer. This focus on self and others is understandable as *Homo sapiens* is a hypersocial species and a subject like astrology, with its windows into personality, destiny, and relationships, is extremely attractive, even if limited to only Sun signs. The same is true for "Dear Abby," celebrity entertainment shows, and reality TV. While astrometeorology and other branches of natural astrology attract only a very few, nearly all astrologers, from students to professional practitioners, study the planetary positions and patterns found in natal charts that generate the information needed for interpreting a personality and life trajectory. At most astrology conferences it is practice and techniques, such as how to work with outer planet transits, planetary midpoints, secondary progressions, essential dignities, and tips to improve counseling skills and manage relationships, that are the primary subjects of the astrological lectures or workshops offered to attendees. Mundane (or natural) and medical astrology are a distant second in both astrological literature and lecture topics at conferences. Astrological theory as a topic in itself is rarely presented, but that topic does have a long history and it is still developing.

THE HISTORY OF ASTROLOGICAL THEORY

Recall that astrological theory was addressed by Ptolemy who, building on the cosmology of Aristotle from a few hundred years earlier, assumed the heavens to be a series of concentric spheres with Earth at the center. The influence of these higher levels on Earth was possible because motion, action, or so-called energy, from the outer celestial sphere was transferred downward toward the Earth (the sublunar zone) via a refined medium, the ether, which permeated space.[1]

---

1. A more extensive account is given in another text by Ptolemy, *On the Hypotheses of the Planets,* a translation of which is contained as Appendix I in *Tetrabiblos,* (Ptolemy 1994, 50–57).

Theories of stellar influences can be found in the ninth-century writings of a few of the great Arab astrologers. Abu Ma'shar, who regarded astrology as a natural science, described astrological causation by referring to Aristotle's model of concentric spheres and attributing a heat-based generative power to the planets that bring form and matter together to make the things of the sublunar world. Al-Kindi, whose metaphysics was Neoplatonic, in the sense that the cosmos was seen more as a living hierarchy of refinement rather than a series of bounded spheres, offered something different: a single mechanism involving rays propagated along straight lines from the planets to Earth and its inhabitants. These rays he saw as the causes that drive the combining of the fundamental components of nature, these being the four elements. Al-Kindi wrote that rays from other stars or planets are also possible as when two planets are in conjunction, or in aspect to each other; their rays will blend producing a unique cause. Further, rays from stars or planets vary in strength according to the obliquity of their angle to the horizon and can be fortified by the rays of other planets or stars. This makes astrology a kind of vector analysis, and Al-Kindi's theory is consistent with historical astrological methodologies in which planetary strength is related to brightness, latitude, declination, angularity, and so on (Al-Kindi 1993, 7–10).

In the late fifteenth century, classical astrological theory, which hadn't really progressed much since the Arabs, was shaken when Pico della Mirandola ripped into astrology and, among other things, raised the problem of exactly how judicial astrology worked.[2] Kepler, who was reading and commenting on Pico over a century after his attack was published, was the next major contributor to astrological theory. His theory of astrology, previously discussed in Chapter 4 and again in Chapter 10, was rooted in ancient ideas, especially those of Pythagoras and Plato, but also Aristotle and Ptolemy. In 1602 he suggested that the Sun, Moon, and planets affected the Earth through two primary forces

---

2. Pico della Mirandola's attack on astrology was broad, but a major argument in the work was the failure of astrologers to account for how the planets, though not the Sun or Moon, could be the causes of anything.

that could transmit two qualities: warming and humidification (Kepler 1984, 238). The first force was light itself, the carrier of heat. From his work on optics, he knew light transported colors and he connected this with the actual colors of the planets. The second force was reflected light, strongest in the Moon, which was the carrier of moisture and humidity. He thought that the planets both radiated and reflected different colors, what we would call today wavelengths, and this accounted for their varied heating and humidifying properties on Earth and its inhabitants (Kepler 2008, T-29–32, 84–86).[3] Kepler's later astrological theory, which is found in his 1619 work *Harmonics of the World,* focused on the astrological aspects.

The angular separations of the planets do establish Pythagorean-Platonic geometric ratios, which Kepler compared to musical tones. His unique proposal was that astrological effects occurred because of the existence of an animal-like faculty by which humans can sense subtle, inaudible, and invisible changes in the sky. When aspects were exact, they produced a non-material tone that could be essentially heard by a responder. The effects of the astrological aspects are then not caused by the planetary bodies themselves, but by the reactions of the Earth soul and individual souls to very specific proportions or harmonies based on numbers that are produced by the planets in their movements; in other words, the music of the spheres, or in modern terms, a wave theory of astrology (Hand, 1982).

Kepler's astrological theory was a cornerstone of his Pythagorean and Platonist theory of everything, which he explained in *Harmonics of the World.*

Goad, a follower of Kepler's astrology who was discussed earlier in this book, didn't have much to say about astrological theory in his *Astro-Meteorologica* (Goad 1686). He thought all planetary effects were ultimately reducible to the interaction of heat and moisture, the Sun being the principal warming agent. But unlike Ptolemy, who presented the planets as capable of producing varying proportions of heat and

---

3. Galileo's telescopic observations of Venus in 1613, revealing its phases, caused Kepler to change his mind in regard to radiation emanating from the planets.

moisture according to their nature, Goad argued that the Sun, Moon, and planets only transmit varying amounts of heat via light, moisture being an effect that is pulled from the Earth where it originates.

Astrology's downfall in the seventeenth century meant theoretical speculation was, for the most part, abandoned. It was practice (and antiquarianism) that kept the subject alive. After staying more or less underground for roughly two hundred years, the subject re-emerged from its cultural marginalization in the late nineteenth century, and new publications, not just reprints of older writings, began to appear. William Frederick Alan (1860–1917), who renamed himself Alan Leo (he was born with both Sun and Ascendant in Leo), was a major figure in this revival. Leo was involved with the Theosophical Society, a revival of ideas from ancient and Renaissance naturalism, mixed with some Hindu philosophy and benign conspiracy theory (advanced beings guide humanity), where he founded a special interest group focused on astrology. The subsequent influence of Theosophy on twentieth-century astrology that came through his teachings and published works cannot be underestimated. Leo's perspective was basically a Neoplatonic cosmology, which has always supported astrology, plus ideas on reincarnation, a combination that has met the psychological and spiritual needs of many people disenchanted with materialist scientific culture—and it continues to be an influence today. More broadly, the late twentieth-century New Age movement is built largely on late nineteenth-century Theosophy, and also ideas from the American New Thought movement, both of which have fascinating histories worth exploring.

In addition to publishing volumes on practical natal chart interpretation, Leo explained astrology as a tool or means of understanding the inner or esoteric laws of nature and offered some metaphysical tidbits on how astrology actually worked. He described the cosmos as one massive organism where the energy of the First Cause is stepped down through individual stars. From this Aristotelian beginning, the Solar Logos, or the Sun, recapitulates the general pattern of the cosmos. It is the local source of cosmic energy that is then distributed via rays by the planetary vehicles, which are intermediaries, much

s in Neoplatonism. The zodiac, which is the Earth's aura
longitudinally from the poles like an orange, is the modu-
d through which the vibrations from the planets, Sun, and
ust pass. Leo's view of the cosmos incorporates the concept
of holons, a term coined by the writer Arthur Koestler, which posits
that individual units make up wholes that are then individual units of
larger wholes. Leo, writing before the term came into use, explained
that "any higher plane or world is in unity when compared with one
below it" (Leo 1918, 11). Of course, most occultists and astrologers
know this principle as the ancient Hermetic axiom "as above, so
below." Leo also stated that the energies of the cosmos tend to be
classifiable in threes, fours, sevens, and twelves, these being the most
cosmic numbers in the Western esoteric tradition. The ancient Maya
would differ, however, as they believed five, nine, thirteen, and twenty
to be the most cosmic numbers.

A few decades after Leo, Dane Rudhyar (1895–1985) elaborated
further on how astrology worked and what it was good for. Rudhyar
was a transplant from Paris to the United States who did much more
than astrology; he also achieved recognition as an influential modern
composer, artist, writer, and philosopher. His thinking was strongly
influenced by Carl Jung, particularly in regard to the process of indi-
viduation, and he introduced Jungian concepts in his modern interpretation of astrology. Perhaps the most significant thing Rudhyar
did for astrology was to raise the level of discussion beginning with
his comprehensive 1936 work *The Astrology of Personality* and later
through a series of highly respected books on astrological practice and
philosophy. These were read by many astrologers in the 1960s and
1970s, especially in California, which was at the center of a new youth
culture. The main theme that runs through his writings is the value
of astrology for the individual. The birth chart is seen as a map of
potentialities, not certainties, and the knowledge obtained from the
chart was to be used as a tool for self-actualization. In this way, astrology provides meaning to a human life. Rudhyar was not a technical
astrologer, and he thought predictions of specific events were misleading because the knowledge obtained by astrology is only useful in

understanding process, not outcome. He was far more a psychological astrologer; he was a friend of Roberto Assagioli and was influenced by Abraham Maslow, Carl Rogers, and others. In 1969, soon after the founding of humanistic psychology by Maslow, he founded the International Committee for Humanistic Astrology. A few years later he took his ideas further by pioneering transpersonal astrology, an astro-psychological perspective where the free will of the individual, operating within the realities of cosmic rhythms, is the central theme. Relative to some of the trends in psychology at the time, Rudhyar made astrology very current.

A major influence on Rudhyar's thinking was the book *Holism and Evolution* by Jan Smuts, the author who coined the words *holistic* and *holism* and saw evolution as a process of making new wholes. Like Leo, Rudhyar also envisioned the cosmos as consisting of greater and lesser wholes, but he emphasized relativity and pointed out that perspective differs depending on what direction one is looking. From the standpoint of a lesser whole, such as an individual, the larger cosmos seems chaotic. But from the position of a greater whole, for example a societal view of the individual, things would appear stable, but also cyclic and following more or less predictable patterns. The relativistic properties of *holons,* meaning "something that is both a whole and a part at the same time," also imply that duality is inherent in any explanation of the cosmos; there is both the individual and the group, the centrality of each depending on where consciousness happens to be focused. The unity that people seem to be driven to seek is then elusive given our mental hardware and software. Rudhyar didn't limit his thinking to philosophical and metaphysical issues; he also thought that actual planetary influences might in a sense be biophysical—electrical and magnetic currents capable of producing changes that react on the vision-making centers (Rudhyar 1936, 102). This approach, taken up by a few scientific astrologers or investigators of astrology, will be discussed below.

The psychologist Carl Jung (1875–1961), a contemporary of Rudhyar, has had a powerful influence on modern astrology. He studied astrology himself for half a century, discreetly, to say the least.

This was a wise move given the deep hostility and, due to nearly complete ignorance of the subject, what were (and are) essentially superstitious attitudes held by many in the scientific community. Jung borrowed some of astrology's core components as key features of his psychological theories. His four functions, at the base of the popular Myers-Briggs Type Indicator, are the four elements (fire-intuition, earth-sensation, air-intellect, water-feeling) of the astrological signs. Jung also conducted statistical research using the positions of the Sun and Moon in the birth charts of couples. In particular, however, it was his ideas about a collective unconscious, the existence and identification of transcendent archetypes, and his theory of synchronicity that came to be woven into a kind of explanation for astrological effect by himself, a number of modern writers on astrology, and a few theory-motivated transpersonal psychologists.

Jung's collective unconscious is basically a non-local inherited repository of instincts and past experiences that is common to all individuals of a species. The material the collective unconscious organizes is said to exist in forms called archetypes. Originally, Jung called archetypes the primordial images of the psyche and hypothesized that in the deeper layers of the species-wide unconscious, accumulated experiences of the same kind occurring for many generations become organized into structures that serve to navigate the world. These hard-wired structures are not accessible to the working rational mind, but they can reach the conscious mind, and dreams, as symbols. The archetypes appear to be similar in some respects to the forms of Plato, which exist behind appearances, and they also have similarities to the morphic fields that shape development, instinct, and memory that were originated by Alexander G. Gurwitsch and later described by biochemist Rupert Sheldrake. While the details of this hypothesis aren't actually worked out, the archetypes as generally understood are non-material structures that guide and shape psychic life. Cultural historian and astrologer Richard Tarnas regards archetypes as universal principles or forces that affect minds, mold the human psyche, and shape the world of experience. He also argues that the planets have archetypal properties and their positioning correlates with world events (Tarnas 2007, 84). Planetary archetypes have been described as transcendent, enduring a priori structures that are timeless yet evolving. Jung thought them to be projections from the collective unconscious of humans. Archetypes are also unquantifiable and cannot be studied scientifically, at least for now, so these ideas are basically a modern twist on Platonic idealism.

The jump from archetypes to the world of astrological planetary symbolism is really quite evident. If we take the community of serious astrologers as field workers at the self/cosmos interface who are reporting their observations, they might say that the astrological planets give structure, coded in a symbolic language, to both the evolving self, the collective, and the flow of time in general. It would be reported by these field workers that the planets hold consistent universal properties; everyone experiences them in more or less the same way, this being one of the things demonstrated by the Gauquelins' studies. The planets would be described as multidimensional, evidenced by their operation on physical, psychological, social, and bodily levels of individuals, in addition to effects on the larger social structures and natural processes. Further, it might be reported that the planets are multivalent, they can be worked with and present possibilities for a conscious co-creation of future reality, a form of participation with the cosmos. The emerging consensus among these field workers would be that the planets are indeed either primary archetypes themselves or, because of correlations and identifications with other structures deep in the human psyche, they are a collective projection. These findings, if you will, suggest that, in either case, the planets are physical evidence of the existence of archetypes, and their motions can be observed to work on the evolving individual psyche. Further, it is possible that a deep understanding of astrological symbolism would offer a better understanding of human consciousness, or at least supply valuable information, than is now offered by philosophers and neuroscientists. This "report" more or less sums up a position held by some astrologers who think about these things. Although there is no direct connection to God in these field observations, the similarities to metaphysical Platonism and Neoplatonism are evident.

Jung proposed that astrology works by virtue of a strange property of the cosmos—an acausal connecting principle he called "synchronicity." Synchronicity posits that time puts a kind of stamp on coincidental, but not necessarily related or even spatially proximate, events. The correlation between a planet in the sky and an event in a person's life he argued was meaningful but acausal. In this view, not all changes require direct cause and effect linkages from one thing to another. Causality, in which one thing directly affects another, continues to exist, but there is also the phenomenon of a grouping or assemblage of coincident events strictly according to meaning, which is synchronicity. Presumably, it is the archetypes, which could be non-local structures, that would allow for such spontaneously organized connectedness to occur. This notion would explain some of the more bizarre manifestations of astrological transits and progressions as projections, such as when others or other things, not the person experiencing the planetary activity in their chart, behave in ways described precisely by the chart. Projection, of course, is a kind of externalization of uncomfortable feelings, a defense mechanism, described in psychology but here blended with astrology.

Jung did a few scientific studies to confirm the reality of synchronicity. One of these matched Sun and Moon positions between couples in relationships, and it initially produced positive results. Later, however, the study was taken to be inconclusive due to the difficulties of creating an unbiased sample to test. Later, Jung moved away from synchronicity as a model for astrological effects when he learned that planetary configurations around the Sun correlated with changes in radio signals or solar activity, as demonstrated by the work that John Nelson and a few others were doing. This far more scientific information suggested a less mysterious cause-and-effect process behind astrology that will be discussed below (Strimer 2000). It appears to be the case, however, that a very high percentage of those in the astrological community continue to use synchronicity as an explanation for astrological effects.

For the larger community of astrologers, who found themselves in the midst of a wave of scientific materialism and hostility toward astrology, Jung was a lifeline. In him they found a brilliant and respected psychologist who not only thought the subject was of value, he also had some original ideas on how astrology might work. In the 1960s and '70s some astrologers, largely inspired by Jung and Rudhyar, came to embrace depth psychology, synchronicity, and archetypes, and they joined with like-minded psychologists and philosophers like James Hillman and Stanislav Grof. By the turn of the century, mostly in California, a kind of theoretical framework for consciousness studies, based on a synthesis of disciplines, was developed that came to be called archetypal cosmology (LeGrice 2009, 2011). The first journal, *Archai: The Journal of Archetypal Cosmology,* was published in 2009. Archetypal cosmology describes an ensouled and evolving cosmos in which life is a participant in the ongoing creation. It draws from mythology, psychology, astrology, cosmology, shamanism, spirituality, history, philosophy, art, psychedelics, and ecology.

Other cornerstones of archetypal cosmology are the system sciences, modern notions such as Bohm's holonomic theory in which the brain is a storage for holograms, Sheldrake's morphic fields that structure organisms and the strange properties of the quantum realm, such as wave/particle dualism and non-local connectedness (i.e., entanglement or "action at a distance"). (The findings of quantum physics are particularly intriguing and, specifically in regard to the blurred distinctions between objective and subjective and non-physical information transport across the vastness of space, are often cited by astrologers as supportive of their subject. Quantum theory deals with non-living matter, however, and links to human life and consciousness are only speculative.) Central to archetypal cosmology is astrology, which offers a kind of proof for this synthesis of ideas. To be clear, archetypal cosmology is a philosophical metaphysics, propped with some edgy scientific notions, that has practical applications in shaping understanding and healing for individuals, groups, and the planet itself. From the scientific perspective, it is an unproven, and untestable, hypothesis. But it is, more or less, what a substantial part of the astrology subculture, and others in related fields, including some psychologists, are most comfortable with in the way of explanations for the observed effects of astrology.

In many respects, archetypal cosmology is a kind of modern multi-dimensional Platonism (monistic idealism) in which the external world of events is understood to be an effect or reflection of a corresponding individual's or group's inner psyche. Its basic tenets, including the idea that outer form is linked to inner consciousness, can explain much of how astrology appears to work to those of a transpersonal psychological bent. Consciousness is then the driving force of the cosmos, a causal agent in a sense, and is the means by which humans are able to participate with the ongoing creation. Planetary positions offer clues as to how participation with the unfolding present moment should best proceed, which is what the best astrologers generally counsel. Archetypal cosmology is teleological. The cosmos is believed to have intent and purpose, which is to raise the level of consciousness and beingness. Evolution is, therefore, more than just change over time (what biologists and geologists assume); it is the movement toward greater wholeness. This too is in line with Jung, who thought that a person's psychological patterns were revealing of a general striving for integration of self. In archetypal cosmology the astrological planets are seen as multi-dimensional structures that shape matter and mind, much as morphogenetic fields are thought to shape organisms as they undergo development. All of this underscores the teleology of archetypal cosmology—the cosmos is an intelligently designed evolver of consciousness.[4]

Archetypal cosmology is indeed a grand synthesis of most of the metaphysical and holistic ideas of the twentieth century, including some from quantum physics, along with a large serving of Renaissance naturalism and Eastern religion. The similarities to Platonism, Neoplatonism, Stoicism, Taoism, and Buddhism are clear and, because of this, an ancient subject like astrology has found a modern home. But, again, this metaphysical synthesis is a philosophy and cosmology, not a science. While this worldview has been developed in fine detail, it is not a program that lends itself easily to quantification and therefore scientific

---

4. For descriptions and discussions on archetypal cosmology see *Cosmos and Psyche* (Tarnas 2007), "The Birth of a New Discipline: Archetypal Cosmology in Historical Perspective" (Le Grice 2009), and "Archetype and Eternal Object: Jung, Whitehead, and the Return of Formal Causation" (Maxwell 2017).

exploration. It is a set of unproven assumptions collected mostly from the humanities and social sciences. (Hostility toward archetypal cosmology would be expected if it drifted too close to the astronomy or physics departments where this kind of thinking was banished a few centuries ago.) This is not to say that archetypal cosmology is wrong or false. Certainly much of it defies scientific testing in the conventional sense, but it does offer an interpretation of modern scientific findings and an update on Platonism, Renaissance naturalism, and also the archaic cosmologies of certain religions. It should also be said that archetypal cosmology, a synthesis of both ancient and modern ideas, takes a systems view of the universe and is a kind of organicism. And if you don't want to believe in it, there is, of course, always the contrasting alternative of the dominant metaphysics that purist RMM science offers: a dead cosmos, deterministic, devoid of purpose, and composed of discrete material parts in which consciousness is no more than a side effect. But most people don't have the time or inclination to wrestle with metaphysics and ontological matters, and so a mostly effortless belief in a sacred, text-based, generalized all-knowing supernatural deity, and an afterlife, is more the social norm.

Another trend that began in the late twentieth century has been the rise of astrological fundamentalism. One part of this movement was an enthusiastic turn toward classical astrology and its methods that was in large part driven by a revival of William Lilly's seventeenth-century writings, discussed in an earlier chapter of the book. Another part was a massive translating effort, called Project Hindsight, led by three scholarly men in the astrological community named Hand, Schmidt, and Zoller who introduced for the first time in English the works of many Greek, Roman, and Arab writers on astrology. These translations opened a pathway into the surprisingly sophisticated methodologies and practices of ancient astrologers, and they stimulated further exploration of Hellenistic, Medieval, and Renaissance ideas on the subject. Along with this trend came the position, among some writers, that the search for modern scientific explanations for judicial astrology was futile, though mundane (natural) astrology was still fair game for that effort.

The metaphysical implications of this fundamentalist retreat from modernity were brought into focus by Geoffrey Cornelius in his book *The Moment of Astrology,* in which he argued that judicial astrology, especially horary astrology (interrogations), is just plain divination (Cornelius 2003). Using Jung's ideas on symbols, the failures of certain statistical studies of astrology, and curious cases where charts based on wrong dates and/or times appear to deliver right results, he made the case that the act of reading a chart for a question is really a moment of participation with the cosmos, an experience at the subject/object divide. By extension, electional charts and natal charts are likewise interpreted in what he calls a "katarchic" moment of immersion in symbolism, *katarch* being a Greek word meaning "beginning" or "initiative." Cornelius argues there is no real difference between forms of divination such as augury, tarot, I Ching, and astrological chart reading, all of which he sees as the presentation to the interpreter of subjective images of reality that are then recognized, processed, and objectified into words. In other words, astrologers, conditioned by their knowledge of symbolic images, select for certain moments within which they then participate with the cosmos, or at least a part of it. This activity of imaginative associations at the subject/object interface he identifies as divination, though it looks to me a lot like intuition, in which subconscious information bypasses rational thought and triggers decision-making. All of this has more in common with the astral magic of Hermeticism and Neoplatonism and stands in sharp contrast to Ptolemy's doctrine of the methodical analysis of seed, or origin, moments that contain causal and temporal information.

If judicial astrology is a form of divination, then the need to explain it in terms of reductionist-mechanistic-materialist science is impossible and this takes a tremendous burden off astrologers and lets them get on with their work. In addition, while doing this is something of a capitulation to the righteous ravings of Pico della Mirandola, it is also a way to keep astrology secure in its own subuniverse. I understand all too well the frustration astrologers endure in the face of relentless dismissals and attacks coming from the dominant culture, most of them spectacular examples of ignorance. While the pressure to conform to the standards of RMM science are immense, the retreat to a non-explanation is not particularly productive if the goal is to grow and advance the field. If anything, labeling astrology as divination is just asking for more isolation. In my opinion, a far better approach is to define astrology in terms of organicism and system science. In this context astrology is then a set of techniques that allow entrance into the guts of a self-organizing system and a perspective on what might be the subject/object divide as illustrated by the observer effect in quantum physics. The points in time deemed significant in horary and electional astrology, moments of choice or action of will, and therefore points where things shift from one state to another, are bifurcation, tipping, or division points. In horary astrology it is the coming into consciousness of a pressing issue, the maturation and organization of a thought that is shared collectively by the questioner and the astrologer. This is a small, but real, bifurcation point in a larger system, and so is a moment elected ahead of time to initiate an event, and so are conceptions and births. I'll say more about this perspective on systems in Chapter 14.

This leaves the issue of how wrong charts sometimes give right answers, a vexing problem in astrology, particularly in regard to celebrities and politicians who sometimes release contradictory birth data for any number of reasons. It is thought by some astrologers that wrong charts, some of which appear to be uncannily descriptive, are evidence of a higher order pressing on human consciousness, something like divination in reverse. While I acknowledge that a deep dive into astrology does raise issues in regard to the nature of ultimate reality, in this case I would look to other, more immediate possible explanations. One is that most wrong charts (due to calculation errors or incorrect data) are still reasonably close to the right time (within a day) and will preserve much relevant symbolism. Because some astrological methodologies are capable of capturing a sizable amount of information from only general features in a given chart, a wrong chart may appear to be the true one. The chances are usually good that precise alignments will also be found that correspond well with some (but not all) known events. There is a solution to this problem, and it is to find the true time of the chart by doing what's called the rectification of an uncertain birth time. Very few astrologers are capable of this highly complex procedure (much like playing chess backwards) that requires math skills, data synthesis, the insights of a psychologist, and the critical analysis and understanding of events that would be expected from an expert historian. What I'm getting at here is that these problems of false charts appearing to provide accurate descriptions also draws attention to the need for astrologers to use their imagination carefully in such situations, to sort out information properly and not shape it to their expectations and biases. On the other hand, accurate diagnosis with faulty data is also something that practitioners in related fields like medicine and psychology no doubt come across from time to time, so I don't think astrology is alone here.

Lastly, there are some people who call themselves psychic astrologers. They know a little about the symbolism but generally not much about astronomy and the nuts and bolts of planetary dynamics and their placement in the diurnal cycle. These people are not a part of the larger astrological community. They are essentially psychics that use astrology the way they might use tarot cards—as set of symbols that trigger shifts in their awareness toward a more immediate use of their imagination, feelings, and intuition. There may be something to this modern version of astral magic. The unconscious mind or a pre-sensory field, as studied by parapsychology researchers, may be capable of anticipating events and assembling normally hidden information using astrological symbols, and some people are much better at finding this mindset than others. Some might call this use of astrology's symbolism divination, but it is not how most deductive astrological interpretation is conducted.

MODERN SCIENTIFIC THEORIES OF ASTROLOGY

During the twentieth century, elements of a new kind of scientifically oriented astrology began to emerge. In the 1930s, mathematician, radio engineer, and astrologer Edward Johndro (1882–1951) proposed a mechanism for the astrology effect that was based on light waves

(Johndro 1933). Johndro lived during the early years of radio and used basic principles of radio transmission to explain his electrodynamic model. He thought the reflected light waves from the planets, which are much shorter than radio waves and therefore not subject to deterioration at distance, could act as carrier waves that bring with them information, in the form of much longer waves, about that planet. Johndro thought each planet has a unique set of oscillating electrical axes that account for its own individualized signal (like a crystal radio). Earth's magnetic field and its organisms (which have electrical properties in cells, organs, and nervous systems) are receivers that sort out the information transmitted in that signal. Electromagnetism, modulated by the geometry of planetary aspects (as in a crystal) and picked up by an organism, is then a cause of the astrological effect. Johndro was somewhat active in the astrological community, and he made contributions to both natal interpretation and the practice of locational astrology, the analysis of one's position on the globe. He identified a point on the celestial sphere that his observations led him to believe was particularly sensitive to electromagnetic forces, and therefore of consequence in astrological charts. This point, named the Vertex, is the western intersection of the ecliptic and the prime vertical and is thought to signify particularly intense interactions with others. It was co-discovered by astrologer Charles Jayne, and it came to be used as a sensitive point in natal charts by many astrologers during the later part of the twentieth century.

At the same time that Jung was developing his notions of how astrology works, John Nelson, mentioned in Chapter 2, developed in the 1940s and '50s a methodology for predicting solar storms (for the tech company RCA) using the angular separations of planets relative to the Sun. His observations convinced him that complexes of planetary alignments concurred with solar storms, flares, and sunspots, and that these disturbed shortwave signals. With this information he developed a method for predicting radio transmission disruptions that was successful enough to keep him employed by RCA for many years. What makes his work clearly fall into the domain of astrology is the fact that the heliocentric planetary alignments he observed were basically the same as those used by astrologers. These were the aspects, specific phase angles in synodic cycles, that were at the core of Kepler's reformed astrology. Other scientists, some working for the government like Jane B. Blizard, noted similar correlations between planetary alignments and solar activity. As mentioned above, the findings of Nelson (who sometimes spoke at astrology conferences) and others in the 1950s and '60s regarding heliocentric planetary aspects and solar activity was something of a wake-up call for Jung, and he apparently backed off from publishing his ideas on synchronicity after hearing of them (Strimer 2000).

Planetary aspects are also central in a wave theory of astrology outlined by Robert Hand that is reminiscent of Kepler's aspect resonance theory (Hand 1982). Building on John Addey's work with harmonics in astrological data (see Chapter 11), Hand suggested that it is the angular relationships between planets and their overtones that produce an astrological signal in the plasma of space. As the planets move through the zodiac they form aspects with each other that, when exact, are able to cut through a background planetary noise. The aspects are then information carrying signals that humans have some kind of receptors for and can be utilized in various ways. This is consistent with Addey's Pythagorean-Platonic ideas on harmonics in astrology and is basically a modern version of Kepler's thought on aspects in astrology where each harmonic has special qualities and therefore interpretive constraints. In Hand's wave theory, the zodiac is primarily a place marker, and it is the ceaseless formation and dissolution of angles between planets, similar in principle to harmonic resonances in music, that explains how astrology might work dynamically, such as through time, hence the music of the spheres.

Theodor Landscheit (1927–2004) was a German astrologer and research climatologist who, in the late 1960s, embraced cybernetics and chaos theory as an approach to understanding astrology. He studied correlations between climate patterns and the movement of the solar system's barycenter (center of mass), which itself is modulated by planetary cycles (described in Chapter 2), and by the early 1970s he was making climate and weather forecasts based on his unique style of heliocentric astrology. Landscheidt, a former high court judge in Germany, moved between social worlds and had a number of scientific papers published in reputable journals and anthologies, but he also spoke at astrology conferences. He referred to astrology as cosmobiology, called astrological charts individual cosmograms, and had his own theory of how astrology worked called the cosmobiological information model. In his view, the Sun and its planets function like an intricate organism, regulated by complex feedback loops, and gravitational fields function as information transmission channels (Landscheidt 1973). Landscheidt's cybernetic view of the solar system extended to the galaxy, and he considered the galactic center to be of great importance in understanding processes in the solar system and on Earth. His was a view of holons; (galaxy, solar system, Earth) or like the title of one of his books: *Sun, Earth, Man.* Unfortunately, Landscheidt was not associated with a major institution, which made his credibility questionable in the scientific world, and his ideas were far too complex for most astrologers to follow. He was also an early critic of global warming, predicting that an inactive Sun during the early twenty-first century would cause global cooling.[5] His ideas on this matter are still cited by climate change deniers, which has not helped his reputation. But he has left a small body of work, some of which I suspect will be regarded as pioneering in the future.

Landscheidt's cosmic heliocentric astrology falls completely within the range of known force fields—gravity and electromagnetism. He thought that at least some astrological effects may be the result of a transfer of planet-driven angular momentum to conditions in the solar surface (or deeper), with effects that include solar energy blasted outward that is intercepted by Earth, a possible mechanism for weather and climate modulation. His work was focused on cybernetics and the behavior of dynamic systems, which was something like a subculture in

---

5. Landscheidt was right about solar activity in the early twenty-first century—it was much lower than the previous seven decades, particularly solar cycle twenty-four. This is consistent with the Gleissburg cycle of approximately eighty-eight years. Any cooling due to this was more than compensated for by anthropogenic greenhouse gas forcing, however, which he failed to account for.

mainstream science at the time, and he didn't invoke any mechanisms that weren't already known to modern science. Landscheidt wasn't entirely alone; he had connections with geoscientists Rhodes Fairbridge and John Sanders, who also stressed the importance of planetary patterns in determining the Sun's motion with respect to the barycenter and the corresponding effects of the solar wind and solar storms on the Earth (Fairbridge and Sanders 1987).

The center of the solar system is the center of mass of the entire solar system. The gravity of the planets orbiting the Sun causes it to move about the barycenter, this phenomenon being the basis of the radial velocity method used to detect extrasolar planets. Landscheidt drew attention to the mutual gravitational interactions of the Sun and planets, specifically where the respective gravitational fields of these bodies meet. In his view it is these boundary layers between attractors, where influences from either attractor approach zero, that are of greatest interest (Landscheidt 1989, 13). It is in this zone that extremely weak signals can have disproportionate effects. The surfaces of spheres (which is where the terrestrial biosphere is located) he saw as places of sensitive balance where symmetry breaking occurs and new structures can emerge. Taking all this into consideration, at least some parts of natural astrology can then be explained as follows: The gravitational forces of planets and the Sun work on the barycenter between them. The shorter-term irregular motions of this point modulate the surface of the Sun, in certain cases resulting in solar eruptions, and these discharges of high-energy particles then blast Earth's magnetic field and atmosphere. Humans, living on the Earth's surface, may then experience weather changes and disruptions in the electromagnetic fields that the biophysicist Chizhevsky and chronobiologist Halberg have correlated with collective behaviors like wars and even personal events like heart attacks. In some ways this model is a modern quantitative version of Ptolemy's and Aristotle's cosmological model in which forces from the planetary spheres bring changes down a causal chain to the sublunar zone.

MAGNETO-TIDAL RESONANCE

Another late-twentieth-century hypothesis of astrological effect is similar to Landscheidt's but involves magnetism and people. Recall from
Chapter 2 that during the 1980s and '90s English astronomer Percy
Seymour proposed a possible sequence of events that could account for
astrological effects, particularly those of the Gauquelin studies. In his
theory of magneto-tidal resonance, Seymour suggested that very small
gravitational forces from planets can lead to very large effects on the
magnetic fields that organisms are receptive to. His multi-link theory
begins with gravitational forces from the planets modulating the low-
density solar surface by a tidal effect. In articulating his theory, Seymour
cited astronomer and mathematician George Biddell Airy's (1801–1892)
theory of ocean tides in an equator-spanning canal that has no continental shorelines to block it. A moving wave, raised by an orbiting body,
circles the globe in this canal. Over time, as the wave follows the tidal
pull of the body, it grows. Seymour argued that when two or more bodies are in *resonance,* that is "traveling together or at specific angles to
each other," their combined gravitational effects are much stronger than
the sum of the two and the wave is greatly amplified or diminished,
depending on the angle between them.

Seymour applied the above process to the evolving magnetic field of
the Sun (i.e., the solar cycle), where pole-to-pole magnetic lines become
twisted due to differential solar rotation, that is lower latitudes rotate
faster than higher ones. This causes a series of magnetic canals to form
parallel to the solar equator. So far this is the standard Babcock model,
but Seymour argued that these solar magnetic canals are responsive to
the tidal effects of planets orbiting the Sun. Two or more planets in
conjunction moving together will significantly amplify the tidal effects
allowing hot gases of the Sun to move to its surface. The idea here is
that planetary gravitational effects become a contributing factor to the
solar cycle itself. The planet-modulated solar cycle affects the magnetic
properties of the Sun, which then, via the solar wind or solar storms,
affects Earth's magnetic field. And because the planets have regular
orbits (Jupiter's is 11.86 years), the solar cycle is kept within a range

centered around 11.1 years. Information is contained in the solar wind, which is modulated by the solar cycle and displays periodicities that are then expressed as patterns in magnetic fields, rates of incoming solar radiation, weather events, and climate trends. This is a heliocentric model, but Seymour thought planetary gravitation could also operate more directly on Earth's geomagnetic field causing rhythmic disruptions that are sensible to the biosphere.

Seymour made a case for magnetism being the primary force behind natal astrology. He first pointed to navigational sensory systems in certain animals, now understood as multiple integrated redundant systems: visual terrestrial, visual stellar, and magnetic. Magnetic navigation, used when visual cues of land or stars are not available, is accomplished with magnetite in some animals and quantum electron pairs in some birds. In other organisms it is apparently the pineal gland that registers the field. If animals are able to sense the magnetic field, what else besides direction can be found in it, and how else might it be used by organisms? Seymour noted several variables, each of which have signals that may provide information. One is the solar daily magnetic variation, produced by the distortion of Earth's magnetic field by the solar wind and modulated by the rotation of Earth within the field. Another is the annual variation of the geomagnetic field that is due to the seasons, and there is also the lunar daily magnetic variation produced by tides in the upper atmosphere, an effect that has electric properties. The amplitude of the lunar daily magnetic variation also varies according to the solar cycle. The final piece in Seymour's theory is that life can do things with this information. At the cellular level, organisms are regulated by direct current electric potential differences that are frequency sensitive. Similarly, there is also the circadian cycle mechanism that runs off electrical gradients in the membrane of some single-celled organisms as mentioned in Chapter 1. This is reasonable as life evolved within the geomagnetic field as both a physical and a temporal environment. In regard to humans, Seymour leaned on the Gauquelins' findings, one of which is that their planetary heredity findings were stronger on magnetically disturbed days, according to the international magnetic field indices.

Earth has a dipole magnetic field that is created by the effects of axial rotation on electrical currents in the outer core, which is liquid. The actual process of field creation is not fully understood, but its form extends out beyond the atmosphere to about five Earth diameters facing the Sun. As the steady solar wind of charged particles hits the magnetic field, a bow shock is formed and the solar wind is pushed around Earth where it merges in a long tail opposite the Sun. Variations in the magnetic field experienced on Earth, as it spins inside this comet-shaped energy and particle structure, are measured in various ways including by intensity, direction, declination (direction relative to the poles), and inclination (up or down) at a number of stations located in places far from electrical disturbances. Earth's magnetic field is not at all static and exhibits a wide range of variations on scales of seconds to centuries. In addition to the Moon and the seasons, variations are also caused by many other factors including solar events, lightning bolts, and possibly even planetary alignments (Payne 1998, 69–76). A charge exists between the ionosphere and Earth's surface; between them is an electrostatic field that ionizes molecules in the atmosphere. Obviously, Earth's surface is bombarded by electrical and magnetic signals, which raises the question of what information is contained in it that life can sense.

Life on Earth evolved within the cover of the geomagnetic field, which has shielded it from cosmic and solar radiation and the interference of other magnetic fields. Adaptations to the field are deeply built into life, and there is evidence that separation from it (such as during time spent in space) can cause mutations as cells try to adapt to the change. An entire subfield of inquiry, called bioelectromagnetics, studies the interactions between magnetic fields and life, including the electrical properties of cells. Consider that cells maintain an electrical charge by regulating the concentration and passage of ions through the cell membrane. The nervous system itself, which "wires" the body, runs on electricity and chemistry. In vertebrates, nerve impulses move through cells by regulating ions from electrolytes (potassium, sodium, magnesium, etc.) inside and outside the cell which, when reversing, produces an electrical charge causing neurotransmitters to jump the gap between nerve cells. Organisms are apparently more electrified than just the nervous system. Robert O. Becker found that organisms have a body surface DC current that is related to the arrangement of their nervous system, perhaps a hint of how acupuncture might work. These electrical properties of organisms may be best described as *biofields,* which are biologically generated forces that are able to receive, encode, and synchronize with even very weak electro or geomagnetic fields. These are known to exist and are being researched (Becker and Seldon 1985; Hammerschlag et al. 2015).

Many studies have demonstrated magnetoreception, the use of the magnetic field in navigation. Recall that this was the area that biologist Frank A. Brown pioneered and was later picked up by geologist Joseph Kirschvink. Magnetic fields have been shown in studies to influence life in many ways, including modulating hormone production in the pineal gland and the growth rate of *Escherichia coli* in the intestinal microbiome. Magnetism and electricity are interchangeable (magnetic fields induce electric fields and electric fields reciprocate), and the full implications of the life-electric-magnetic connections are as yet to be worked out. The emerging subdiscipline called biofield physiology is looking into these matters, however. The questions being asked here in regard to astrology are (1) are planetary signals captured in a consistent way by Earth's magnetic field, and (2), do organisms extract these signals from the noise and use them, and for what purposes? If something like this is the case, then it might explain at least some parts of astrology. Studies in biofields and bioelectromagnetics have the advantage of being related to human health, and that need will produce funding and direct investigations. Possible connections between the planets and the body, however, are another matter. As described in Chapters 1 and 2, funded scientific efforts to sort out the solar system influences on climate and life are extremely complicated, and it can take decades to reach conclusions.

The question relevant to natal astrology here is whether Earth's magnetic field, shaped by solar events and lunar gravity on multiple time scales, carries other subtle but complex solar system signals, possibly planetary, that humans can register and utilize. For example, are Shumann resonance signals, Earth-ionosphere electromagnetic extremely low frequencies that match human brainwaves and are impacted by solar activity, utilized in any way by humans? At present, questions like this one may not be fully answerable, but knee-jerk rejection may be short-sighted.

While magnetotactic bacteria and navigating animals like fish, bees, and birds can read the magnetic field for directional information, there is still an argument over whether there exists innate human navigational abilities, though there is some evidence that they do exist (Baker 1982). Then there are the Gauquelin findings that astrological effects were stronger during times of high magnetic activity and therefore, based on that evidence, information from at least some of the planets can be transmitted via the magnetic field. The work needed to test, clarify, quantify, and interpret these forces, linkages, and receptors would be immense, however, and not likely to be funded. It would seem easier to ignore or even deny any astrological effect at all and avoid the tedious and potentially controversial work on a scientific research project that is not urgent and pays next to nothing. Certainly, those who identify as astrologers are not prepared to do this work and so must wait for insight from scientists into these possible explanations—whenever they get the time and money to do it.

How astrology might work (or at least parts of it) is a question that will require a lot more knowledge to answer adequately. As Einstein said, the universe is likely to be stranger than we can imagine. Approaching the problem in terms of what is presently known in physics and biology brings one quickly to the edges of pure speculation. Perhaps organisms possess a weak biofield that is sensitive to weak electromagnetic signals. Maybe there are fields or forces other than electro-magnetism or gravity yet to be discovered, ones capable of conveying information that organisms can decipher. And then there are the findings of Lilly Kolisko, that inorganic chemical reactions change when they occur at the same time as planetary aspects—if this is the case, how is this possible? Are planets a factor in the modulation of molecular arrangements?

Many people are drawn to non-material explanations, and not a few contemporary science writers have suggested that the cosmos is actually conscious, this position being called panpsychism. The idea that nature is mind (idealism) does solve the mind-matter dualism problem and would appear to offer more to a unified general philosophy than materialism (physicalism). It also has a long history and is found in the thought of a number of ancient Greek philosophers, and modern philosophers such as Spinoza, Leibniz, James, Whitehead, and Bateson, to name a few. The peculiar nature of the quantum world especially has inspired many publications that build the case for panpsychism. Just the facts of non-locality alone, which have been absorbed into the eclectic assemblage of ideas called archetypal cosmology discussed earlier, are enough to disturb certain modern assumptions about nature. It is always possible that today's magic will become tomorrow's science. Being a methodological naturalist with strong agnostic leanings, a high tolerance for ambiguity and paradox, and no compelling need to adopt beliefs, I will only say we just don't know what is really going on and that there appears to be plenty of learning ahead if we are open to it. This practical attitude I inherited from my father, who was an electrical engineer, and his father, who was a farmer. In the spirit of science as I conceive it, we should always be willing to make adjustments to our understanding as more group-confirmed information becomes available—and we should be honest about what we don't know.

THE DEVELOPMENTAL PLANETARY IMPRINT MODEL

Many organisms, including humans, undergo developmental changes before reaching adulthood, and these appear to occur in discrete stages on the physical, emotional, and mental dimensions. People who work with children generally organize these periods in terms of age or by simply describing milestones to be achieved by a certain age. A common scheme has the following labels: infancy or baby (up to two years old), toddler (one and a half to three years old), preschooler (three to six years), middle childhood (six to twelve) and adolescence (eleven to eighteen). More precise stages, or sequences, in human development have been named by a number of developmental theorists, and a review of this literature points to a consensus of about four main stages. At each of these stages what may be entrainment attractors build templates that are used to manage the self and navigate the social environment. In spite of the fact that each theorist has focused on a specific aspect of development, such as psycho-sexual, cognitive, and the like, a comparison of the major schools of thought on this subject, including those of Freud, Erikson, Piaget, Steiner, Wilbur, Leary/Wilson, and others, shows a more or less standard pattern of stages that could be summarized as follows:

FOUR STAGES IN HUMAN DEVELOPMENT

Age	Stage	Function	Structures	Body	Planet
0–2	Bio-survival	Attachment	Security, trust, belonging	Metabolic	Moon
2–4	Ranking-territorial	Identity	Autonomy, position in social hierarchy	Immune	Mars
3.5–13	Language-learning	Social coordination	Communication, problem-solving, navigation	Nervous	Mercury
8–16	Socio-sexual	Mating	Relationships, courtship, creativity	Reproductive	Venus

These four distinct stages, or sequences, given here with approximate ages, are based on the generalized observations of several noteworthy psychologists. There is both supportive and unsupportive evidence for the existence of these stages, so they must be taken as hypotheses, not fact. In contrast to the apparent rigidity of these stage theories, the tendency today among those working with children is to allow for individual variations; stages are seen as highly flexible and overlapping. This is a tolerant and practical approach and may have benefits in calming parents' expectations. But leaving this contemporary perspective aside, I've argued in a number of publications that these developmental periods coincide in a remarkable way with the symbolism of the inner planets in astrology, the planets that are traditionally associated with the most personal of human traits and characteristics (Scofield 1987; 2000).

There are also correlations between these stages and planetary cycles. Using the positions of the planets at birth as starting points, the completion of these planets' cycles when they are coincident with the solar return or its opposite are good matches with the ages of these four developmental stages. The solar return is an event that occurs every year at the birthday, when the Sun passes through the same degree of the zodiac it was located in at birth. Each day of the year has a specific, though latitude-dependent, day/night ratio, or photoperiod, and there is abundant evidence (as shown in Chapter 1) that, by measuring the length of daylight, organisms can track the seasons accurately. For reproductive purposes this is extremely important. Plants need to know when to flower and animals when to mate. Modern humans, being domesticated primates, likely have remnants of similar processes that evolved over millions of years. These may be buried in the nervous and sensory system and, at least hypothetically, can sense when the birthday (or its opposite, which is the reciprocal of the day/night ratio of birth) comes around. If the birthday, which is a key point in the annual solar cycle for the individual, happens to coincide with the return cycle of another planet, the photoperiod signal may then contain additional information. If so, this signal could be used as a cue for switching on or off developmental stages that are driven by hormones or other body-regulating chemistry. All of this, of course, is speculation, but emerging subfields like bioelectromagnetics and biofield physiology are investigating the existence in organisms of extreme sensitivities to light, magnetism and electricity, out of which emerges information used for whole-organism self regulation. Discoveries in these sub-disciplines may fill in gaps in our understanding of solar system dynamics and organism behaviors.

Developmental psychology (and ethology) has shown that there are periods of imprint vulnerability when external experiences have extra power to shape the growing self. These are the critical or sensitive periods that occur during the appropriate developmental stages, periods when exploration of new territory (consistent with sequential maturing capabilities) is occurring. Going back to the list above and comparing it to the traditional descriptions of the planets, it is obvious to anyone knowledgeable of astrology that these four stages of development correspond very closely to the symbolism of the Moon, Mars, Mercury, and Venus, in that order. Based on this observation I proposed a model called the developmental planetary imprint hypothesis (DPI), which links stages with planets. Birth initiates the Moon (lunar) stage of attachment, which extends to roughly age two. During this time instinctive responses, cognitive qualities that operate rapidly and emotionally, are being shaped (System 1 in Daniel Kahneman's model as described in his book *Thinking, Fast and Slow*) (Kahneman 2011). It turns out that at the second solar return the Moon will be located roughly 90 degrees from its birth position. Also, at the second birthday, Mars will be very close to its birth position, having completed roughly one cycle since birth. These two events, possibly received by the endocrine system as signals embedded in the geomagnetic field, could be what shuts down, or at least mutes, the bio-survival sensitive period of the Moon and opens up the autonomous ranking-territorial window of Mars. At the fourth birthday, Mars is again near its birth position and the Moon is found roughly 180 degrees from its birth position. This may mark the ending of the Mars stage, but it also suggests that stages may overlap to some extent and that some imprinting of the Moon stage has continued.

The half (demi-) solar return at 3.5, the solar return at either the sixth or seventh birthdays (these vary, but which one may say something about learning progress), and the solar return at the thirteenth birthday (in all cases) occur when Mercury is locked into phase with its birth position, these being times when learning is accelerating and lifetime interests may be imprinted. (Mercury's type of cognition can be correlated with Kahneman's System 2 thinking.) At the eighth birthday Venus returns to its birth position precisely, and the sensitive period for socio-sexual matters may be switched on. The same double return (Sun to Sun, Venus to Venus) occurs at the sixteenth birthday, which may mark the decline of socio-sexual imprinting; the period between age eight and sixteen is the period when mating patterns become established. At the eighteenth birthday both Sun and Moon return to their birth positions within a few hours, this being the well-known Metonic cycle, and it may possibly be a signal for completion of the developmental process for humans (at this stage in evolution). By this time cognitive development and the personal and social identity will (in most cases) have become established, and the individual is then ready to navigate the world, though this will vary from culture to culture.

What I'm suggesting with this hypothetical model is that during these periods of planetary resonance with the Sun, centered on specific birthdays (photoperiod being information that is recognized and used by many organisms), it is possible that the coincident reception of a planetary signal triggers hormonal processes that initiate periods of change, growth, and imprint vulnerability. These signals may also be involved with physical developments as well, but here I am focusing on personal and social development. Once a developmental stage is opened, sensitivity to certain kinds of imprints may be greatly increased and, through a kind of entrainment, these can be used to build the structures or frameworks of the psyche on which the evolving personality is built. This astro-developmental model, if it actually works, is like the four developmental stages proposed by psychologists, a generalization or ideal pattern that not all individuals will follow. If it does work, and there is only anecdotal evidence for it, it may be an artifact from earlier times, possibly before civilization, when populations were smaller and people were more directly exposed to the cosmic environment.

Planetary returns coincident with the solar return on the birthday vary in distance (measured in degrees of celestial longitude) from the Sun by a small amount. How precise these phase correlations turn out to be for a particular individual may say something about their development and account for variations among individuals. For example, if at the second solar return the Moon is widely square its birth position, off by 15 degrees, but at the fourth solar return it forms a very close opposition with its birth position, say within 2 degrees, this could suggest a longer period of development in regard to the establishment of attachments and security needs, but also a quicker ending. A longer period of attachment may be a good thing, or not, depending on the circumstances of life during this time. Once the basic self and identity is established and shaped by imprints, perhaps by the eighteenth birthday when both Sun and Moon return at the same time, planetary positions in the future passing over these specific planetary positions at birth (called transits) may no longer leave imprints. But, the imprints taken during development may then function as organized nervous system templates. When these templates are activated by transiting planets, a person may find themselves drawn toward situations that replicate in some ways the events that took place during the periods of imprint vulnerability. From this perspective, layers of accumulated experiences timed by planetary recurrences become the basic framework of the psychic self that is constructed by adulthood. The events of adult life continually challenge the strength and value of these fundamental imprints. This may be the basis of at least some interpretations and predictions made for individuals by practicing astrologers.

Some of the above ideas should be briefly expanded on here. Imprints are defined as when sensory information gleaned from an external event/stimulus is embedded or internalized somehow in the developing organism's nervous system as a memory. Exactly how these neural networks are laid down at these times and exactly where in the brain or body memories are located is not well understood. In the DPI model, specific imprint vulnerabilities (critical or sensitive periods) are likely to be activated at solar returns (birthdays) that occur simultaneously with planet returns. This photoperiod information (light) and the aspect (phase) between the Sun and a planet may activate parts of the brain to accept imprints. In a sense this could be action at a distance through some as yet unknown media, similar to

Kepler's ideas on resonance and how an organism "hears" the planets. Or it could be that the sensory system of the organism is picking up fluctuations in the magnetic field generated by planetary gravitational fields, or registering solar activity that is also modulated by the gravity of the planets. What goes on in an organism at the quantum level is a field of study in itself, called quantum biology, that includes magnetoreception in the sensory system.[6] Regardless of a precisely known agent of causation, once a developmental stage is triggered, the actual events and circumstances that are experienced in the external world during a period of imprint vulnerability shapes the framework of a newly established neural/memory network and serves as a structure for identity.

Let's suppose the events that correlate with early triggers (e.g., transits) to the natal astrological chart leave imprints on the developing neurological circuits. This may explain how a transit in later life works: The transit activates the imprinted memories that are then used to select information from the present (including surrogate actors reminiscent of past significant others, as well as other circumstances) and assemble that information into a pattern that can be comprehended and acted on. In this recapitulation can be seen a mechanism of sorts: a response to planetary positioning that activates imprint memories that then generate thoughts and release specific hormones. Such a process would be a quick way to solve problems, and if it had good survival value in the distant past it would have been retained. The activation of an imprint circuit then leads to choices, made mostly unconsciously, and to an observer it may appear to be destiny at work. It all sounds fatalistic, and it may have been so in the distant past, but today the process may be less precise and allow for plenty of wiggle room. Once the self-identity system of a person is up and running (at the Metonic return around age eighteen), and

---

6. Links between mind and matter have been studied by a few scientists (e.g., Brian Josephson) and some findings in parapsychology (psychokinesis) are intriguing, but there has been considerable resistance from skeptics to such knowledge, largely because acceptance of them implies rejection of RMM science as absolute.

the personal and social environment is being successfully navigated, then the system (person) may become self-conscious and thus begin the long, hard work of growing and cultivating personal awareness and exercising true free will. Self-knowledge and consciousness-raising could then be seen as a kind of unraveling of past imprints in the subconscious that are moved into the conscious mind to be redigested. But to be perfectly clear, this is just a speculative hypothesis with only observational anecdotal evidence derived from a limited number of cases.

In summary, I'm suggesting that at least a part of astrology may have a biological basis, one shared with other life forms. Life has evolved in an environment of photo, tidal, and magnetic signals, and life has used these as structures and grids to run biological processes and also to build a self. The self is a composite of behaviors that functions like a system and, like all self-organizing systems operating far from equilibrium, it is sensitive to very subtle influences. The development of individual identity in humans may then be a byproduct of events and social interactions during childhood, which were internalized during periods of imprint vulnerability on a schedule clocked by photoperiod and phase information that was transmitted via electromagnetic signals that were modulated by the Sun, Moon, and planets. The results of this multi-link causal chain make for a variety of individual types shaped by the astronomical and social environment possible, and in the context of evolution by natural selection, this serves to better adapt the species to its environment over time. While all of this may sound mechanistic, it is certainly not a variation of the behaviorist blank slate view, which is. This is a hypothetical organicist model of the internalization of the periodicities of the temporal environment and its use by the organism to function in the world in its own unique way. This model is likely not limited to humans. It's possible that other primates (and other organisms) with different developmental periods may utilize correlations between planet cycles and photoperiod in similar ways, and as evolution brings changes over long periods of time, these timings would shift or be used differently.

The DPI model is a perspective on personal identity development that describes in modern terms possible interconnections between the macrocosm and microcosm. The science is, of course, in the details, but if it turns out that there is something to this model, it should add to knowledge that life internalizes the sky.[7]

---

7. On the other hand, this DPI hypothesis hasn't been of much interest to astrologers who, for the most part, regard it as an inadequate explanation when considering the full scope of astrology. Given that only five points in the astrological palette are used in this model, as it is presented here, it does leave unexplained the roles of the other planets and points commonly in use (I have addressed this issue in my writing (Scofield 2001)). I think the hypothesis is worth considering, that explaining all of astrology in one stroke is asking a lot, and it may be more productive, initially that is, to tackle it in parts.

13

MARGINALIZED

Astrology is either an ancient and valuable system of understanding the natural world and our place in it with roots in early Mesopotamia, China, Egypt and Greece, or complete rubbish, depending on whom you ask.

LINDA RODRIGUEZ MCROBBIE,
IN *SMITHSONIAN* MAGAZINE

Society is very forgiving when the error is on the side of the reigning metaphysics; a virtuous cycle that tendentiously maintains its ruling status.

BERNARDO KASTRUP, *BRIEF PEEKS BEYOND*

By now anyone who has made it this far will be in a position to think in a reasonably informed way about both factual and possible solar system influences on terrestrial systems and people. Most of the material covered in Chapters 1 and 2 is found in the modern sciences, though the information is scattered about in subdisciplines, most of them closely related to the geosciences and biology. This multidisciplinary collected body of knowledge also uses modern terminology and modern physical models of how things work. But closely related is the tradition of Western astrology, which also tracks natural phenomena such as photoperiod (zodiac), diurnal cycles (the houses or domiciles), synodic phases (aspects), tidal maximums and minimums, seasonal cycles in solar radiation, and more. A serious inquiry into the subject will reveal that astrology is a systematized and rational methodology that appears to be designed for assessing solar system influences on terrestrial systems (and people). It has persisted in Western culture in spite of fierce opposition from the defenders of divine authority and unnuanced human free will and modern RMM science.

The results of centuries of anti-astrology campaigning are apparent. Any astute observer knows that dominant institutions of the Western world—that is the media, the government, educational institutions, and the scientific establishment—all regard astrology as a bogus practice that is not to be taken seriously. The subject is marginalized, rejected, or treated as entertainment by these pillars of culture. I've argued in this book that the reasons for this state of affairs is best seen in the terms of history. Rejection by entrenched religious institutions and modern liberal beliefs of human autonomy are mostly about issues of free will that go back millennia. The RMM scientific viewpoint rejects astrology for its apparent lack of evidence and acceptable mechanism. And then there is the tradition of badly behaving rogue astrologers and poor representation of the field by astrological spokespersons. These problems, which are both historical and current, are important, and they deserve consideration in any realistic discussion of astrology's outsider status and what might be done about it. But it is one thing to stand outside and point at the subject with a stick and another to actually enter the world of astrology itself and see what's going on in it; an apparently dangerous exercise because it is never attempted by the critics of astrology. To compensate for this missing data in the public record, in this chapter I will put on my anthropologist hat and, doing my best to be fair and balanced, report on the subculture.

INSIDE THE ASTROLOGY WORLD

While millions of people follow astrology online, I would estimate (very roughly) that there may be at least one hundred thousand people in the United States who identify as astrologers. This is only a guess on my part as this kind of data doesn't seem to exist or is very hard to find. However many there are, these people don't use the term astrologist to describe themselves, though some outsiders continue to use it. (I've never known a psychologist to use the word psychologer to describe themselves, either.) It's likely that the adoption of the term astrologer parallels the astronomy/astronomer convention. That being said, the vast majority of people who identify as astrologers are amateurs, that is they love the subject and make it a lifetime learning project. Some keep to themselves, learning from books and the internet, while others learn from teachers and socialize with those who share their interest. During the twentieth century, most astrology teachers held classes at their homes, or sometimes an office, and some taught community night classes held at local schools. With the internet now a part of modern life, more and more teaching typically takes place online, and many do this using group meeting software. After a few classes, most amateurs will attempt to do some chart readings, usually for friends and family, and will find that "astrologer" has become a part of their identity. Many will become involved with the offerings of online astrology websites, and a small percentage will attend astrology conferences where they can meet with others who speak the symbolic language of astrology fluently. Some may become confident enough of their abilities to charge money for readings. There are also some professionals in other fields, many holding advanced degrees, that study astrology on a very technical level and use it for personal purposes including self-knowledge and decision-making. These people are generally very discreet about their interest in the subject. I would guess that of those who actually study astrology seriously, maybe twenty thousand practice it and a quarter of those make a full-time living at it. These rough estimates are probably rising steeply as astrology continues to spread via the internet.

Professional astrologers are those who manage to earn steady money from doing astrology, a challenging prospect as it turns out, and this selects for personalities that are mentally focused, resistant to criticism, can work for themselves, and are comfortable with self-promotion. Mastering the use of complicated astrological software and maintaining a website and social media pages are now necessary skills, in addition to knowledge of the subject itself. Income is generated by offering various kinds of chart readings at rates that are similar to those of psychotherapists, though the astrologer does most of the talking while synthesizing a considerable amount of data. There is also time spent in preparation, although it is less today compared to the past, since the required astronomical calculations are now done far more quickly with software. Some professionals augment their income through teaching, publication royalties, and speaker fees, all of which also serve to advertise their practice. A few write or sell software, offer apps to download, publish their own books, host a podcast or other media show, or sell their predictions in an annual publication. Jobs for astrologers are almost non-existent, a factor that selects for highly independent personalities capable of running a sole proprietorship. While most professional astrologers work at success in business, those with natural charisma and good communication skills are also able to rise to higher social levels and attract followers. A best-selling book or wide internet presence may be all they need to become established.

There are a number of professional astrologers who can hold their own with professionals in other fields. Some are excellent consultants who can deliver to clients respectable and very nuanced forecasts, assessments of their life path, and suggestions for timing events they wish to initiate. Their accuracy is their best advertisement, and word-of-mouth drives their business. Others are counselors, some with psychology or social work degrees, who can pull relevant information out of a birth chart, often going far deeper and much faster than a psychotherapist, and both guide and coach clients through life-changing events. (It has been my experience that many psychotherapists see astrologers when they need to deal with their own problems.) Some have gone deeply into the history of astrology. Robert Hand, author, lecturer, software developer, and practitioner in astrology, who we met earlier in the book, earned a doctorate in Medieval studies and wrote his dissertation on Bonatti's use of astrology in warfare. He has attended academic conferences, and unlike the highly specialized professors that make up the very small community of those who study the history of astrology, he actually knows how to do the subject. English scholar and astrologer

Nicholas Campion created an academic postgraduate program on cultural astronomy at the University of Wales (at the Sophia Centre). A master's degree can be earned by taking courses on the astronomical and astrological components of archaeological sites, historical figures and movements, and also modern culture. Related topics are presented at regular conferences that bring to the Sophia Centre academics from other countries. Lee Lehman is an exemplary astrological practitioner who has also taught and lectured on astrology at a high level. She has a doctorate in botany and has been reconstructing the traditional Western medical system (the European equivalent of Ayurvedic and traditional Chinese medicine), which involves the use of herbs and other remedies and practices, but is informed by astrology. Richard Tarnas is a college professor, cultural historian, and astrologer who authored a popular college textbook on European intellectual history and also a comprehensive book that considers astrology as a legitimate historical tool. He is an advocate of archetypal astrology as a proper astrological metaphysics.

There are a few national and international astrological organizations in existence. Most were established in Western countries during the twentieth century, the earliest ones in England. In 1938 the first major American organization, the American Federation of Astrologers (AFA), was founded in Washington, D.C., and later moved to Tempe, Arizona, where it has remained. For many years the AFA has published a member's journal, maintained a curriculum with exams and certificates, and held conferences. Today it is more of a publisher of books on astrology but also continues its certification program, one that requires basic astrological chart calculations. In the 1970s, two research-oriented organizations were formed in the United States: the International Society for Astrological Research (ISAR) and the National Council for Geocosmic Research (NCGR). Research in astrology, such as the studies of Michel and Francoise Gauquelin, was publicized and encouraged by these organizations, and astrology began to receive more attention from people outside the field. From a broader perspective, astrology was a part of the consciousness-raising trends of the 1960s (along with yoga and eastern spirituality, progressive political awareness, Native

American practices, etc.), and its integration with Jungian and human-istic psychology moved it closer to mainstream subject matter.

The reaction to this rapid increase of interest in the subject was a hostile condemnation by militant skeptics, including the signing of an anti-astrology statement by over one hundred "leading" scientists and the formation of the Committee for the Scientific Investigation of Claims of the Paranormal (CSICOP). In particular, the validity of the Gauquelins' studies (see Chapter 10) became the focus of a lengthy debate, one that went well beyond the level of scientific understand-ing (and interest) of most astrologers and one that continues to the present in highly specialized journals. (The orthodox reductionist-materialist worldview of CSICOP would suffer enormous damage should the Gauquelins' work be deemed legitimate.) Both ISAR and NCGR began with sincere efforts to improve astrology, but the skeptic crackdown had big effects. To survive, these organizations shifted the focus from research to subject matter that was of interest to amateur and professional consulting astrologers—basically natal chart inter-pretation. Other groups were formed since the 1960s, some local or regional and also a few more national or international groups. Every few years a large conference, the United Astrology Conference (UAC) is sponsored by a group of organizations and attracts up to twenty-five hundred attendees. I should note that the gender distribution at most astrology conferences favors women over men; I would estimate that females make up roughly 80 percent of those attending conferences. This ratio (similar to that in the medical and health fields, educa-tion, and social sciences, as determined by academic degrees granted) may also be reflected in the larger population of people who iden-tify as astrologers or hold a serious interest in the subject (American Academy of Arts and Sciences).

During the twentieth century a number of small astrology schools, almost always based around a single personality known for their compe-tence, came and went. Around the turn of the century the first online schools of astrology started up. One of them, Kepler College, attempted to make a real subject out of astrology by developing a curriculum that taught its history and philosophy, the astronomy and mathematical calculations that underlie the astrological chart, and the astrological methodologies of other cultures. For several years Kepler College, operating online but with a week of in-person classes in Seattle each semester, was very much a legitimate college, more academic than anything previously seen in the field. But enthusiastic support from within the astrological community, which is nearly all practicing or aspiring chart readers, was lukewarm. In addition, there was heavy skepticism from accreditation boards who were ignorant of the subject and not interested in looking into what was actually being taught. Nevertheless, Kepler College managed to hold accreditation in the state of Washington for a few years, something no other astrology school has done. But with a U.S. government crackdown on the rapid proliferation of online colleges, and the outrage of some in the academic community over its very existence, Kepler was forced to retreat into becoming more of a trade school that offers mostly classes on natal chart interpretation, and also courses on metaphysical and esoteric studies, to make ends meet. Kepler College and the International Academy of Astrology (IAA) are today the largest online schools that offer a wide variety of classes and, for those who complete a prescribed program, they both offer certificates of completion.

Certification in astrology is a controversial topic, and most who identify as astrologers ignore it or are unaware of it. The larger organizations offer certification exams based on what they believe constitutes a proper curriculum in the subject. Most of these fixate almost entirely on natal astrology and two (AFA and NCGR) also have chart calculation requirements. Until the advent of the home computer astrologers had to convert the time, date, and place of an event or birth into an astrological chart and other data relevant for interpretation, a process that could take hours. For centuries it was trigonometry that was required to do this, and as I've said before, it could be argued that it was the needs of the astrologer that was a driving force in the development of this branch of mathematics. During the seventeenth century logarithmic tables began to replace tedious trigonometry calculations, simplifying the work, and by the twentieth century the construction of an astrological chart had been reduced to mostly addition and subtraction using interpolation tables, a process that greatly diminished any real understanding of the process. For a decade hand calculators sped up the work, but with the coming of astrological software in the late 1970s a chart could be produced in less than a second. This change led to a dilemma. Should chart calculations be eliminated in certification exams or should they be retained in some form? Elimination has been supported by probably a majority of contemporary astrologers who are either math-phobic or just don't want to be bothered with the work. Retention is supported by a minority who believe there is value in understanding the astronomy of the solar system and the movements of the celestial sphere. This same problem has been faced in surveying and nursing: both fields have accepted modern technology that eliminates previous calculations, but both chose to retain at least some elements of their traditional working knowledge.

Regulation in astrology today has no teeth. Distinguishing between amateur and professional astrologers is not easy as anyone can print up business cards or build a website and call themselves an astrologer. When someone in the media wants to do a story on astrology, they often settle on interviewing people who have a strong internet presence but are unknown to members of the certifying organizations. Professionalism in astrology is then largely based on self-promotion, not exam-based credentials. I would describe the situation as a free-for-all that selects for high-energy independent types who don't always cooperate well with others in a group (except as leader). Their strategy is to develop a brand for their particular style of astrology, and then market it. This is to be expected in a capitalist culture, and it has been the case with other self-employed professionals like lawyers. Self-promotion, sometimes shameless, will lower the level of discourse, and this can work against consistent standards in any field.

It is not easy to make a living in astrology. Many professional astrologers do chart readings and earn a modest income but also have a spouse who works, earns a bigger salary, and has job-related benefits. Few professional astrologers make a full-time living from the subject, and even fewer earn more than the average psychotherapist. The internet has opened the potential for earning through astrology, but it also has mostly, but not always, lowered the level of dialogue on the subject. At the same time, it has allowed astrology to be visible to many, many more people than ever before, and the younger internet generations are far more knowledgeable about the subject than their parents and grandparents. While there are some very good astrology websites, many more are built around Sun sign astrology, along with a few tidbits like knowing when Mercury is retrograde and what the latest planetary configurations might mean. Market-driven anything seeks the lowest common denominator. We see similar results in politics.

The actual nuts and bolts of astrology became both simplified and more diverse over the past century. During the Middle Ages and the Renaissance, the form of astrology that developed in Hellenistic times grew to become extremely complex. A look at some of the texts from those times reveals not only very sophisticated mathematical understanding, but also a mastery of hundreds, if not thousands, of astrological configurations and their blendings that were developed to pull out significant details when delineating an astrological chart. These were all based on the five visible planets plus the Sun and Moon. Then, beginning with the discovery of Uranus in 1781, more planets, and also asteroids, were discovered and began to be tested by practitioners to extract information useful for analysis. (Consensus on the astrological meaning of a new astronomical point comes relatively quickly and then remains stable, though refinements may continue to be proposed.) During the twentieth century the procedures for chart interpretation began to change and many older rules were dropped in favor of a more easily absorbed set of methodologies that incorporated some of the findings of psychology. In the 1990s this modernization of astrological methodology was challenged when a few leading astrologers took on the formidable task of translating the many ancient and Medieval astrological texts that existed only in Latin or Greek, all of this being done without institutional support. Today, a sizeable percentage of astrologers now have some grounding in Hellenistic and Medieval/Renaissance (classical) astrology, which is surprisingly complex, and some specialize in it. Another modern trend has been the introduction of the astrological tradition of India into the Western world. At first this tradition (mostly unchanged in its fatalism and methodologies over time) was referred to as Indian or Hindu astrology; toward the end of the twentieth century it was rebranded as Vedic astrology. Its Sanskrit name is Jyotiṣa.

In the first half of the twentieth century a very different kind of astrology was developed in Germany. Alfred Witte, a surveyor, put together a system of astrological analysis based almost entirely on symmetry, that is the symmetrical positioning of planets along the ecliptic. Some of this was based on ideas from Kepler and ancient astrologers, and some of it was original. In Germany, Witte's concepts and techniques, along with those of his followers, are known as the Hamburg school; in the United States, the names Uranian astrology or symmetrical astrology are more commonly used. This system of astrology, which does share important features with traditional Western astrology, is as complete in itself as is Hellenistic or Hindu astrology. In general, this new method of chart analysis investigates planetary positions along the ecliptic (with the zodiac signs serving mostly as place markers) and locates axes (or diameters) around which planetary symmetry is found. A very simple example would be two planets spaced 100 degrees from each other on the ecliptic. In between the two, 50 degrees from each planet is their midpoint, which is seen as a point on an axis that extends to the opposite point in the zodiac. Another pair of planets may be found to also share this midpoint, or a planet may be positioned directly on the axis, such configurations being called planetary pictures. A synthesis of the planetary symbolism, essentially the same as in traditional astrology, is then attempted, and the information is translated into potentialities of expression and also future trends. The methodology of symmetry also attempts to locate in time when an axis may be triggered by a planet passing over it (a transit), which may designate observable events in a life. Very few astrologers practice this kind of astrology exclusively, which requires using specialized graphic tools and software, but it continues to influence the field and parts of it have been absorbed into mainstream modern Western astrology.

A PERSONAL PERSPECTIVE

My personal experience with astrology has been mostly positive. I became interested in it in my late teens after noticing that nearly all my closest friends (males and females) were born within a day of October 10, though some in different years. At the time I was a science major at Rutgers and was skeptical, but having been interested in astronomy as a youth, I realized that this date marked a specific point in Earth's orbit. I then took a look at some astrology books a girlfriend had, along with some I picked up at a used bookstore and, unlike any other skeptic I've ever met, I actually studied the subject. After some experiments with the timing of events I became convinced that astrology was not a superstition, fantasy, or scam; it was a powerful methodology that used a symbolic language to map the temporal environment both subjectively and objectively. It also raised many questions about the nature of reality. Over the next few years I used astrology to study my life and also that of friends and acquaintances (in the process gaining a lot more self-knowledge than my peers). The charts of partners and the astrology of relationships was of particular interest to me. I kept notebooks, observed patterns, and mastered the math required to do more complex analyses of events and trends. I met Ken Negus, head of the German department at Rutgers, who was an expert on Goethe. While writing a book on German Moon poetry, he had researched astrology in order to better understand his subject and came to the same conclusions about it as I had. Negus got involved in national astrology organizations, founded the Astrological Society of Princeton, New Jersey, and believed astrology was in desperate need of a standardized curriculum. In 1979 he hosted a national conference on astrological education (I was a delegate) that produced a tentative framework. The results led to the creation of the NCGR educational curriculum that today, in the form of the Professional Astrologers Alliance (PAA), offers certification at four levels of competency.

From about 1975 to 2008 I had a full-time private practice and managed to make a living from astrology, allowing for a family, house, car, vacations, and no debt except a mortgage. I worked with a wide range of clients that included a few very wealthy people, teachers, housewives, middle-class business owners, starving artists and musicians, corporate white-collar workers, and quite a few psychotherapists. I wrote hundreds of articles and some books, worked on astrological software with Astrolabe Inc., one of the earliest software companies, taught classes at my home and spoke at conferences. There was always pressure on me from clients, and sometimes the media, to make predictions. It has been my experience that astrology is quite good at assessing the future by producing useful descriptive and timing information. Predictions, however, are another matter, and every astrologer knows that exact predictions are impossible. Even getting close to a reasonable sketch of a future event requires knowledge of other non-astrological variables and conditions including probabilities and scientific details, in addition to the complexities of astrological calculations needed to establish the parameters and timing. Unfortunately, the demand from the public for a prediction, and also the need for a self-employed astrologer to boost their business, is so high that many astrological predictions of low quality become widely circulated. A reasonably accurate forecast also depends on accurate data and the ability to properly analyze it. A vast amount of information is contained in a single astrological chart, and synthesizing it requires much knowledge and years of experience. The perfect storm happens when the media asks a relatively inexperienced astrologer to quickly pick a winner in an election, or some other contest, where the birth times of the competitors are unknown. Given that an error of four minutes in a birth time can amount to an error of a year in the timing of an event in one's life, the probability that any forecasts will be spot-on with an uncertain birth chart drops steeply. But the pressure to perform is constant and many astrologers will take the bait. I learned early on to keep my mouth shut if I didn't have confidence in my data or was asked for outcomes in regard to very complex situations that would require many unpaid hours to properly analyze.

I also studied history and non-Western astrology in order to better understand how the subject had developed in various cultures and what were the differences and commonalities. Mesoamerican astrology, the astrology of the Maya and Aztecs, was a special interest, and I began to speak on it at conferences. Once, at a conference in Dallas, I gave a talk titled "Introduction to Mesoamerican Astrology." Only a few people attended my talk, and afterward one of the organizers of the conference (known for a series of instructional books titled *The Only Way to Learn Astrology*) reamed me out for the title of my talk. "It is too boring," she said. I replied (with tongue in cheek), "OK, how's this: Sexual Secrets of the Aztec Astrologers?" She said that was a great title—and I did use it for the next several years and it did draw more people. It also made me more cynical about hype and commercialization in astrology, a necessary pathway when institutional support is almost totally nonexistent. An effect of this is that most astrology conferences are not very academic, if at all, and that guarantees a better turnout for the event. In general, featured topics, many dramatized with exciting titles, are directed toward the improvement of natal chart interpretation skills, understanding relationships with astrology, current political events, and how to do forecasting.

In the social world of astrology today there is little interest in criticism, theory, or research. I once gave a talk at a conference titled "What's Wrong with Astrology?" that riled a few people, and was never asked to give it again. Some years later I put together a talk called "What is Astrology?" that looked deeply into the subject from both philosophical and scientific perspectives. The first time I gave it was for a philosophy class at Northeastern University, and the response was good—I was asked a lot of very thoughtful questions. Another time I gave it at a national astrology conference. The organizers of the conference put me in the least desirable time slot and I consequently had few attendees (though they were also attentive and had good questions), something that counts for invitations to speak at future conferences. A few years later, at a major conference attended by about two thousand people, my talk on the history of astrological chart calculations drew only two attendees! This is some track record, but I don't feel rejected because I also speak on popular topics that draw respectable numbers. Along the way I have made many friends in the world of astrology, a cultural bubble that is full of very intelligent people, is largely female, leans politically progressive, often avoids internal disagreements, is protective of colleagues, tends to be philosophically vitalistic and Neoplatonic, values intuition, and is not all that interested in science except where it might support astrology.

In my mid thirties, I attended Montclair University part time and earned a master's degree with a focus on the history of science. I liked learning and wanted to go further, but family life and my equally strong interests in mountain hiking and playing in rock bands kept me busy for the next fifteen years. But at around age fifty I was recruited by the late microbiologist and evolutionist Lynn Margulis who wanted me to teach her class Environmental Evolution when she traveled, though doing this required me to get a doctorate in her department—geosciences. For a dozen years or so I worked with her at the University of Massachusetts, first as a graduate student and her teaching assistant, then briefly as an adjunct. For my Ph.D. I focused on paleoclimatology, which I found very interesting, and I got along well with the professors and other graduate students. While at the university I was discreet about my long history and continued interest in astrology and rarely discussed it, except with a couple friends I had made there.

A few years into my program, Lynn, who knew very little about astrology except that there were a disproportionate number of Scorpios in her extended family, outed me in the department. She told me that for a couple years she had been bringing up the subject to individual faculty members at UMass and other scientists she knew in order to gauge their response. She said she found that nearly all of them would become agitated (in some cases apoplectic) at the mere mention of the word *astrology*. She would then grill them on what they actually knew about the subject and found that all were completely ignorant. Interestingly, it turned out that the head of the geosciences department at the time had a mother who was an astrologer, and he was clearly a bit embarrassed about it. At one point he called me into his office and let me have it for bringing up astrology in a presentation on astronomical cycles I gave in a weekly seminar on evolution. I responded by accusing him of being ignorant of the subject and completely reliant on assumptions. I seem to remember saying he should listen to his mother. (To his credit, he did apologize to me the next day in an email.) During the entire time I worked at UMass (mostly in geosciences but also in biology and one semester in the astronomy department) no one, except for my two friends in the Margulis lab, ever asked me a single thoughtful question about astrology. Apparently, curiosity is blunted where everyone believes that they already know all they need to about the offending topic—or somebody they trust knows somebody they trust, and so on. Lynn told me she thought I was among the most scientific people she knew because I was curious about nature, patient, and methodical and would withhold judgment on topics until I thought I knew enough to have an informed opinion. In the end, I didn't really fit in that well with academia. And at the same time, I was an odd case in the astrological community. This never mattered much to me as my identity was diverse—hiking, backpacking, skiing, playing in rock bands, and building and fixing things are activities that have always been satisfying for me personally and socially, though I had to curtail my outdoor sports interests after a bad accident in 2008 on my second Saturn return (at the age of fifty-nine).

THE WAR ON ASTROLOGY

Astrology as a subject is stuck between a rock and a hard place. The public is not very clear on what astrology is and what astrologers do because the media treats it as either entertainment or delusion. One thing that keeps astrology marginalized is its dumbed-down and highly visible popular feature, ubiquitous in print and internet media—the "horoscope" columns. As I've said before, these columns are non-threatening generalizations for publishers who wish to stimulate their readers' self-interest. They are to the real meat and potatoes of astrology as the "Dear Abby" column is to the field of psychology—but they are the dominant face of astrology in mass media and an easy target for attacks on the subject. Unlike "Dear Abby's" advice, which is also non-threatening, there is real concern among those in the culture striving to preserve their version of rationality and their sacred, though superficial, beliefs in free will.

Every so often a study is published that measures the percentage of people who "believe" in astrology. Recent surveys indicate that roughly 30 percent of Americans believe in astrology, with higher percentages among women than with men. The problem here is that *belief* is a loaded word and to use it reveals that both the authors and the participants of such studies are colluding in sustaining a kind of ignorance built around their own notions of what is rational and what is not. Whenever belief in astrology is on the rise, there is alarm that irrationality is increasing; a fear that doesn't seem to be equally applied to religion with its own very strange, imaginative, and sometimes dangerous notions. The National Science Foundation's 2014 Science and Engineering Indicators study found a decline in the percentage of Americans who consider astrology to be unscientific, which it is, if you only consider it as a practice. However, the same is true for many brands of psychotherapy that use methods that haven't been thoroughly studied but get a pass because the results are thought to be positive. Nobody seems to be studying the results of astrological consultations.

Asking people if a subject is scientific or not may not be the best question. Most people don't know the difference between science, technology, or engineering and don't realize the extent to which medicine and psychology base many of their products and methods on studies that don't hold a candle to those of Michel and Francoise Gauquelin. Critical thinking is not to be expected from a culture in which a large percentage of people can't distinguish fake from factual news in media, buy into conspiracy theories, have embraced truly ridiculous beliefs about human exceptionalism, vote for delusional con artists, and believe they know more than virologists and climate scientists.

Consistency in critical thinking is not necessarily found in those self-appointed defenders of science that have a large media presence either. Bill Nye, the Science Guy, is a mechanical engineer who comes down hard on astrology, but along the way he displays extraordinary ignorance. In one of his shows he points out that, while people may think they are born under one of the zodiac constellations, the astrologers are actually off by one whole sign because of precession. But, as I've said elsewhere in this book, nearly all Western astrologers don't use constellations, they use signs that are 30-degree sections of the ecliptic measured from the equinoxes and solstices, and most of them know what precession is. Yet his argument, based on a major misunderstanding, is delivered with confidence. He then argues it is stupid to think there are just twelve kinds of people, which is true, but this also shows how very little he really knows about the subject as no competent astrologer believes that either.

Neil deGrasse Tyson is an astrophysicist who also throws punches at astrology. In addition to routinely bringing up the tired "zodiac signs are wrong" argument, he also draws on a statement attributed to Carl Sagan, that the gravity of the obstetrician at a birth is much greater than that of Mars.[1] The implication is that he knows how astrology works (gravity), yet elsewhere he and others argue that astrology's problem is its lack of a mechanism. Apparently, he can have it both ways. More often deGrasse Tyson criticizes astrologers themselves. In one interview he rightly pointed out some errors in an astrologer's assumptions about lunar eclipses occurring at the deaths of the Kennedys. I would agree with him that the astrologer was making wild assumptions, but I would also see this as a judgment of the entire subject based on a single person who, for one reason or another, managed to represent the entire field at that moment. I've already pointed out that it is mostly the astrologers with connections or the loudest media presence, not the really qualified ones, that get selected to be in a position like that of this person. DeGrasse Tyson also refers to the fact that most people can't distinguish between a "horoscope" written for their sign or for one of the other eleven, which, due to the low level of self-knowledge in our culture, is true. But competent astrologers also know about the Barnum effect and keep that issue in mind when doing counseling, educating the client along the way. He also says that the Babylonians projected their culture onto the stars and people today shouldn't be taking these myths seriously, which implies astrology is basically beliefs, like a

---

1. Apparently, the gravity of Jupiter is a bit higher than that of the obstetrician, according to Shaun Dychko on the online discussion on the forum "College Physics Answers." Chapter 6, Question 39PE. (accessed 2021).

religion. While I respect Nye and deGrasse Tyson because they promote reason and science, they are way out of their element when it comes to understanding astrology. But they have the podium.

While it is mostly science popularizers that attack the subject, a few big names in science have made their views on astrology known. One often quoted is evolutionary biologist Richard Dawkins, best know for his reductionist hypothesis of the so-called selfish gene. Dawkins, like other critics, not only shows his ignorance of the difference between signs and constellations by assuming they are one and the same, but he also criticizes their naming, which occurred millennia ago, this being a topic in archaeoastronomy, not astrology. His advice can be found all over the internet on quotation websites: "We should take astrology seriously. No, I don't mean we should believe in it. I am talking about fighting it seriously instead of humouring it as a piece of harmless fun." This is a statement of absolute intolerance delivered sideways. Here's another one: "Astrology not only demeans astronomy, shriveling and cheapening the universe with its pre-Copernican dabblings. It is also an insult to the science of psychology and the richness of human personality."[2] Dawkins, himself apparently immune to criticism, broadcasts his uninformed opinions with supreme confidence.

The same attitude is found in the humanities. Dorsey Armstrong is an English professor who recorded a series of excellent lectures on the Black Death for The Great Courses, an online lifelong learning community (formerly branded as The Teaching Company). When she comes to a consideration of what the leading intellectuals of the time thought were the causes behind this devastating event, she notes that they pointed to a conjunction of planets that were thought to produce corrupted air (miasma). Then she gets very serious, looks straight at the camera, and authoritatively lectures her listeners on the wrongheadedness of astrology, calling it "bunk." Because doctors used

---

2. Anti-astrology quotes from Dawkins are found all over the internet in websites that offer inspirational quotes but don't specify where they came from. I believe some of these were taken from interviews, however.

leeches in the Middle Ages should the entire field of medicine also be considered bunk? With authoritative pronouncements on astrology, like these coming from the best academics, it takes a lot to buck the tide.

One more example of a scientist who gets away with uncritical thinking is "Everyday Einstein" Sabrina Stierwalt, Ph.D. This astrophysicist and host of the *Everyday Einstein* podcast called *Quick and Dirty Tips,* makes all the usual false assumptions about astrology and preaches her beliefs to the public, in her case aimed at the eleventh-grade level. Right off the top she states that the signs are off by 30 degrees, claims studies show astrological predictions to be as accurate as random guesses, and explains that there is no science to support astrology. Apparently, she doesn't seem to know astrologers don't use constellations and doesn't say what studies have tested predictions (I know of none that are rigorously scientific). As for science that supports astrology, she doesn't mention the Gauquelins, but she does cite Carlson, which, as described in Chapter 11, is hardly an authoritative study because it was a biased experiment with many design and execution problems.

Stierwalt suggests that belief in astrology leads to authoritarian values. She takes this notion from anti-fascist philosopher Theodor Adorno who linked newspaper "horoscopes" with psychological dependency (Kumar 2012). I find this assertion hard to believe given that few who read these take them that seriously, at least in my experience. It also contradicts the fact that the astrological community is predominately liberal politically, some of whom need to write pop astrology "horoscopes" to keep bread on the table. And, staying on Stierwalt's level of understanding, I wonder if fortune cookies have the same power over people who eat regularly at Chinese restaurants. Stierwalt also seems to know with certainty, and this judgement is delivered authoritatively but without evidence, that astrology fails to predict outcomes any better than random guesses. Her last arrow is the accusation that astrology has no mechanism, which, at present, is true. But ideas on this matter exist as discussed in Chapter 12. Hasn't this problem been true of most processes in natural science, many of which have been discovered only after decades of well-funded research carried out by the best minds? The mid-nineteenth-century challenge of figuring out exactly what heat is and how it relates to work is just one example. Presumably, heat existed before it was measured and explained.

All of the above pronouncements on astrology continue a modern anti-astrology tradition that can be traced to the late 1960s and 1970s when astrology was growing, improving, and gaining more respect, at least among those in the counter-culture. As was previously described in Chapter 11, a statement signed by 186 "leading" scientists (including 18 Nobel Prize winners) publicly made it known that they objected to astrology. This statement, sent to every newspaper in the United States and Canada, was followed up with a number of articles in various magazines aimed at debunking the subject by labeling it pseudoscience.[3] The statement itself is authoritarian and not well-informed. Philosopher of science Paul Feyerabend compared it to the *Malleus Maleficarum,* the sixteenth-century book on how to spot a witch, and said that these objections really show that scientists are willing to pose as authorities on subjects they know nothing about. Other skeptics, stimulated by the witch hunt of the moment, worked on articulating demarcations between science and pseudoscience. Karl Popper was, and is, often cited as having solved this problem by defining a scientific theory as one that is falsifiable, or refutable. Every test of a theory is, indeed, an attempt to falsify it. Theories such as those of Marx and Freud, which have great explanatory power, are not falsifiable because they cannot be tested, so they cannot be considered scientific theories. The same is true of astrology, that is astrology as a practice, and also of psychoanalysis and some forms of medical practice. Astrology and medicine may both use scientific data in their work, but actual practice is loaded with variables that are hard to pin down and mathematically model. Popper's definition also serves to reinforce a definition of science as exclusively reduction-

---

3. This event coincided with the formation of the Committee for Scientific Investigation of Claims of the Paranormal (CSICOP), which continues under a new name (Committee for Skeptical Inquiry [CSI]) to fight selected forms of irrationality. The statement was published in the American Humanist Association's newsletter and is found in "The Philosophy of Science and the Occult" (Grim 1982).

ist, and this implies that it is also materialist and mechanistic, as those qualities are needed for the definition to work. Without measured units of a substance placed in a narrow context, testing that can falsify is not possible.

The attack on astrology by 186 leading scientists was successful. Although the battle over astrological "irrationality" was won in part by cheating and ignorance (see Chapter 11 on the Gauquelins' Mars effect), the subject's rise during the late 1960s and 1970s was nipped in the bud. To this day astrology remains stigmatized, and science popularizers and militant skeptics diligently keep it that way.

Accusations of deception and gullibility are routinely aimed at astrology. Self-fulfilling prophecy, where someone makes a prediction or an interpretation from a birth chart and then it somehow happens, is believed by critics to explain astrology. The idea here is that prediction becomes true because of wishful thinking and the Barnum effect. The mechanism of astrology is then explained as the power of suggestion on the subconscious, so anything that starts with a birth chart and turns out to be true is really just a case of self-deception. This probably happens in some measure with a small number people who dabble with astrology, and for professionals, suggestion can sometimes be an issue in regard to dealings with clients. However, it is very easy to bypass this frequently used putdown by observing others who are completely unaware of astrology. A reasonable way to see for yourself if astrology works or not is to closely observe the lives of prominent public personalities (if their birth data is known and accurate) or (with their permission) friends or relatives. With objective observation the Barnum effect is effectively canceled. I do think there are many people who are very susceptible to suggestion and probably should limit their involvement with astrology—and likewise with religion, politics, conspiracy theories, and a few other subjects. Others who are more experienced, knowledgeable, and detached are, if at all, far less affected by this problem.

Another tactic of astrology skeptics is to focus on failed predictions. To be fair, we should first acknowledge that economists, pollsters, and also weather, stock market, finance, sports, and fashion forecasters have far from perfect records. But these are, for the most part, well-funded business, corporate, or governmental organizations that collect and use data and modeling techniques. Their results are usually given as probabilities and estimates. Meanwhile, competent astrologers continue to offer prognostications, many of which have proven to be quite accurate in a way that is uniquely astrological. For example, at astrology conferences in New Jersey, Baltimore, and Boston in the summer of 2019, many months before the Covid-19 pandemic began, I gave presentations that described the year 2020 as a particularly complex year, possibly even a pivotal point in world history. I suggested that a black swan–type event would occur during the first quarter of 2020 and trigger a cascade of problems that would last for several years. Exactly what these problems would be I declined to specify and only suggested that wars, natural disasters, and major social and economic crises were good candidates. However, I said the larger issue was government control over the people. Among other things I predicted Trump would not be re-elected, a woman would play a prominent role in the election, China would continue to rise in power and reach, and so would authoritarianism in general. This presentation was not a research project, just a quick take on the future based on the Saturn-Jupiter-Pluto conjunction in Capricorn that was exact in early 2020. I was far from the only astrologer who made such a call; the Jupiter-Saturn-Pluto conjunction of 2020 and its possible effects were common knowledge in the astrological community and had been discussed for years. This information, if developed in collaboration with experts in world affairs and sponsored by a reputable institution, may have proven useful.

Another way astrology is "proven" false in popular media is by referral to statistics, chance, and coincidence. This usually takes the form where you have a group of people, say U.S. presidents, organized by their Sun signs. It turns out that more presidents were born under Aquarius and Scorpio (eleven) than Aries and Virgo (four). Since the sample size (n = 46) is small, the difference is not statistically significant and is deemed due to chance. Because astrology is nothing more than zodiac signs to the average ignorant critic, "studies" like these conveniently allow the entire subject to be discredited in full public view in spite of serious flaws in design, misuse of statistics, and slanted interpretation. This is why the Gauquelins' findings are so interesting—they test the planets, not the signs, and the results are statistically significant, that is if you take the time to unwrap the mischief perpetrated by stonewalling skeptics that is passed off as critical thinking. So given the vast number of variables to be considered when doing astrology, and the difficulty of reducing these to units, attributing a correlation to chance is by far the easiest way to be done with the wretched subject. And if a correlation is found, there are other ways to kill it. One is with a meta-analysis that will pick up scores of poorly designed studies and sink the good ones. Another more recent, also indirect, critique of astrology is that it is like machine learning in which a known pattern, say a historical cycle of something, is matched with some data and the correlation found is extrapolated into the future. This apparently makes astrology irrelevant because we now have big data and computers and we can get results in that way and not have to deal with planets, even if their astronomical periods are actually buried in the numbers (Boxer 2019).

Scientists are skeptical by nature and this is a good thing because it is the basis of critical thinking. A ground rule for purist RMM science is that correlation is not causation and only plausible causation pathways (like billiard balls) lead to something like certainty in science (which itself remains uncertain and subject to change). This is how the chemical and petroleum industries can often get away with emissions of deadly pollution. Good correlative studies between a chemical and a health problem are often not enough; a causative link between the two needs to be shown decisively to get some attention from lawmakers. Meanwhile, as data slowly accumulates, weird, toxic, and even deadly effects on people and other life forms become more pronounced.

Of course, there are many other correlations in our world today that are taken for granted and even celebrated (like medical studies that match a treatment with an outcome) but lack a completely understood mechanism or significant evidence of causation. Causation can be elusive, and statisticians, social scientists, and philosophers often can't agree on what is certain.[4]

But unlike astrology, most of these don't challenge assumptions about self and individual destiny or mess with the fundamental Western notions of free will in a cosmos presumed to be directed by an incomprehensible supernatural being. Astrology, at least some major parts of it, is stuck between a rock and a hard place, that is, between the demand for an elusive mechanistic causation and the beliefs of Western anthropocentric monotheism.

ASTROLOGY'S INTERNAL PROBLEMS

So astrology is in a situation where it is labeled as pseudoscience, denied serious coverage by the mainstream media, and is routinely attacked by uninformed scientists and militant skeptics. What should advocates of astrology do about this? Probably the first thing would be to stop letting the critics define the subject. It should be made clear that astrology is a subject that includes history, theory, research, and a practice. The assumption that astrology is only a practice, held by both astrologers and critics, is counterproductive to the field. A practice is not a science, so it shouldn't be judged as such. Scientific criticisms should be directed to issues of theory and research, even if these areas are woefully undeveloped and there are only a handful of astrologers who can offer informed responses. In this way the subject itself will be challenged properly and over time should benefit from better directed criticism.

Then there is the matter of what kind of a subject astrology is. Given that astrology has been applied to humans, animals, agriculture, and weather, the definition must be very broad. My suggestion for decades has been that astrology is a subject that develops and works with mapping and timing techniques for self-organizing systems. Ilya Prigogine (1917–2003) called such systems dissipative structures, these being organized phenomena that are operating on the edge of a gradient, cheating

---

4. It is the existence of uncertainty that Big Oil has used to discredit global warming science.

entropy, extremely volatile, and subject to minute forces. These would include a living cell, a multicellular body, the Earth's biosphere and surface (Gaia), a group, team or collective behaviors, the stock market, a hurricane or tornado, neurological activity that organizes itself into a mind, or a set of behaviors operating together (synchronized) in such a way as to create an emergent integrated personality. I'll say more on this in the next chapter.

The mapping techniques of astrology include the astrological chart itself, often inaccurately called the horoscope. This chart is essentially a time-slice from a specific location calculated for time of a phase transition or bifurcation event, such as birth. Astrology produces tools to analyze solar system dynamics, such as plots that represent planetary cycles, and the information gathered is then transferred to the subject material whether it be an event or a person. In other words, astrology uses geometry and graphic techniques to describe and analyze, or map, the behavior and trajectories of what are now referred to as the emergent phenomena of systems operating far from equilibrium. Defining astrology as a branch of systems science could do a lot for the subject conceptually and in regard to its scientific treatment. But many who identify as astrologers (most being practitioners) seem to want to believe that astrology is fine the way it is, even if it is defined as a form of divination. Some astrological fundamentalists would have it reset to its original condition of some two-thousand years ago. While knowing the past can be of benefit, failure to advance the field will only keep astrology out of touch (and out of collaboration) with related academic disciplines like psychology, economics, meteorology, and astronomy. These subjects long ago opted for non-supernatural, data-driven, and science-based explanations for how things work, and now these subjects fill most of the needs that astrology once fulfilled in society.

Rejection of astrology is generally rendered by those who see the subject as failing to conform to scientific standards in a RMM context, believed by many to be the only truth there is. Astrologers, for the most part, subscribe to a holistic-vitalistic worldview, which they regard as truth also. This is a great divide. Recall from Chapter 8 that mechanistic-reductionism and holism-organicism are considered to be mutually exclusive worldviews. Many, if not most, of the working scientists I've met appear to accept reductionist science as the standard model, one that is critical to getting grants and keeping one's job. Not much attention is paid to the fact that this view has some serious ontological problems given the uncertainties of quantum mechanics, the matter of self-consciousness, and the origin of life. There are also its larger negative effects on society like the environmental and social destruction it has sponsored, along with its ally, capitalism. Arguments against materialism have been made by many writers, and I claim no philosophical expertise in these matters, but it appears that not a few thinkers regard the RMM worldview as, at least, problematic (see the writings of Whitehead, Capra, Berman, László, Varela, and Katstrup).

For the most part, astrologers hold to a completely different worldview than that of RMM science. Doing astrological chart interpretation selects for a certain kind of thinking, one that favors intuition, imagination, generalization, and pattern perception. These qualities of thought may also favor beliefs that are impossible in the context of RMM science. For example, reincarnation is held to be a reality by many in the field, but this implies belief in the existence of a disembodied soul. A high percentage of practicing astrologers appear to accept what amounts to a teleological-vitalistic metaphysics. In this view, one shared with archetypal cosmology, the cosmos is animated by a soul-like life force that has the purpose of growing consciousness. What astrology does, in this view, is to detect, or construct, meaning of things. When I first encountered astrology in 1967 it came with a full dose of Theosophy, reincarnation, and soul evolution that was called by authors and practitioners simply "metaphysics." Even today astrology is often described as a spiritual science (which means non-materialistic), and this implies an ontology of monistic idealism.

After a decade or so of being exposed to popular astrological metaphysics I found there were just too many unanswered questions and not enough data to support assumptions, and also that alternate hypotheses or explanations for questionable phenomena like reincarnation existed. I saw that most astrologers were in certain ways anti-science and content with a loose metaphysical ideology built on beliefs that are not testable because there are no units of anything to test. And if you can't test anything you are left with only assumptions and beliefs that are not tightly integrated, as is the case with religion. I was tolerant of this situation because I recognized the basic human need to understand the world and the convenience of belief in achieving that goal, but I wasn't on the same page with most in the field.

By the 1980s my thinking had turned more toward a methodological naturalism with a high tolerance for uncertainty. I learned a great deal about myself and others by studying astrology, and I practiced it as a consultant, interpreting astrological charts for clients as maps of conditions to manage and insights to consider. I used astrology as a way of making life meaningful, but this is not meaning as given from or controlled by the cosmos, it is meaning constructed from the analysis of a focused (time-slice) view of solar system structural dynamics. Astrology applied this way, and the best practitioners do this, provides a working explanatory framework for life navigation, but, as with any wisdom tradition, the user needs to be consciously engaged with the process. Years of experience have taught me that for people seeking self-knowledge, interpretive astrology can produce a far better life script than chance or decisions based on fears, obligations, urges, or other factors. (This use of astrology as a way to create meaning can be extremely valuable and is perhaps the most important thing it can contribute to human society.) And given that astrology also maps the flow of time, it can deliver information that psychotherapy cannot.

My astrological practice was geared toward solving problems and assessing relationships, though at times I had to resist being manipulated by clients seeking to use astrology as a tool for control over others. In this way I was doing the same thing as others in the field, and I also continued to participate in the larger astrological community. At conferences, I was a speaker, but I was also something of a freelance anthropologist, an observer of that social world, engaging in conversation with some very bright and thoughtful individuals all the while respecting others' rights to their beliefs. At no point did I find any evidence that astrology was not an operative force in people and in the world, that astrology only worked if you believed it did, or that I was deceiving myself in any way. Understanding why this was so became my problem, hence this book.

It's one thing to live in a society and hold to an alternative metaphysical ideology; it is another to do business in a strange language. While it is true that many highly specialized subjects like law or the social sciences have produced complex vocabularies to clarify subtleties (and also sustain a knowledge hierarchy), these are simply words used to enhance precision, not symbols that group together a range of ideas. Because astrology uses an ancient symbolic language, insights discovered through it are not easily transferable to other disciplines. An important question asks if the symbolic language of astrology, and its role in what is essentially a discipline of pattern processing and synthesis (the activity of doing astrology), is absolutely necessary for measuring, timing, and analyzing the self-organizing systems to which it is applied. Astrologers would say yes, it is the right tool for the work. They also argue there is beauty in the symbolism and that because symbols are multivalent (operate consistently on different levels or perspectives) they allow for synthesis of concepts, something that cannot be accomplished easily with linear thought. I am in agreement; astrological symbolism is a magnificent artifact of the ancient world that continues to be functional and relevant, at least to those who use it. Symbol use in the practice of astrology as a counseling technique is not the problem. But would newer, more precise names that retain the power of symbols be useful in both doing astrology and communicating with other fields? Yes or no, the planetary symbolism and terminology issue has kept astrology isolated from other related subjects for millennia.

The primacy of symbolism in the practice of astrology is a reason that it has been considered to be an occult subject and associated with tarot, divination, and natural magic. The ability to make sense out of what are regarded as cryptic symbols, it seems to me, is not that different from learning a language. If you start early on, you do much better than if you start later in life when your brain plasticity is diminished. I picked up the symbolism quickly at around age eighteen, not uncommon if you came of age during the 1960s and were drawn to astrology. Most of the best astrologers I know started about that age or in their twenties. In contrast are people I've met later in life who, when introduced to astrology, struggle hard to get it, if they can at all, and this seems to have no correlation with intelligence. Some of the most brilliantly scientific people I know are hopeless when it comes to understanding the symbolism because it is a kind of thinking that is completely foreign to them. This seems to be especially true for those with a physics or an academic philosophy background, and it puts them in a situation where their considerable knowledge is useless. Perhaps this is because such people are inculcated into the reductionist way of seeing things beginning in early childhood. Another possible explanation is that the extremely rational and logical mind lacks the flexibility needed to interpret and blend astrological symbols, which may explain why so many good astrologers are also artists or musicians. Contrast the mental processes of a jazz improviser with that of a scientist. The former is engaged in a constant process of creative synthesis, while the latter is following a set of logical procedures. This suggests that doing astrology is facilitated by a certain kind of talent for pattern processing, what Foucault would say is the ability to see resemblances. So, if doing astrology hinges on the interpretation of a specific set of symbols, then communicating with other subject areas (e.g., social sciences) becomes problematic. As mentioned above, this creates a serious kind of cultural marginalization. The field of astrology will need to make some language adjustments if it is to contribute to the making of knowledge in modern society. In regards to consulting and counseling, this is probably not so important.

The public associates astrology with prediction. Let's assume some information about the future from astrology is possible, then how could such knowledge be improved and made useful to the public? This would require considerable cooperation from astrologers who are used to working independently and protecting their brand. Then there are the technical subtleties and social responsibilities involved in forecasting and timing probable events. In addition, critics hold astrologers to a style of prediction that they (the critics) are familiar with. They fail to appreciate that, while astrology approaches prediction in its own way, it does have the ability to see the larger economic, cultural, and political trends and the ability to place reasonably accurate time constraints on future events. And again, the astrological community is not organized in ways that can regulate predictions produced by individual astrologers, many unqualified, who operate with few restraints. One rogue astrologer with a failed prediction is all a militant skeptic needs to prove astrology wrong once and for all. Further, testing general predictions made by astrologers is no simple matter. What is more important in a prediction: the timing, a thoughtful interpretation of the underlying trends, or the naming of a single specific event?

In response to its marginalization, the social world of astrology is closed and isolated, a culture bubble. In this regard the astrological community would be defined by sociologists as an outsider group. Like most outsider groups that hold to a set of values and ideas not held by the dominant society, astrologers make unique knowledge claims but fail to communicate effectively with the educated non-astrological world. This means that those in the dominant society (the insiders) have no easy way to evaluate the subject. And since the majority of those who identify as astrologers believe astrology is only a practice, those who may actually know something about theory or research are hard to find. In addition, loose standards, questionable distinctions of competence, and disagreements within the field tend to drive factionalism. All of the above reinforces cultural isolation. In contrast, subjects like psychology, economics, meteorology, or astronomy are well-established and require years of formalized study and examination programs for those who seek to go beyond amateur status. Additionally, institutions exist to support career professionalism in these fields. In astrology, which lacks institutions, professionalism mostly means success in the business of building and maintaining a private consulting practice—jobs for doing theory or research don't exist. And in regard to building a practice, the rewards for passing certification exams are minor and this diminishes regulation of the field. There is a circularity here in which an outsider group, astrology, is frustrated by the dominant insider society that controls requirements for acceptance, respect, and participation. This rejection causes the outsider group to rebel against the norms of the insiders, including maintaining a solid system of ranking professionals, and doing so guarantees more rejection.

People who identify as astrologers find themselves in an impossible quandary. Some are concerned with the problems of larger cultural relevance and seek out graduate degrees in other subjects in order to deepen their understanding, but they are a very small minority. The majority mostly work at improving their chart-reading skills and socialize with the like-minded. By ignoring the larger social and cultural situation, the field only sustains a libertarian free-for-all that fuels astrology's drift away from the mainstream academic subjects while continuing to set loose upon the world waves of unregulated practitioners. Some of these astrologers will be brilliant, able to do more in one session with a client than a psychotherapist might do over the course of a year. But most probably won't, and some may further damage the reputation of the field. It happened before in the sixteenth and seventeenth centuries when unregulated astrology, propelled by the printing of almanacs containing prognostications, lowered the reputation of the subject. Even when astrologers produce quality work, there is trouble. The burst of astrology publications during the 1960s and '70s, which promoted astrology as a serious subject, got pushback from militant skeptics who were not, for the most part, actual scientists. The dominant culture took this well-publicized rejection to imply that astrology was bogus, and its association with psychics, tarot cards, and palm-reading was reinforced. Not much has changed since then.

The cultural perception of astrology as entertainment, or fortune-telling, has led to government intervention. I remember when the state of New Jersey attempted to pass a law taxing astrologers by placing the subject in the same category as pony rides and canoe rentals. It took a major letter-writing campaign to get astrology taken off the list. Some cities, like Las Vegas, require astrologers (and card readers, psychics, palm and crystal ball readers, etc.) to be licensed. While most local or state governments consider astrology only as fortunetelling, and may use steep licensing fees as a regulatory tool for what they don't want to encourage, the city of Atlanta does specify three categories (astrology, fortunetelling and handwriting analysis) and currently charges $500 a year for a license to practice. Unless the field of astrology can explain itself in a way that gathers respect in the dominant culture, more of this external regulation is likely to occur. One way for astrology to get some respect would be for a few prominent astrologers, possessing great social and communication skills and the ability to disarm the skeptics, to somehow become celebrities and be seen routinely discussing the subject intelligently on the major networks. This sort of thing usually gets the interest of the pubic and can lead to changes of opinion based on understanding and familiarity.

The obvious, and realistic, alternative to the forms of marginalization described above is to make astrology more rigorous by raising the standards of the field. If this road is taken, astrologers need an agreed-upon definition of astrology as a subject with a history, theories, research, and practice. Limiting the field to the techniques and methodologies required for birth chart interpretation is myopic, like saying the field of medicine is limited to taking and evaluating X-ray, MRI, and CT scans. Another thing that should be done is to pay more attention to certifications and encourage research, which some of the organizations have been doing in small ways. General certification exams should be thorough and be a means of establishing high standards and enhancing professionalism, and these should be followed by further required professional training. There does seem to be a general agreement on what constitutes a core curriculum, which has already been accomplished by some of the organizations and several schools. But there also needs to be wider community support for the upgrading process and an effort to reduce and consolidate the number of general certification exams (specialty exams that measure interpretive skills would be add-ons as they are in other fields). Establishing standards would help to grow educational and professional institutions.

In regard to research in astrology, interest and support are desperately needed. While the 1970s was probably the peak of modern research into astrology, mostly the statistical testing inspired by the Gauquelin's work, this trend was not sustained. Since the 1900's, some academic-quality work on astrology's history has been done and several astrology journals that publish peer-reviewed papers exist, but these trends are not well-known to the field; the typical person identifying as an astrologer reads articles that are on the magazine level. More recently there has been a renewed interest in researching and testing parts of astrology statistically, and a few conferences have been held where findings were presented. There are now also software packages available, specifically written for astrologers, that can handle modest research projects. It is the case, however, that most astrologers are practitioners and their interests are focused on practice. Only a very few of them have an understanding of how to properly research the subject. The main reason that this situation exists is that serious astrology research simply does not pay the bills. Of course, astrology as a practice is important, but while it is true that one must know how to read natal charts to make a living as an astrologer today, it is also true that there is more to astrology and these other sides of the subject will need a lot of attention if marginalization is to be overcome.

What's very sad about the rejection of astrology by the mainstream cultural institutions is that astrology, as I see it, has something valuable to contribute that is more than self-knowledge, counseling, healing wounded personalities, making life meaningful, and predicting trends in useful ways, these alone being an impressive program. Astrology could prove revolutionary in that its mechanisms or processes, once found, may revolutionize our scientific understanding of how nature works on the systems level. Or maybe our understanding of how nature works will have to change first and then astrology will seem more reasonable. I'm sure that the issues of free choice and determinism in philosophy, which do not now include astrology, will be reinvigorated if astrology once again is able to enter the discussion. Will astrology be the wedge that drives Renaissance naturalism back to center stage and ushers in a new "magical" cosmos? I doubt that and agree more with Morris Berman, who argued that a "re-enchantment" of the world will only come when the social implications of systems science are better understood and seen to be a necessary complement to RMM science (Berman 1981). I also don't doubt that in the future a deeper understanding of astrology will drive discoveries, perhaps some suggestive of Platonic notions that cross disciplines and others pointing to processes that add to our understanding of systems, organic and inorganic, and even the cosmos itself.

It is possible that in the future there will be an opportunity for astrology as a subject to rejoin the scientific and academic communities, but this can only happen if it becomes a rigorous and truly professional field with both educated researchers and talented practitioners that coexist, with clear boundaries, along with many more amateurs, much as astronomy does. Or astrology can continue to sustain group norms where everyone is an astrologer, a situation that is probably a reaction to the marginalization described in this chapter. To be fair, most marginalized groups suffer from a kind of societal rejection and adopt detached and defensive coping behaviors. Individuals in such outsider groups find that their well-being is promoted by identifying with their group's unconventional norms, and this then serves to further solidify isolation from the larger society. The astrological community, like most other marginalized outsider groups, is challenged by a range of problems not faced by the larger society, and adaptations to this exclusion often reinforce isolation and misunderstandings. In spite of these obstacles, the field of astrology, which I believe has a great deal to offer the larger society that represses it, has survived, has protected its history, and has evolved. This is an accomplishment in itself.

14

Time-Mapping a System

In the systems concept of mind, mentation is characterized not only of individual organisms, but also of social and ecological systems.

Fritjof Capra, *The Turning Point*

In the reductionist philosophy of science it is commonly assumed that nature is compositional, that small parts build larger structures, living or not living. In that view the notion of emergent properties is controversial if not rejected. Emergence, a somewhat mysterious process that suggests certain phenomena amount to more than the sum of their parts, is deemed not possible perhaps because this immediately reduces reductionism to a method limited to a certain class of applications.

Both the physical and social sciences always have areas of inquiry that are not well-understood. Hypotheses to fill knowledge gaps are routinely proposed, but not until enough evidence piles up does a theory emerge that is better able to integrate the field. Geology had competing theories on mountain building until plate tectonics, developed in the 1960s, convincingly explained that process. Paleoclimatology was similarly transformed by the Milankovich cycles, variations in Earth's orbit around the Sun, that were finally proven with hard evidence from marine cores. Suddenly, climate in prehistory and the cycles of more recent ice ages were explained with one theory. Biology was mostly field observations and descriptions based on morphology until Darwin (and Wallace) made sense of how organisms change over time with a modern theory of evolution by natural selection that has only gotten broader and stronger, even spilling into other disciplines. Grand theories such as these give their subjects integrating principles that create a stable platform for understanding, but they also serve as a hub for more expansive research. A general theory of astrology would go a long way toward pulling the field together, but it can't be one that is based entirely on esoteric and evidence-poor ideas that keep it disconnected from other subject areas. A good place to start, in my opinion, would be to first define astrology as a special case in the field of systems science, one with established practical applications.

It should be obvious by now that astrology does not lend itself to reductionism because it does not study easily reducible phenomena. Things like electrical currents, the reproductive patterns of fruit flies, chemical reactions, and the light from a star can be measured precisely and reduced to units and therefore make great subject matter for reductionist science. Wherever this detail-driven, bottom-up method of doing science can be applied it will ultimately reduce its subject matter to physics. This in turn supports the notion that if it can't be measured, then it can't exist, that measurement and equations serve as explanations, and that physics is complete or will soon be. Problems exist, however, and this approach doesn't work so well in all fields of inquiry. For example, fully-funded, physics-informed behaviorist psychology has successfully mapped out only a tiny portion of the subject matter the founders of psychology originally intended to study. But all can agree that the reductionist approach has been incredibly powerful and transformational in many other ways. During the past four centuries a vast amount of knowledge about the natural world has been made using this method, and engineers have applied the findings to produce useful labor-saving technology, better living through chemistry, the internal combustion engine, pocket-sized computers, and robots.

Not all things in nature are so easily reducible to a formula. Separating one component or another from a subject under investigation, that is starting with the parts, is not always a productive, or even possible, research program for a certain class of phenomena. Ecosystems, as well as cells, individual organisms, and weather systems like hurricanes are phenomena that are not static or easily broken into units. They are energized systems that are recursive, have boundaries, are self-sustaining, and display emergent properties. The study of this less compliant subject material can be found in the domains of multiple subjects, including biology, climatology, economics, and the social sciences. These disciplines have developed complicated specialty areas that grapple with systems and must therefore confront the limitations of reductionist science. This need drives the devising of methods that attempt to quantify and model phenomena in ways that allow for prediction. But the interdependence of the system and its environment, with both in flux, presents problems that are not easily reducible to parts and absolute boundaries or equations. This means standard reductionist scientific analysis of systems, these being phenomena composed of innumerable variables, will be incomplete—unless you believe its only a matter of time before it won't.

Self-organizing systems are phenomena in which a whole appears to be greater than the sum of its parts. They form spontaneously out of disorder, manage entropy by slowing it down, and are sustained by external inputs and internal feedback loops. A self-organizing system is hard to quantify and is the reason that daily weather forecasting is limited, why economists struggle to predict market behaviors, why pollsters are often wrong, why scientists can't create life in the lab, and why medical science can't prevent death. Russian medical scientist Alexander Bogdanov (1873–1928) may have been the earliest to propose a multidisciplinary approach to the study of system-like phenomena, a science he called "tektology." A few decades later biologists (e.g., Ludwig von Bertalanffy) were calling for a new organismic top-down approach to these problems. Scientists from many other disciplines, including anthropologists (e.g., Gregory Bateson), and especially economists and engineers (e.g., Norbert Wiener), also contributed to developing rational frameworks and methodologies appropriate to the study of complex and self-regulating systems. By the mid-twentieth century these new scientific approaches were often referred to as systems science, a generic label used for a group of theoretical and investigative fields that includes cybernetics, emergence, complexity, and chaos.

At roughly the same time that Bogdanov was expressing key concepts of system science in his writings on tektology, physics was being revolutionized by relativity and quantum mechanics. Systems science, which slowly developed as a form of scientific inquiry, was eclipsed by the new physics with its dramatic implications and narrow focus on specifics and equations. Today, systems science is still not a fully developed and widely understood set of scientific methodologies, and its ideas are spread over multiple subject areas and subdisciplines. One reason for this may be that the kind of thinking and style of problem-solving required to do this work is not commonly taught, or worth staking an entire career on. RMM science, which in our time has the podium and the funding, selects for a certain type of mind that compartmentalizes and handles details well. We see this plainly in the culture of specialists that has been produced by the medical field, a culture that has had great success in fixing broken parts but also great failures in grasping how the body works as a whole. In contrast, system science favors a wide-ranging knowledge base and the ability to generalize, think holistically, and synthesize.[1] System science is a broad perspective and top-down approach to making knowledge that considers process to be as significant as parts, recognizes organized patterns and recurrent features, seeks out self-regulating feedback loops, and investigates relations with other systems. The argument of this book is that astrology studies, with its own

---

1. Systems science has been described thoroughly in many publications over the past half century, and I would direct interested readers to other authors who have dedicated volumes to explaining how this alternative and complementary form of science works and why it is vital to the future of humanity, in particular *The Systems View of the World* (Laszlo 1972), *The Turning Point: Science, Society, and the Rising Culture* (Capra 1982), and *The Web of Life* Capra 1997).

methodologies, self-organizing systems. It has produced a set of mapping, timing, and interpretive techniques for such systems that other disciplines cannot match. These techniques are mostly based around the traditional astrological time-slice chart calculated for the time of a recalibration, phase transition, bifurcation, or critical event, but there are also various graphic depictions of planetary cycles that have long been used in practice.

SYSTEMS CONCEPTS APPLICABLE TO ASTROLOGY

Self-organizing systems are responsive to very weak signals in the environment. This is because a system is held together by multiple feedback loops that, over many iterations, can amplify a weak signal. As noted in Chapter 2, some climate and meteorological phenomena are driven by very subtle astronomical factors that include lunar gravity, solar cycles that vary only slightly in radiation output over time, barely measurable orbital cycles that modulate gravitational strength, and quite possibly cycles of galactic cosmic ray flux variations, both short and long-scale. This is the cosmic environment that, in its rhythmic persistence, establishes a subtle, but quite real, temporal framework that floods the terrestrial environment. Darwinian natural selection works on this level as ecosystems and organisms must adjust to the externally generated temporal environment that, like tides and the changing rhythm of day and night, drives physical changes in environmental conditions over time. Organisms must remember these, or they fail. Those that are best adapted to their environment, both spatial and temporal, will tend to increase in number over time compared with those who are not adapted.

The fluctuations of the astronomically driven temporal environment and its physical consequences in weather and climate cycles are also drivers of group behaviors such as migrations, market fluctuations, and politics. These are cases of the entrainment of collectives to the time-environment, not just individuals exercising free choice. That the Sun and Moon move life on Earth is not at issue, but response to planetary motions is. Centuries of anecdotal evidence amassed by the astrologers of the past, and also a few scientific studies, say planets also play a significant role in modulating certain types of terrestrial phenomena. It is this factor that is contested as known forces from distant planets are thought to be too weak to have effects. But when planetary influence is demonstrated, even as a correlation (i.e., the Gauquelins' studies), the contemporary understanding of how life operates in time and within the context of the next order self-organizing system, the solar system, will be forced to make some adjustments.

There are two other points related to the sensitivity of self-organizing systems to relatively small signals or forces that are relevant to astrology. Edward Lorenz, a meteorologist and mathematician who worked at MIT, studied the general circulation of the atmosphere. While testing a few weather variables (temperature, pressure, wind, etc.) in an early computer model that ran simulations of atmospheric conditions (this was an attempt at making long-range forecasts), he found that where he started the program, that is his starting values, made a huge difference in the outcome. Completely different outcomes could be due to very small (in the thousandth place of a decimal) differences. What this implied was that the initial conditions of a test run in the computer model greatly affected the long-range outcome of the calculations, and this was not a good sign for the goal of making long-range forecasts. Lorenz called this "the butterfly effect"—a butterfly flapping its wings in Brazil could set off a tornado in Texas—and it was a cornerstone in the development of chaos theory.

Bifurcation theory is the mathematical study and modeling of changes in dynamical systems, these being physical systems that are studied by mathematicians. What's known is that a small parameter change made to the mathematical framework established for a system has the potential to cause a sudden and much larger set of changes. These would include instability, breakdown, or movement toward another kind of stability, either of these occurring at what's called a bifurcation point. The mathematics behind this process is complex, and the plotting of curves involves a mixture of linear and nonlinear differential equations, the former substituting for the latter as approximations are sought for.

A simple example is called the logistic map, which is built from an equation that shows how a small change in a constant will suddenly disrupt a normal pattern, which then breaks into a complex and seemingly, but not really, chaotic pattern. The logistic map is considered a classic demonstration of non-linear mathematics in describing the evolution of a system, and it demonstrates in graphic form how sudden destabilizing effects, often compared to a heart attack in which a normal regularly beating heart suddenly goes into fibrillation, are triggered by only small changes. Bifurcation simply means branching or splitting in two, and a bifurcation point marks a critical point of instability in a system that can lead to chaos or another kind of stability. Other terms for this kind of change are phase transition or tipping point. The idea here is that self-organizing systems are not linear and predictable, and sensitivity to very small changes can trigger a boundary crossing that leads to branching or chaos.

Ervin Laszlo describes three types of systems: suborganic, organic, and supraorganic (Laszlo 1972). The first includes atoms, which are stable entities having properties that are not the same as the sum of protons, neutrons, and electrons would predict. The second type, organic, incudes cells, organs, and bodies, each being self-regulating open systems with permeable boundaries that allow for exchanges with the environment. Medical astrology, or iatromathematics as it was known centuries ago, is applied to these systems. The activity of the nervous system with all its feedbacks has emergent properties: the mind, the sense of self, and a personality. If the body is subject to planetary influences, by extension, the individual personality, being an emergent property of the body (nervous system), is subject to them as well. Astrology is most often used to map these biological and consciousness-related systems, analyzing and classifying them in various ways using a sophisticated, though antique, typology. Laszlo's third type of system is supraorganic, the networks and structures of human social reality that are built on communications that coordinate individuals. On this level the properties of many individuals—their beliefs, values, culture, and so on—coalesce and become organized and synchronized, forming a boundary and displaying an identity. This type of system (e.g., stock market, a nation state, etc.) is analyzed by what is today called mundane astrology.

The existence of a social reality that is its own self-organizing system has support from scholars in both the life and social sciences. An understanding of such an intangible entity begins with what it emerges from, that is life, and an important concept in the understanding of life, *autopoiesis*. This term, which means "self-making," was coined by biologists Humberto Maturana and Francisco Varela (Maturana and Varela 1992). It is used to describe how a living system (cell, body), which is composed of multiple feedback loops circling back in on themselves, constantly self-organizes in a process that over time displays a coherent pattern. This process of living they regard as equivalent to cognition because the system, being recursive but structurally open, interacts with the environment and selects from it specific information and materials needed to maintain itself. Evolution has produced a number of collective cognitive networks, bee and ant colonies are good examples, but human language makes for highly efficient exchanges of ideas and information used to coordinate group behavior. Social networks, which are made of a shared body of knowledge activated by communication exchanges, also develop an identity and sense of belonging, as well as boundaries in regard to meanings and expectations. Such networks might also be viewed as a synchronization of minds.

Maturana and Varela were concerned with the issue of whether human social systems are autopoietic in themselves, or are they just a medium in which biological systems interact and become linked. Sociologist Niklas Luhmann also looked at this question and concluded social systems do have properties of living systems. He developed a social systems theory based on the idea of communications between people forming a distinct autopoietic system, one that processes meaning and is separate from the environment.[2] The social system is then built from communications, not the physical people, and examples are the stock market, social media, and the nation. Organizations (e.g., political parties, sports teams) are a class of social

---

2. For a review of Luhmann's ideas, see "Niklas Luhmann's Theory of Social Systems and Journalism Research" in the journal *Journalism Studies* (Görke and Scholl 2006).

systems that exist over time, manage to coordinate the interactions of their members, and do not always require physical presence, yet they sustain an identity. In his book *Gaian Systems,* Bruce Clarke classifies this subject material as falling into the domain of neocybernetic theory, a second-order type of cybernetics that transcends the limited structures of conventional cybernetic systems: feedback control, homeostasis, and self-correction.

Neocybernetic systems theory expands on the concept of autopoiesis, which divides living from non-living systems, by emphasizing autonomy, cognition, and system identity that emerges from the circular causality of the system. Some systems are structured couplings of both biotic and abiotic factors that Clarke places in a special class he calls metabiotic systems. This category includes human (biotic) social systems, which today rely on electronic communications (abiotic), and also Gaia, as described by Lovelock and Margulis, in which the biosphere is coupled with its physical environment (Clarke 2020). Luhmann elaborated on social systems including a theory of social evolution, but here I wish only to draw attention to the fact that a number of social science thinkers have described human society as the producer of a network of communications that at the very least has autonomy and identity and transcends the individuals that built it. (Recall from Chapter 1 that chronobiologist Franz Halberg included the socio-psycho-physiological realm in his scheme of biological and environmental rhythms that he called chronomics.) Traditional and modern applications of astrology to this phenomenon in mundane astrology suggest that society may indeed operate as a self-organizing system and as with other systems is likely to be responsive to very weak signals.[3]

Complex systems contain subsystems. Arthur Koestler, author of the novel *Darkness at Noon* and many other books on a wide range of topics, including a biography of Johannes Kepler (*The Watershed*), came up with some names for this pattern in nature. In his book *The Ghost*

---

3. This topic has its share of debates to which astrology, if it were more rigorous, could contribute. For more on this, see the article "Can Social Systems be Autopoietic? Bhaskar's and Giddens' Social Theories" in the *Journal for the Theory of Social Behaviour* (Mingers 2004).

in the Machine, Koestler argued against Cartesian mind-body dualism and its influence on psychology. He developed the concept of the holon as a way to eliminate duality in thought and made reference to the Roman God Janus to illustrate his concept. Janus had two heads, looking in opposite directions from each other. Looking in one direction, a holon is itself complete; but looking in the other, it is subordinate to a much larger entity, which itself is a holon. Any holon is then simultaneously independent and dependent, and this pattern extends in both directions. For example, an atom is part of a molecule that is a part of a cell, which is part of an organ, which is part of a body, and so on. Koestler saw organisms as containing a hierarchy of self-regulating holons that are in one sense autonomous, but also dependent on and coordinated with their greater environment. This sequence he called a "holarchy." What this implies is the existence of multiple layers of self-organizing systems in more or less the same place and time, each both containing and being subsumed by other systems, and all are responsive to signals from their environment. Should one of these systems select for an external signal, others up or down the line may then be affected by it.[4] An example of transference of signal within an ecosystem would be aquatic insects stirred by lunar tides that then stimulate the feeding behaviors of fish.

How do the above ideas apply to astrology? One of the key notions in astrology is the importance of the initiating moment. Astrological charts are most often calculated for a biologically critical point, most common being the shift between water and air breathing that occurs at the separation from the mother, at birth. This point in time is regarded as highly sensitive and capable of registering, and apparently imprinting, a vast amount of astronomical information that is carried through the course of a lifetime. This astronomical data is then used by an astrologer to construct a map of fundamental tendencies and future potentials.

---

4. It should be noted that a downward transmission of information in a holarchy produces a situation where the origins of a signal may appear mysterious to holons below it and possibly suggest teleology is at work. This is demonstrated by the power of the placebo effect in which the mind acts on the body. Astrology, interpreted to be fatalistic, may be another example of this directional process.

Birth might be seen as both a seed moment in time where there is sensitivity to initial conditions but also as a phase transition, or bifurcation point. The initial conditions of the cosmic environment at birth (equivalent to Lorenz's starting values) establish a framework (and trajectory) that is unique as it branches off from that of the mother of which it was once a part. Other moments, where initial conditions are selected consciously (i.e., by doing electional astrology), also mark branching points in time where unique trajectories begin. Such might be the launching of a ship, the coronation of a queen, the opening of a business, or the vows in a marriage ceremony. All of these mark points of change, bifurcation points perhaps, where the flow of time that is experienced by a collective of persons is altered in some significant way, one that has implications for the future. These consciously created moments would then have special properties analogous to that of a birth.

The application of astrology to a sports team, a city, the stock market, or a nation (as is routinely done in mundane astrology) presumes that such intangible entities exist. Exactly how these systems, built on the more or less synchronized communications of many people and the material means by which these communications take place, should be categorized is an open question. As noted above, some biologists and many social scientists have studied social systems that operate over long periods of time, beyond the lifespan of their founding members, and find these hold to a set of values and maintain a kind of boundary. Astrologers who work with this metabiotic subject material, which they regard to be quite real, report that they respond to the motions of the planets. In many cases an initial starting or bifurcation point (i.e., opening of a store, the launching of a spacecraft, the incorporation of a business, etc.) can be located in time. An astronomical map of that transition point then supplies information as to how the system develops through time and when critical points, including those that may produce another bifurcation, are likely to occur. In other cases planetary cycles will correlate with the evolution of a social system as it changes over time. Some astrologers work exclusively with the stock market, a social activity that translates the communications of millions of individuals into measurable units. A number of techniques have been employed to forecast its trajectory, including the cycles of the planets relative to themselves or relative to a point in space such as the crossing of the celestial equator and ecliptic (the vernal point). An example of this would be the market trader W. D. Gann (1878–1955), who developed a unique methodology that utilized planetary cycles to track prices. Another approach has been to use the astrological chart for the origin of the New York Stock Exchange and also charts for the incorporation of companies, or for their first trade on the market. The best results are usually achieved using a combination of techniques that have proven themselves when compared with historical data. Since markets are driven by a wide range of information coming from other social systems, accurate future forecasts are challenging and require considerable expertise. I know a few astrologers who do this work and some of them have been consistently rated among the best market forecasters in the business, though most avoid using the word "astrology" if possible. While most of these practitioners have done well for themselves financially, assuming that if they are not rich then astrology doesn't work fails to take into account the differences between studying a subject and being a participant. These don't always overlap, and such things are not limited to astrology—sports gamblers are almost never great athletes and film critics aren't movie stars.

ASTROLOGY IN SYSTEMS

The subject matter of astrology, that is self-organizing systems, can be depicted in terms of a holarchy. At the lower end of the scale would be the behavior of certain chemicals in a liquid state. Lilly Kolisko's experiments (see Chapter 10) with metals in solution are suggestive of a few things. One is that the reaction rate of certain chemical mixtures (that produce a pattern as they crystallize on filter paper) were shown to vary during very close planetary conjunctions (Moon and Saturn) or during an eclipse. A second point is that these experiments appear to support the traditional correspondences between planets and certain metals. In the process of drying, the metals in solution self-organize and, apparently, this transition can be affected by events in the solar system. These experiments have been replicated, but no units have been applied to the results, only forms subjectively interpreted, and the work has been done in the context of Rudolf Steiner's holistic brand of science. Should there be something to these findings, sensitivity to solar system dynamics may be found at the chemical level and play a role in the self-organization of an organism's membrane, metabolic, neurological, and other chemical systems.

Cells in the body combine to form structures and specialized organs that then play specialized roles in the larger system they are a part of. Many bodily systems run on circadian cycles and probably lunar cycles as well, though these are not as well understood. Astrology, however, has a long tradition of correspondences between body parts and the planets that is based on function within the larger system of the body, these being at the core of traditional medical astrology. To give a few highly simplified examples, Mercury is associated with communication and movement of information; in the body it is connected with the nervous system. The Moon is said to be nurturing, and it rules the begining stages of metabolism, including the mouth and stomach. Venus is associated with balance, and it correlates with the kidneys (among a few other parts), which regulate water balance in the body. Mars is the planet of force and power, and it correlates with the muscles and the immune system. Each planet is linked with the body parts that its function describes; in the case of Saturn, which is the planet of structure and boundaries, correlation is found with the skin and the skeleton. Consider the heart as an organ. You can live without a lot of other body parts, but you can't live without a heart; it is central to life and associated with vitality. Its ruler is said to be the Sun, center of the solar system. A deep knowledge of anatomy and the functions of the many systems in the body is a prerequisite before applying the traditional taxonomy of astrology. The parts of the body, each fulfilling specific functions, are subsystems in the larger organism; holons themselves, but dependent on the next larger holon.

The parts of the body, linked together and communicating by feedback loops, sustain the next level self-organizing system—the neocybernetic systems of mind and the individual self. The sense of self is an emergent property, something that rises out of the activities of the nervous, endocrine, and metabolic systems. The understanding of the self is incomplete because of its complexity. It was discussed early in the history of psychology by William James and since then the psychology of self has become a major topic, one that has many subsets including self-awareness, self-esteem, self-concept, and self-image. Thought appears to arise from chemical messaging and the firing of neurons in the brain, and out of this emerges a stream of self-consciousness, at least in humans who have language. This activity gives rise to not only a sense of self, but it also integrates behaviors and attitudes that form a personality. It is on this system level that astrology has produced a body of knowledge, refined over at least two millennia, that offers a way of modeling who a person is. This is not a twelve-type cookie-cutter method of personality types. Astrology offers a wide range of qualitative measurements that are capable of describing an infinite number of types. Some parts of this method have been captured in the Gauquelins' studies and perhaps some in a few of the zodiac studies, but because reductionist science cannot measure a synthesis of meanings, convincing studies are limited. The emergent properties of the body called the personality and the sense of self are themselves a holon that encompasses and directs the body, but looking in the other direction, it is also a part of the larger social systems that contain it.

Social systems are sustained by the continuing activity of many individuals. Synchronized flight formations of birds and the coordinated hunting behaviors of wolves are non-human examples. Far more complex is the combined communications of multiple humans, that is the actions of language-based cognition that organizes and coordinates social behaviors. This emergent property of human groups maintains an identity, marks boundaries, and can persist long after its original founders have left—institutions, corporations, and nation-states being good examples.[5] With the advent of technology (printing, telegraph,

---

5. This view of social entities having properties similar to individuals appears to support those who have legalized the notion that corporations are people, though I think these distinctions need to be nuanced in order to rein in the power of corporate capitalism.

telephone, email, internet, smart phone) the power of metabiotic social systems has grown. This is mundane astrology, the astrology of the behavior of groups and the impact of the environment on them. The behavior of a nation may change with the rise and fall of different political persuasions, and destructive events such as an earthquakes, storms, or plagues may decimate the population, but the nation still maintains a distinctive identity. The multivalent qualitative taxonomy of astrology is applicable to such social systems, as it is to the holons below it, and the standard techniques apply. While the Sun correlates with strength in the individual psyche, it correlates with leadership in a social system. The Moon correlates with the public, which is usually a receptive part of the social body. Again, given here in simplistic form, Mars is the military, Venus the arts and culture, Mercury the media, Saturn the government and laws, and Jupiter the protective institutions, such as hospitals, schools, and churches. Analysis of social systems by astrology involves multiple approaches to the subject matter, requires a deep understanding of it (often missing—most astrologers don't also hold degrees in history or other social sciences), and is far more complex than can be described here. Since social systems can be enormous, astrological techniques employed will vary but generally include long planetary cycles in addition to charts for phase shifts or bifurcation points in the history of the system.

Social systems as an emergent property of human groups may explain the astrological techniques called interrogations (horary) and elections. An interrogation is a chart calculated for the time a pressing question is asked, and an election is a time selected in advance to commence an activity. As noted in earlier chapters, these practices are so strange that many astrologers regard them as a form of divination or astrological magic. Social systems, being self-organizing, will undergo adjustments to their environment as well as acting on it. Sometimes critical moments will occur that lead to branching or bifurcation points. In the case of interrogations, an individual mind (experiencing concern and strong emotions) may be temporarily in a kind of extreme resonance with the social system (next level holon). An analysis of this moment, using the time-slice technique of the astrological chart, may reveal something about the information circulating within the social system at the time, and in this way, connections may be discerned. Using this data, a competent astrologer may detect pathways in the flow of time that have interpretive and predictive value. With electional astrology, the astrologer scans future planetary positions and locates a moment (based on the study of past experiences) where the pathways of the social system will be most receptive to a specific type of initiative. Elections are then a method of predicting a class of future conditions and then acting on them, something that since Hellenistic times has been considered to be an advanced practice of free will. It can be seen that the Stoic notion of cosmic sympathy linking all parts of a living cosmos makes perfect sense for the practice of these branches of astrology.

The next level holon is that of Earth itself, which has a biosphere occupying a region roughly ten kilometers above and below sea level, an atmosphere, hydrosphere, lithosphere, and a core of molten and semi-molten rock and metals that drive internal convection currents. Gaia, as described by Lovelock and Margulis, is the complex self-organizing metabiotic system that is the biosphere and its physical environment (atmosphere, hydrosphere, lithosphere), the two entwined. Here billions of organisms (mostly microbes) modulate global temperatures, the composition of atmospheric gases, and the pH of the oceans and soils, among other things, through their many kinds of metabolism. As Earth rotates digitally under solar radiation (night and day), the atmosphere is heated and moved, creating winds, while the tilt of the axis drives the reception of solar radiation—all of which then self-organizes into what we call the weather. Weather, traditionally mapped by astrometeorology, is a complex system. Climate is similar, but its self-organization occurs on a much larger timescale and is affected by planet-induced orbital variations that also impact the biosphere, though the biosphere can respond by driving changes in atmospheric gases that modulate climate as well. It is from the coupling of life with its environment that self-organization emerges and long-term environmental stability is achieved. This is the subject material of natural astrology, which studies correlations between solar system dynamics and Earth phenomena, such as weather, climate, and the flow of heat from the interior (probably influenced by spin-rate and tidal forces) that is a factor in earthquakes and volcanic eruptions. As was made clear in Chapter 4, the standard astrological techniques of the time-slice (astrological chart) and the cycles of planets (revolutions) have long been employed in analyzing the weather as well as the timing of geological processes. Since weather controls agriculture and plays a major role in regard to population dynamics of individual organisms (i.e, bacteria, locusts), astrology has had a long association with farmers and has also attempted to map outbreaks of plagues and pestilence. There are interesting precedents here but, for reasons noted in previous chapters, astrology has made little progress in these matters since the rise of modern science.

As a holon, Earth encompasses the self-organizing networks that sustain the biosphere and contain the individual and social holons of organisms on its surface. But Earth is subsumed by the dynamics of the solar system, which consists of varying gravitational resonances, fluctuating electromagnetic signals, and the motions of objects that occasionally impact the biosphere. In a holarchy there is a top-to-down process, the larger holon containing and modulating the lesser holons that are its parts. In seeing the solar system as the next largest system, it is possible that the planets act like nodes enveloped within the medium of charged particles that make up what used to be thought of as empty space. The constant gravitational shifts (tidal forces) of the planets relative to each other as they orbit the Sun, the small variations in Earth's magnetic field caused by solar activity, and the bombardment of a wide range of high-energy particles that exhibit periodicities, all stepped down through the holarchy, may ultimately be what astrology can measure qualitatively. Or maybe there are other linkages, transfers of energy, signals, or forces that are at present unknown. For now, observed correlations and patterns in the context of a systems science model should be a completely acceptable scientific program to anyone who is curious, a good observer, unbiased, has funding, and is not infected by the belief called scientism. While it is true that there have been strong objections to systems theories from critics, and that proponents of systems theories are still more of a subculture in science than the rank and file, I argue that astrology has something of great value to offer this broad explanatory trend in science.

A FUTURE FOR ASTROLOGY

Let's assume for a moment that over the next century or so (barring a global descent into idiocracy or shameless authoritarianism) some explanations and mechanisms for astrology will be found and the subject is somehow resuscitated. What would it look like and what will it offer? Given that natal astrology as a practice has survived mostly intact over the past four hundred years under difficult conditions, we can assume that this branch will continue to exist and will probably experience significant advances in technique and application, much of this based on technological advances in computing and artificial intelligence, and from the findings of scientific studies. Better tools and new and more verifiable knowledge will complement the activity of astrological consulting and counseling which, at its best, is even today producing insightful information about the human situation at levels comparable and often exceeding that of contemporary economic advisors and psychotherapists.[6] It will be these kinds of changes in the actual data used and how it is presented that could make natal astrology a century or two from now an even more powerful professional service.

Changes are also likely to occur in the methodologies normally employed by astrologers. At present the circular astrological chart form, the conventional time-slice map, is the standard tool. Most consulting astrologers today use this map as the primary reference grid and, with considerable mental gymnastics applied to pattern processing, synthesize the data it contains and translate their insights into words a client can use. There are some shortcuts to sorting out chart data. A number of information sorting methods, or weighting systems, have been developed in the past, and many are in use. With advances in astrological

---

6. The best astrologers are really quite good but, as I've stated many times, the lack of widely accepted certifications and the economic pressure of selling one's services has maintained extreme variations in the quality of astrological practice.

software, these methodologies might be presented in forms other than simple lists, perhaps in three-dimensional graphics that can simultaneously merge chart data with dynamic images of the celestial sphere that a practitioner could then interpret. In other words, progress in computer technology could assist in what is probably the most difficult part of interpreting an astrological chart—sorting through immense amounts of data and focusing on only those points that emerge from the background noise and are most relevant to the problem at hand. Presumably, such advances would be built on actual findings from large-scale funded research projects and lead to a role for astrology in artificial intelligence projects that could revolutionize the field.

On the societal level, better education for astrologers and more rigorous requirements for those intending to practice could clarify in the public's mind who is really an astrologer. Every other profession has advanced itself by adhering to standards. With better-educated and better-prepared people offering astrology-based consulting services the field would begin the long process of rehabilitating its reputation. Possibly, courses on astrology may be taught at colleges and universities, though more likely, a few schools dedicated to teaching astrology will focus the subject just as schools have in fields such as chiropractic, acupuncture, natural medicine, and other forms of alternative healing. These schools, institutions that astrology as a subject needs, may also play a role in encouraging research of the social-sciences variety in astrology, and also support technical publications. With higher standards, collaboration and mixing with the more progressive schools of psychotherapy will likely occur. The fact that astrology brings an immense amount of information to the process of working with individuals in need, and is by far the most sophisticated system for generating self-knowledge, suggests to me that astrology will become a major contributor to psychotherapy. It is even possible that training programs in psychotherapy may eventually require some course work in astrology. Likewise, financial and business consulting using astrology will increase and become presented more in line with already established models used by financial advisors and economists. My thinking, then, is that if standards can be maintained and institutions be established, the counseling and consulting part of astrology will probably continue to improve in content, method, and reputation.

The projected success of practicing astrology above will be largely conditioned by developments yet to be seen. First, and most crucial, would be the publication of scientific studies on the elite level that reveal strong correlations between planetary positions and human life and some sort of explanation for an astrological effect. Once it becomes apparent that the correlations found are no more threatening in regard to human freedom than genetic lines and early childhood family environments, rejection from faith-based institutions may diminish to some extent or at least be checked. Given that religions are among the most conservative institutions in all human societies (except when initially "divinely" produced), some pushback from them would be expected. It is remotely possible, however, that some of the smaller and more progressive religions or congregations will find ways to work with astrology (a few are doing this now) and possibly incorporate some of its ideas into doctrines. At the same time there will likely be pushback from some segments of society where astrological advice, or the use of astrology in artificial intelligence, may be seen as too much of an advantage for those who apply it (as in gambling), or too much information for situations where evenhandedness is crucial (as in job applications). A potentially disruptive use of astrology may result in problems requiring legal solutions.

Second, findings that support astrology as legitimate knowledge will drive those in other disciplines to better understand the subject. The social sciences, particularly psychology and economics, may be the first to utilize such findings. Biologists and system scientists may also find useful information in astrology that is consistent with findings in chronobiology and, with funding and institutional support, begin to study more thoroughly the linkages between astronomical patterns and the behavior mapping of self-organizing systems. Physics, which must have a mechanism that can be explained in a formula, could be the last of the sciences to take astrology seriously. Eventually, mechanisms may be found by astrophysicists as the nature of space, plasma, particles, gravity, dark matter, and energy become better understood. Or perhaps some other force or forces will be discovered that show how information moves within the solar system. I have long thought that because the sciences as we know them today don't take into account the implications of astrology, they are left with explanatory power that is necessarily limited. With astrology fortified by new findings, that subject could develop a comprehensive theory more in line with what is known through astrophysics. Finally, the acceptance and use of astrology by people in high places will cause many to follow their lead and become more open to the subject. Human nature drives the formation of pecking orders that are usually expressed as fashion, trend, or social orginazation. Like other primates, a few lead, many more others follow.

The applications of astrology to the study of self-organizing systems may force greater cooperation between disciplines. When a planetary cycle is found to be correlated with both a biological and meteorological process, it will be difficult to keep these correlations separate and distinct. As a subject of its own, astrology will probably emerge as a multidisciplinary field, much like how mathematics is spread among the sciences, and quite possibly without its traditional name. If this should happen and the subject finds itself more securely established as a respectable area of inquiry, it will tend to attract generalists comfortable working in multiple disciplines, rather than reductionist thinkers. Over time this could shift some of the emphasis in scientific thinking away from reductionism as the agenda-setting mode of science and, along with similar trends in subjects like organismic biology, economics, and climatology, allow systems thinking to take a leading role where appropriate. An informed reductionism would then be limited to solving specific linear problems in the larger context of a systems view—and not dictate absolutes.

Existing subdisciplines could potentially benefit from some traditional techniques in astrology. An example in chronobiology might involve dividing the annual photoperiod wave into discrete sections. This technique would not necessarily be limited to the twelve sections of the zodiac; other harmonics may turn out to produce more reliable information in certain situations than in others. One would think that anchoring phase transitions at the equinoxes and solstices, which limits the range of divisions, makes good sense. For example, division by eight, twelve, sixteen, twenty, twenty-four, and so on allows each quadrant the same number of sections. Division by seven can only be anchored at one of these points. The same approach could be applied to the houses, which is what the Gauquelins did when they divided the diurnal cycle into twelve, eighteen, or thirty-six segments. The long tradition in astrology of transforming wave-like phenomena (photoperiod, diurnal cycle) into discrete (digital) sections not only facilitates the application of mathematics but is also interesting in itself. Ideas like this were behind Kepler's theory of everything, in which he reduced planetary orbital motions into discrete phase angles that formed geometric solids, musical tone ratios, and astrological aspects.

Economics in the broadest sense is another study that would be able to absorb and improve astrological techniques. Astrology has long been applied to market cycles, particularly stocks and commodities, these being self-organizing social systems that are translatable into units (e.g., price points) for study. The success records of the astrologers who are dedicated to this work are quite good and recognized to some extent, often reluctantly, by their peers in market forecasting. As was mentioned in Chapter 2, since the nineteenth century economists have recognized a business cycle of roughly eighteen years. It's long been known among astrologers that the real estate cycle roughly follows the 18.6-year lunar declination cycle, registered as the movement of the lunar nodes through the zodiac. While each individual cycle can vary from the mean, most of the time prices begin to fall when the north node of the Moon approaches the vernal equinox. With some attention given to the subject, it may be possible to better understand the variations between successive cycles, a variation that is not unlike that of Schwabe solar cycle lengths, which are not fully explained at present and are taken as a near periodicity. The origin of this business cycle, which may involve weather and agriculture and how it can be manipulated by governmental policies and laws (an example of collective free will?), is a topic that could be explored. The field of economics is full of information that could be much better organized and coordinated with planetary cycles, single and multiple. An important point is that experienced astrologers know that an isolated planetary cycle rarely has strong explanatory power—instead, it is the synthesis of multiple combinations and higher order recurrence cycles that better correlate with phenomena.

Another natural application for astrology in the social sciences would be in history, the subject that studies the self-organizing dynamics of human societies over time. This has been attempted previously and is found in the works of astrologers from Medieval times to the present. For most of this time it was the cycles (formerly called revolutions) of Jupiter and Saturn that were thought to produce the largest outline of history. Over the last century these ideas have been greatly expanded not only due to the discovery of additional outer planets, but also the perfection of astrological techniques that better model the observed correlations between planetary cycles and historical phenomena. I'm not implying that these techniques are 100 percent accurate, but with a good understanding of history, an experienced astrologer is able to discern and describe patterns in historical trends and make estimates of future trends as well. Timing historical changes and cyclic patterns is one thing, interpretation is another. Astrology's multilayered and non-repeating matrix of planetary motion gives, I would argue, a nuanced understanding of the unfolding of the now and, consequently, a guide to the possible meanings of historical events. Because planet movements are cyclic, better comparisons between historical periods could be made. Given this close fit between history and astrology I would expect that a great deal of what has been called mundane astrology would be assimilated by historians and used in the same way as they already use information from sociology, anthropology, archaeology, and other disciplines. Probably a name for these techniques other than astrology would be found. My suggestion would be "solar system dynamics."

An astrology of sorts is already used as an organizing principle in the geosciences. The Milankovich (orbital) cycles, described in detail in Chapter 2, are able to map out and explain the periodicities of ice ages and climate cycles over vast periods of time, as far back as two hundred million years or more. The Earth system is a self-organizing system composed of both physical and biological forces linked by a wide range of complex feedback loops. Over deep time the elements themselves are recycled and contribute to the composition of the soils, seas, and atmosphere, which has attained a relatively stable configuration—this is in essence Gaia theory. On smaller timescales solar cycles have been correlated with weather, climate, and the biosphere, though the precise mechanisms behind solar cycles, which may involve planets, are still being worked out. One area where the methodologies of astrology may play a role is in regard to earthquakes and volcanic eruptions. A few correlations, explained by gravity, between these events and the Moon have been reported, but because of the many variables involved in the shifting of large blocks or the release of pressures from the mantle in an eruption, there is no precise formula. Earthquake prediction today is largely based on knowledge of faults, specifically the amount of accumulated stress. It is in the triggering of these geological conditions that a more sophisticated astrological methodology may prove useful.

The medical field, which works with the self-organizing system of the body, may benefit considerably from astrological methodologies as it has in the past. What astrology brings to medicine is detailed information on human variations. Modern medicine is mostly reductionist and generally treats people as if one size fits all. A wide range of human individual differences in metabolism, for example, exist, and few medical practitioners have the time to work these out with any single patient and so refer them to nutritionists who, over some time, may discover what helps and what doesn't. This slow process of discovery could be sped up using what is already known in astrology, but probably much more could be learned if the subject was studied seriously. A second contribution from astrology to medicine, one that dates back to Hippocrates, is in regard to timing. Accurate information along these lines could help make better estimates of healing times, locate when interruptions in the healing process may occur, and designate the dates and times that surgeries should be attempted or medications should be taken for best results. The latter is already a little-known subdiscipline called chronopharmacology, one of the ideas promoted by chronobiologist Franz Halberg.

The branch of astrology called elections, where a future time is calculated for the time of an initiative, will probably draw the attention of competitive individuals and corporations. Over time, studies may show that after all other factors are considered, the configurations of the planets and their positions relative to the horizon do make a difference in the unfolding of an activity. Timing the launch of something, for example a business, marriage, or other project, with astrology may come to be shown to have some value, but it is not as easy as it sounds. The inconveniences of controlling and adjusting schedules may be seen to outweigh the benefits. While electional astrology is employed today, discreetly, I have some concerns about its use in the future because it could be exploited. Some situations, such as public sporting events, could be timed to favor one side over the other. Initiatives such as launching a campaign, selling insurance, or enacting a public policy are possible applications that either enhance a legitimate need to compete or minimize negative repercussions. None of this is new; Ronald Reagan is known to have used electional astrology in scheduling important events, including ones that led to a deal with the Soviet Union.

If my assumptions are correct, the subject of astrology will progress over the next centuries and regain some of its former status. Much of the subject will be caught up in the activity of practice, basically serving the needs of individuals. Other parts may be used in other disciplines, probably in conjunction with artificial intelligence, to map and measure growth and crisis in self-organizing systems. A larger question is, how much of a contribution to general knowledge will astrology actually make? In the world of physics, which is populated mostly by reductionists, unless some mechanism is found to explain astrology that is based on known forces it will continue to be kept on the sidelines. If enough data accumulates that suggests something entirely new is going on in nature that has not been previously considered, perhaps in this way astrology may prove to be truly revolutionary. I wouldn't count astrology out as a big game changer. The subject has been there before.

SUMMARY OF POSITIONS

I fully expect *The Nature of Astrology,* which is both a science book on astrology and an astrology book on science, to provoke reactions from both the scientific and astrological communities. In order to be perfectly clear about my views, which I believe are likely to be misinterpreted, I will now summarize my positions on a few of the points I've been making. The first is science itself. I'm not anti-science in the least—I just think holding all of nature 100 percent accountable to reductionist, materialist, and mechanistic science, the brand favored by the average scientist and held as sacred and worthy of worship by militant skeptics, can be counterproductive, if not pathological. RMM science is a great tool, but applied bluntly can lead to assessments that are narrow in scope and possibly dangerous when it comes to bodily health and the environment, both of which are systems. Saying that "science" has shown something to be false because it fails to meet the demands of reductionism may be reasonable in most situations, but such a narrow and limiting definition of supposed "truth" may, in other situations, be short-sighted.

In my view, science is a democratic process of making knowledge that has been evolving for millennia. It now uses a wide range of instrumentation and techniques such as observation, measurement, comparison, critical thinking, and mathematical modeling to sort out what is being studied. I have been very clear that the scientific study of self-organizing systems requires different approaches than the study of a rock or a simple chemical reaction. Life, consciousness, and certain complex environmental processes (weather, climate) require a scientific methodology that begins with the whole system, or a model of it, and then, with knowledge of process, patterns, and structures, parts can be located where RMM science might be applied. This approach could be called informed and targeted reductionism. What I am suggesting is that using RMM science randomly on a self-organizing system, which is how skeptics "test" astrology, may produce isolated pockets of detailed information that lack integration, focus on static factors rather than processes, and may create more problems than they solve. My point is also relevant to environmental systems and the body and mind.

My position on astrology is that it is a real phenomenon that needs more study and that defining it as only a practice is misleading and needs to stop. Explaining how it works in the context of modern RMM science is very difficult, and studies are few and most poorly designed, but the work of the Gauquelins should at least give open-minded critics something to think about. I am of the opinion that astrology operates on multiple levels, but there is a general sharing of fundamental principles (or laws), mostly in terms of geometry and resonance. How a Sun-Saturn opposition can modulate weather may be explained by gravitational resonances, but we don't know for sure because it hasn't been studied, only a correlation has been shown. How a Sun-Saturn opposition can correlate with personal matters is something else. I've suggested looking for answers in subdisciplines like biogeomagnetics and biofield physiology; these research areas have found that environmental patterns in light and other electromagnetic information can be utilized by organisms even if these signals are very weak. At present, we don't know enough about these things and I am only speculating and offering ways to think about them. I am also suggesting that looking for correlations between planetary patterns and self-organizing systems in the context of known forces is a start, a beginning in scientific analysis, but not proof of anything. Because astrology appears to operate on several levels some entirely new linkages or forces may need to be considered or discovered. The unanimous judgement by generations of astrologers that remote and minuscule bodies like Pluto have powerful effects on people and society will probably require a lot more knowledge about solar system dynamics, the nature of space, and the reception of signals by systems than is available now. Our present understanding of nature is far from complete, and we should keep in mind that today's science was yesterday's magic. As for divination as an explanation for astrology, I don't think it is helpful.

Regarding astrologers, it appears to me that, while there is a great interest in improving interpretation and predictive techniques as applied to consultations, there is little interest in astrology as a broader subject. In fact, most people who identify as astrologers think of it as only a counseling or consulting practice, and their interests are the technicalities of the craft. While astrology as a practice has become quite sophisticated, and I support a lot of what is being done by some very competent practitioners, it needs to be recognized that, as a subject, astrology is very weak on theory and philosophy. Research is attempted, but with virtually no funding or institutional support, and with few astrologers who are knowledgeable of research design and statistics, not much gets done. Meanwhile, outsiders to the field don't understand the subject well enough to design a proper study, and they are generally hostile and likely to skew studies toward what they expect to find. One area where astrologers have made progress is in regard to the history of the subject. This has involved translations from many languages and critical scholarship. Along with this interest in history has come a tendency to elevate subjects like Hellenistic or Hindu (Vedic) astrology over modern astrology. While this may serve to better acquaint modern astrologers with ancient techniques that can then be tested, and it may fortify a practitioner's identity in a competitive marketplace or social scene, it also can work against any true evolution of the subject in a modern framework. My opinion, however, is that this reclamation of history is a necessary step toward the larger goal of making astrology relevant and reliable.

There are two final observations I wish to make about judicial astrology. The first is that it may work for everyone, but it may not be good for everyone. Some people have the ability to be detached from themselves and from social situations. For whatever reasons, which most likely involve genetics, parenting, friendships, and educational background, they can stand apart from a situation and actually study and learn from it. Much like an anthropologist or other social scientist these people will have a much better appreciation of astrological effects, will benefit from having their chart read, and may even become interested enough in the subject to become a self-practitioner. In contrast are those who fall into two general categories. One is the hyper-rational types who are so focused on details that they fail to see the larger picture. These people I've observed to either become agitated over an astrological detail or make assumptions that are counterproductive to understanding, or even to their well-being. The other category are people who are so emotionally charged, possibly the result of childhood trauma or abuse, that the mere suggestion of an upcoming stressful transit to their birth chart is enough to trigger high anxiety. These people should be very careful with astrology as they may misread situations and even create self-fulfilling prophecies. They probably also need to be careful with psychotherapy or any life consulting practice as well. Most people are somewhere between these extremes and may only benefit from astrology from time to time.

The second point is that curious people may want to test out astrology for themselves. What I recommend to them is that, using the birth data (month, day, time, place) of someone they know well, like a spouse or child, they obtain the natal chart (online or with software) and follow transits to it over the course of a year or more. Of course, they can do this with their own natal chart as well. A way to start is to follow the daily and monthly motions of the slower-moving planets, particularly Mars, Jupiter, Saturn, and Uranus, through the zodiac (all of this is available online or with software) as they pass over, oppose, and square the planets in the birth chart—these are called transits by astrologers. As one of these planets approaches to within a degree of the natal chart's sensitive points (the planetary positions and the Ascendant and Midheaven), observe and make notes of behaviors and event patterns. Younger people will tend to display planetary effects in the form of behavior and social interaction, and older people may display them more in the form of bodily issues. After a few months, not only will it be apparent that astrology is working, but a more nuanced understanding of how it works will be gained. Since the culture has rejected astrology, thinking about it properly requires an exercise like this. Of course, much can be learned from books, but a hands-on project is a good way to start.

For those who have finished the book, I thank you for your patience. I hope you have found it to be interesting and, having been exposed to some unfamiliar facts and concepts, have expanded your thinking patterns. The arguments I have made are supported by evidence, except in those cases where I have been clear that a given statement is speculative. At the most fundamental level there is abundant evidence that the solar system environment impacts systems of all kinds on Earth and that life evolves in both spatial and temporal environments. Astrology is a multi-millennial project that explores and measures this cosmic environment using a technical methodology that incorporates geometry, semiotics and pattern perception. My assertion that astrology is real and not a delusion is based on a life-time of exploration of that subject—and also the stability of its symbolic language over the millennia as demonstrated by the Gauquelin studies. How Western astrology became marginalized during the 16th and 17th centuries is interesting history and not as simple as one would assume. And because astrology works strictly with self-organizing systems, it makes sense that an understanding of the subject should be not be framed by the methods of RMM science—investigation and study needs to be applied in a context established by system science. There is so much more to know about astrology, a subject with a practice. Those who choose to ignore the uninformed and biased warnings of the thought police of scientism and the fear-based judgments of fundamentalist religions will find it well-worth investigating.

References

Adderley, E. E., Bowen, E. G. (1962) "Lunar Component in Precipitation Data." *Science,* 137 (3532). 749-750.

Adel, Miah M., Saiyeeda F. Hossain, and Hannah Johnson. 2014. "Favored Zodiac for Celebrity Births." *Journal of Social Sciences* 9(4):164–72.

Alabdulgader, Abdullah, Rollin McCraty, Michael Atkinson, York Dobyns, Alfonsas Vainoras, Minvydas Ragulskis, and Viktor Stolc. 2018. "Long-Term Study of Heart Rate Variability Responses to Changes in the Solar and Geomagnetic Environment." *Scientific Reports* 8, no. 2663.

Alexandrinus, Paulus. 1993. *Introductory Matters.* Translated by Robert Schmidt. Cumberland, Maryland: Golden Hind Press.

Al-Kindi. 1993. *On the Stellar Rays.* Translated by Robert Zoller. Cumberland, Maryland: Golden Hind Press.

Allen, Don Cameron. 1964. *The Star-Crossed Renaissance.* New York: Octagon Books.

Alley, Richard B. 1998. "Icing the North Atlantic." *Nature* 392:335–37.

Alley, Richard. B., Clark, P.U., Keigwin, L.D., and Webb, R.S. 1999. "Making Sense of Millennial-Scale Climate Change." pp. 385–394. In Clark, Peter U., Robert S. Webb, Lloyd D. Keigwin, eds. *Mechanisms of Global Climate Change at Millennial Time Scales.* Washington DC: American Geophysical Union.

American Academy of Arts and Sciences n.d. "Gender Distribution of Bachelor's Degrees in the Humanities." AMCAD Website. Accessed May 2020.

Aristotle. 1952. *Meteorologica*. Translated by H.D.P. Lee. Cambridge, Mass.: Harvard University Press.

Aristotle (c. 350 BCE). 1961. *Parts of Animals*. Translated by Peck. A. L. Loeb Classical Library. Cambridge, Massachusetts: Harvard University Press.

Aristotle. (c. 350 BCE). 2017. *Politics 1*, xi. Translated by Benjamin Jowett. Digireads Publishing.

Arny, Thomas T. 1998. *Explorations: An Introduction to Astronomy*. New York: McGraw-Hill.

Augustine. (c. 400) 1963. *Confessions,* Vol. 1. Books 1–8. Translated by W. Watts. Loeb Classical Library. Cambridge, Massachusetts: Harvard University Press.

Augustine. 2009. *The City of God*. Translated by Marcus Dods. Peabody, Massachusetts: Hendrickson Publishers.

Aveni, Anthony. 1980. *Skywatchers of Ancient Mexico*. Austin, Texas: University of Texas Press.

Babcock, H. W. 1961. "The Topology of the Sun's Magnetic Field and the 22-Year Cycle." *Astrophysical Journal* 133:572.

Bacon, Francis. (1623) *Works of Francis Bacon,* collected and edited by Robertson, Ellis and Heath. 1900. The Riverside Press, Cambridge.

Baigent, Michael, Nicholas Campion and Charles Harvey. 1984. *Mundane Astrology*. Detroit: The Aquarian Press.

Baigent, Michael. 1994. *From the Omens of Babylon: Astrology and Ancient Mesopotamia*. Burnaby, British Columbia: Arcana.

Baker, R. Robin. 1982. *Human Navigation and the Sixth Sense*. New York: Simon and Schuster.

Baliunas, Sallei, and Willie Soon. 1996. "The Sun-Climate Connection." *Sky and Telescope* (December): 38–41.

Balling, Robert C. and Randall S. Cerveny. 1995. "Influence of Lunar Phase on Daily Global Temperatures." *Science* 287:1481–82.

Barnett, Audrey. 1966. "A Circadian Rhythm of Mating Type Reversals in *Paramecium multimicronucleatum*, Syngen 2, and its Genetic Control." *Journal of Cellular Physiology* 67:239–70.

Barton, Tamsyn. 1994. *Ancient Astrology*. London: Routledge.

Bazylinski, Dennis A. 1999. "Synthesis of the bacterial magnetosome: the making of a magnetic personality." *International Microbiology* 2:71–80.

Beatty, J. Kelly. 2011. "A Sign of the Times." *Sky and Telescope,* January 20, 2011.

Becker, Robert O. and Gary Seldon. 1985. *The Body Electric: Electromagnetism and the Foundation of Life.* Fort Mill, South Carolina: Quill.

Beer, Arthur and Peter Beer, ed. 1975. *Kepler, Four Hundred Years.* Oxford, United Kingdom: Pergamon Press.

Berger, A., J. Imbrie, J. Hays, G. Kukla, and B. Saltzman. 1984. *Milankovitch and Climate: Understanding the Response to Astronomical Forcing.* Dordrecht, Netherlands: D. Reidel Publishing Co.

Berger, Peter and Thomas Luckmann. 1967. *The Social Construction of Reality.* Norwell, Massachusetts: Anchor.

Berman, Morris. 1981. *The Reenchantment of the World.* Ithaca, New York: Cornell University Press.

Bianchi, G.G., and I.N. Mccave. 1999. "Holocene Periodicity in North Atlantic Climate and Deep-Ocean Flow South of Iceland." *Nature* 397:515–17.

Bigg, E.K. 1963. "A Lunar Influence on Ice Nucleus Concentrations." *Nature* 197:172–73.

Bijma, Jelle, Jonathan Erez, and Christoph Hemleben. 1990. "Lunar and Semi-lunar Reproductive Cycles in some Spinose Planktonic Foraminifers." *Journal of Foraminiferal Research* 20, no. 2: 117–27.

Blakemore, R.P. 1975. "Magnetotactic Bacteria." *Science* 24, no. 190 (4212): 377–79.

Bobrick, Benson. 2005. *The Fated Sky.* New York: Simon & Schuster.

Bonatti, Guido. 1994. *Liber Astronomiae,* Tractate 1, chapter IV. Translated by Robert Zoller. Cumberland, Maryland: The Golden Hind Press.

Bond, Gerard C., Showers, W., Elliot, M., Evans, M., Lotti, R., Hajdas, I., Bonani, G., Johnson, S. 1999. "The North Atlantic's 1-2 k.y.r Climate Rhythm: Relation to Heinrich Events, Dansgaard/Oeschger cycles and the Little Ice Age." In Clark, Peter U., Robert S. Webb, Lloyd D. Keigwin, Eds. *Mechanisms of global Climate Change at Millennial Time Scales.* Geophysical Monograph Series 112, pp. 35–58.

Bono, James J. 1995. *The Word of God and the Languages of Man. Vol. 1: Ficino to Descartes.* Madison, Wisconsin: University of Wisconsin Press.

Boose, E., and E. Gould. 1999. Shaler Meteorological Station (1964–2002).

Harvard Forest Data Archive: HF000. A newer time-series, Fisher Station, contains data since 2002.

Bos, Gerrit and Charles Burnett. 2000. *Scientific Weather Forecasting in the Middle Ages: The Writings of Al-Kindi.* London: Routledge & Kegan Paul.

Bowden, Mary Ellen. 1975. *The Scientific Revolution in Astrology: The English Reformers, 1558-1686.* PhD diss., Yale University.

Boxer, Alexander. 2019. *A Scheme of Heaven.* New York: Norton.

Bradley, Donald A. and Max A. Woodbury. 1962. "Lunar Synodical Period and Widespread Precipitation." *Science* 137:748–50.

Brier and Bradley. 1964. "The Lunar Synodical Period and Precipitation in the United States." Boston: *Journal of Atmospheric Sciences.* Vol. 21.

Britton, John, and Christopher Walker. 1996. "Astronomy and Astrology in Mesopotamia." In *Astronomy Before the Telescope*, Christopher Walker, ed. New York: St. Martin's Press.

Brown, Frank A. 1960. "Subtle Factor and the Clock Problem." *Cold Springs Harbor Symposia on Quantitative Biology* XXV.

Brown, Frank A., J. Woodland Hastings, and John D. Palmer. 1970. *The Biological Clock: Two Views.* Cambridge, Mass: Academic Press.

Bünning, Erwin. 1958. *The Physiological Clock.* Berlin: Springer-Verlag.

Bünning, Erwin. 1960. Opening Address: Biological Clocks. Cold Spring Harbor. Symposis on Quantitative Biology, Vol. XXV *Biological Clocks.* The Biological Laboratory.

Burtt, E.A. 1954. *The Metaphysical Foundations of Modern Science.* New York: Doubleday & Co.

Butterfield, Herbert. 1957. *The Origins of Modern Science 1300–1800.* New York: The Free Press.

Campion, Nicholas. 1989. "Astrological Historiography in the Renaissance." In Kitson, ed., *History and Astrology.* New York: Mandala.

Campion, Nicholas. 2009. *A History of Western Astrology.* 2 vols. London, New York: Continuum International Publishing Group.

Capp, Bernard. 1979. *English Almanacs 1500–1800.* Ithaca, New York: Cornell University Press.

Capra, Fritjof. 1982. *The Turning Point: Science, Society, and the Rising Culture.* London: Bantam.

Capra, Fritjof. 1997. *The Web of Life.* New York: Anchor.

Cardan, Jerome. (1970). "The Choicest Aphorisms of the Seven Segments." In *The Astrologer's Guide.* c. 1570 Tempe, Arizona: American Federation of Astrologers.

Carlson, Shawn. 1985. "A Double-blind Test of Astrology." *Nature* 318 (December 5): 419–25.

Carpenter, Thomas H., Ronald L. Holle, and Jose J. Fernandez-Partagas. 1972. "Observed Relationships Between Lunar Tidal Cycles and Formation of Hurricanes and Tropical Storms." *Monthly Weather Review* 100(6):451–60.

Casiraghi L., Spiousas I., Dunster G.P., et al. 2021. "Moonstruck Sleep: Synchronization of Human Sleep with the Moon Cycle under Field Conditions." *Science Advances* 7, no. 5 (January).

Casper, Max. 1993. *Kepler.* Translated by C. Doris Hellman. Mineola, New York: Dover Publications.

Cassirer, Ernst. 1963. *The Individual and the Cosmos in Renaissance Philosophy.* Translated by Mario Domandi. New York: Harper Torchbooks.

Chapman, Sydney and Richard S. Lindzen. 1970. *Atmospheric Tides: Thermal and Gravitational.* Dordrecht, Netherlands: D. Reidel Publishing.

Childrey, Joshua. 1652. *Indago Astrologica.* Edward Husband.

Chotai, Jayanti, Mattias Lundberg, and Rolf Adolfsson. 2003. "Variations in Personality Traits Among Adolescents and Adults According to Their Season of Birth in the General Population: Further Evidence." *Personality and Individual Differences* 35, no. 4: 897–908.

Christianson, J.R. 1979. "Tycho Brahe's German Treatise on the Comet of 1577." *Isis* 70(251):110–40.

Ciarleglio, Christopher M., John C. Axley, Benjamin R. Strauss, Karen L. Gamble, and Douglas G. McMahon. 2010. "Perinatal Photoperiod Imprints the Circadian Clock." *Nature Neuroscience* 14, no. 1: 25–7.

Cicero. 1933. *De Natura Deorum Academica* ("On the Nature of the Gods"). Translated by H. Rackham. Cambridge, Mass.: Harvard University Press.

Cipolla, Carlo M. 1994. *Before the Industrial Revolution: European Society and Economy, 1000-1700.* New York: W.W. Norton & Co.

Clark, Vernon E. 1960. "An Investigation of the Validity and Reliability of the Astrological Technique (Part I)." *In Search* 2, no. 4 (Winter): 44–69. (Part II) *In Search* Vol. 3 Nos. 1 & 2, Spring/Summer. pp. 25–33.

Clark, Vernon E. 1961. "Experimental Astrology." *In Search* (Vol. not numbered) (Winter/Spring): 102–12.

Clarke, Bruce. 2020. *Gaian Systems*. Minneapolis: University of Minnesota Press.

Clarke, D., T. Gabriels, and J. Barnes. 1996. "Astrological Signs as Determinants of Extroversion and Emotionality: An Emprical Study." *The Journal of Psychology* 130(2):131–40.

Cock, William. 1670. *Meteorlogiae*. Jo. Conyors.

Cole, Lamont C. 1957. "Biological Clock in the Unicorn." *Science* 125, no. 3253: 874–76.

Collingwood, R.G. 1960. *The Idea of Nature*. Oxford, United Kingdom: Oxford University Press.

Cooper, Glen M. 2011. "Galen and Astrology: A Mésalliance?" *Early Science and Medicine* 16(2):120–46.

Copleston, Frederick, S.J. 1962a. *A History of Philosophy, Volume I, Parts I and II Greece and Rome*. New York: Image Books.

Copleston, Frederick, S.J. 1962b. *A History of Philosophy, Volume II, Mediaeval Philosophy, Part I*. New York: Image Books.

Cornelius, Geoffrey. 2003. *The Moment of Astrology*. United Kingdom: The Wessex Astrologer.

Courtillot, V, Y. Gallet J. Le Mouël, F. Fluteau, and A. Genevey. 2007. "Are There Connections Between the Earth's Magnetic Field and Climate?" *Earth and Planetary Science Letters* 253(3–4):328–39.

Crombie, A.C. 1959. *Medieval and Early Modern Science, Volume II*. New York: Doubleday & Co.

Currey, Robert. 2011. "U-Turn in Carlson's Astrology Test?" *Correlation, Journal of Research in Astrology* Vol. 27, no. 2 (July 2011): 7–33.

Currey, Robert. 2017. "Can Extraversion [E] and Neuroticism [N] as Defined by Eysenck Match the Four Astrological Elements." *Correlation, Journal of Research in Astrology,* Vol. 31(1).

Curry, Patrick, ed. 1987. *Astrology, Science and Society: Historical Essays*. Suffolk: The Boydell Press.

Curry, Patrick. 1989. *Prophecy and Power: Astrology in Early Modern*

England. Princeton, New Jersey: Princeton University Press.

Dalton, David. 2004. "John H. Nelson's Theory of Propagation: Is There Anything to It? *Ham Radio on the Net* (blog). July 29, 2004. https://www.eham.net/article/8828.

Dampier, Sir William Cecil. 1943. *A History of Science*. Cambridge University Press.

Dansgaard, W., S.J. Johnsen, H.B. Clausen, D. Dahl-Jensen, N.S. Gundestrup, C.U. Hammer, C.S. Hvidberg, J.P. Steffensen, A.E. Sveinbjornsdottir, J. Jouzel, and G. Bond. 1993. "Evidence for general instability of past climate from a 250-k.y.r ice-core record." *Nature* 364:218–220.

Darwin, Charles and Francis Darwin. 1881. *The Power of Movement in Plants*. D. Appleton & Co.

Davis, Marc, Piet Hut, and Richard A. Muller. 1984. "Extinction of Species by Periodic Comet Showers." *Nature* 308:715–17.

Dean, Geoffrey, and Arthur Mather. 1977. *Recent Advances in Natal Astrology*. The Astrological Association of Great Britain.

Dean, Geoffrey. 1983. "Shortwave Radio Propagation: Non-correlation with Planetary Positions. *Correlation* 3.1:4–37.

Dean, Geoffrey. 1985. "Can Astrology Predict E and N? 2: The Whole Chart." *Correlation* 5(2):2–24.

Dean, Geoffrey. 1986. "Can Astrology Predict E and N? 3: Discussion and Further Research." *Correlation,* 6 (2): 7–52.

DeCoursey, P.J., J.K. Walker, and S.A. Smith. 2000. "A Circadian Pacemaker in Free-living Chipmunks: Essential for Survival?" *Journal of Comparative Physiology A* 186:169–80.

Dee, John. (1558, 1568) 1978. *Propaedeumata Aphoristica*. Edited and translated by Wayne Shumaker. Berkeley: University of California Press.

Devore, Nicholas. 1947. *Encyclopedia of Astrology*. Philosophical Library.

Dewey, E.R., and E.F. Dakin. 1949. *Cycles: The Science of Prediction*. New York: Henry Holt and Company.

Dewey, Edward R. and Og Mandino. 1971. *Cycles: The Mysterious Forces that Trigger Events*. Portland, Oregon: Hawthorne Books.

Digges, Leonard. 1605. *A Prognostication Everlasting*. Felix k.y.ngstone. London

Dreyer, J.L.E., 1963. *Tyco Brahe*. Mineola, New York: Dover Publications.

Dunlap, Jay C. 1999. "Molecular Bases for Circadian Clocks." *Cell* 96:271–290.

Dunlap, Jay C., Jennifer J. Loros, Patricia J. DeCoursey. 2004. *Chronobiology: Biological Timekeeping.* Sinauer Associates, Inc.

Dvornyk, Volodymyr, Oxana Vinogradova, and Eviatar Nevo. 2003. "Origin and evolution of circadian clock genes in prokaryotes." *Proceedings of the National Academy of Sciences* 100(5):2495–2500.

Eddy, John A. 1976. "The Maunder Minimum." *Science* 192(4245):1189–1202.

Elgin Dairy Report: A Weekly Bulletin of Dairy Information, Volume 24, No. 6, July 6, 1914. "Do the Planets Affect Our Weather?"

Eliade, Mircea. 1956. *The Forge and the Crucible.* University of Chicago Press.

Endres, Klaus-Peter and Wolfgang Schad. 1997. *Moon Rhythms in Nature.* Edinburgh, United Kingdom: Floris Books.

Ertel, Suitbert. 1988. "Raising the Hurdle for the Athletes' Mars Effect: Association Co-Varies with Eminence." *Journal of Scientific Exploration* 2(1):53–82.

Ertel, Suitbert. 2009. "Appraisal of Shawn Carlson's Renowned Astrology Tests." *Journal of Scientific Exploration,* 23(2):125–37.

Ertel, Suitbert and Kenneth Irving. 1996. *The Tenacious Mars Effect.* Somerset: Urania Trust.

Ertel, Suitbert and Kenneth Irving. 2000. "The Mars Effect is Genuine: On Kurtz, Nienhuys, and Sandhu's Missing the Evidence." *Journal of Scientific Exploration* 14(3).

Fagan, Cyril. 1950. *1950 Zodiacs Old and New.* Los Angeles: Llewellyn Publications.

Fagan, Cyril. 1971. *Astrological Origins.* Woodbury, Minn.: Llewellyn Worldwide.

Fairbridge, Rhodes W. and John E. Sanders. 1987. "The Sun's Orbit, A.D. 750–2050: Basis for New Perspectives on Planetary Dynamics and Earth-Moon Linkage." In Rampino, Sanders, Newman and Konigsson. *Climate: History, Periodicity, and Predictability.* New York: Van Nostrand Reinhold.

Field, J.V. 1984. "A Lutheran Astrologer." *Archive for the History of the Exact Sciences* 31:189–272.

Filipovic, M. D., J. Horner, R.J. Crawford, N.F.H Tothill, and G.L. White. 2013. "Mass Extinction and the Structure of the Milky Way." *Serbian Astronomical Journal* 187:43–52.

Foster, Russell G. and Leon Kreitzman. 2004. *Rhythms of Life: The Biological Clocks that Control the Daily Lives of Every Living Thing.* Yale University Press.

Foucault, Michel. 1970. *The Order of Things: An Archaeology of the Human Sciences.* New York: Vintage Books.

Frankel, R.B., R.P. Blakemore, and R.S. Wolfe. 1979. "Magnetite in Freshwater Magnetotactic Bacteria." *Science* 203:1355–1356.

French, Peter. 1972. *John Dee. The World of an Elizabethan Magus.* New York: Dorset Press.

Friis-Christensen, E. and K. Lassen. 1991. "Length of the Solar Cycle: An Indicator of Solar Activity Closely Associated with Climate." *Science* 254:698–700.

Fuzeau-Braesch, Suzel. 2007. "An Empirical Study of Some Astrological Factors in Relation to Dog Behaviour Differences by Statistical Analysis and Compared with Human Characteristics." *Journal of Scientific Exploration* 21(2): 281–93.

Gaquelin, Francoise. 1982. *The Psychology of the Planets.* Epping, New Hampshire: Astro Computing Services.

Gauquelin, Michel. 1969. *The Scientific Basis for Astrology.* New York: Stein and Day Publishers.

Gauquelin, Michel. 1983. *Birth-Times: A Scientific Investigation of the Secrets of Astrology.* New York: Hill and Wang.

Gauquelin, Michel. 1991. *Neo-Astrology: A Copernican Revolution.* London: Arkana.

Gardiner, Brian G. 1987. "Linnaeus' Floral Clock." *The Linnean* 3(1): 26–29.

Gardner, W.W. and H.A. Allard. 1920. "Effect of the Relative Length of Day and Night and Other Factors of the Environment on Growth and Reproduction in Plants." *Journal of Agricultural Research* XVIII: 53–606.

Garin, Eugenio. 1990. *Astrology in the Renaissance.* New York: Penguin/Arkana.

Gehlken, Erlend. 2012. *Weather Omens of Enūma Anu Enlil: Thunderstorms, Wind, and Rain (Tablets 44–49).* Leiden, The Netherlands: Brill.

Geneva, Ann. 1995. *Astrology and the Seventeenth Century Mind.* Manchester, United Kingdom: Manchester University Press.

Genovese, J. E. C. 2014. "A Failed Demonstration of Sun Sign Astrology." *Comprehensive Psychology* 3(16).

Gilder, Joshua and Anne-Lee Gilder. 2005. *Heavenly Intrigue: Johannes Kepler, Tycho Brahe, and the Murder Behind One of History's Greatest Scientific Discoveries.* New York: Doubleday.

Gillman, Michael and Hilary Erenler. 2008. "The galactic cycle of extinction." *International Journal of Astrobiology* 7(1):17-26.

Gingerich, Owen. 1973. "Johannes Kepler." In *Dictionary of Scientific Biography*, edited by C.C. Gillispie and M. DeBruhl, vol. 7:289–312.

Ginzburg, Carlo. 1986. "Clues, Myths, and the Historical Method." In *The High and the Low: The Theme of Forbidden Knowledge in the Sixteenth and Seventeenth Centuries.* Johns Hopkins Univ. Press.

Goad, John. 1686. *Astrometeorologica, or Aphorism and Discourses of the Bodies Celestial, their Natures and Influences.* J. Rawlins.

Gonda, Xenia, P. Erdos, M. Ormos, and Z. Rihmer. 2014. "Season of Birth Shows a Significant Impact on the Distribution of Affective Temperaments in a Nonclinical Population. *European Neuropsychopharmacology* 24: S345.

Görke, Alexander and Armin Scholl. 2006. "Niklas Luhmann's Theory of Social Systems and Journalism Research." *Journalism Studies* (August).

Grafton, Anthony. 1991. *Defenders of the Text: The Traditions of Scholarship in the Age of Science, 1450–1800.* Cambridge, Massachusetts: Harvard University Press, 1991.

Grafton, Anthony. 1995. Girolamo Cardano and the Tradition of Classical Astrology. The Rothschild Lecture.

Grafton, Anthony. 1999. *Cardano's Cosmos: The Worlds and Works of a Renaissance Astrologer.* Cambridge, Massachusetts: Harvard University Press.

Grafton, Anthony. 2002. *Magic and Technology in Early Modern Europe.* Washington, D.C.: Smithsonian Institution Libraries.

Grafton, Anthony and William R. Newman, eds. 2001. *Secrets of Nature: Astrology and Alchemy in Early Modern Europe.* MIT Press.

Granada, M.A. 2005. "The Discussion between Kepler and Roeslin on the Nova of 1604, in Supernovae as Cosmological Lighthouses." *ASP Conference Series* 342.

Gribbin, John and Stephen Plagemann. 1974. *The Jupiter Effect: The Planets as Triggers of Devastating Earthquakes.* New York: Random House.

Grim, Patrick. Ed. 1982. *Philosophy of Science and the Occult.* Albany, New York: SUNY Press.

Guinard, Patrice. 2002. "Astrology: The Manifesto, Part 5." *Considerations* XVII(2):69–90.

Hahm, David E. 1977. *The Origins of Stoic Cosmology*. Columbus, Ohio: Ohio State University Press.

Halberg, Franz, Germaine Cornélissen, George Katinas, Elena V. Syutkina, Robert B. Sothern, Rina Zaslavskaya, Francine Halberg, Yoshihiko Watanabe, Othild Schwartzkopff, Kuniaki Otsuka, Roberto Tarquin, Perfetto Frederico, and Jarmila Siggelova. 2003. "Transdisciplinary Unifying Implications of Circadian Findings in the 1950s." *Journal of Circadian Rhythms* 1:2.

Halberg, Franz. 2004. *Solar wind's mimicry can override seasons in unicells and humans*. Unpublished paper received from the author.

Halberg F., Otsuka K., Katinas G., Sonkowsky R., Regal P., Schwartzkopff O., Jozsa R., Olah A., Zeman M., Bakken E. E., and Cornélissen G. 2004. "A chronomic tree of life: ontogenetic and phylogenetic 'memories' of primordial cycles keys to ethics." *Biomedicine & Pharmacotherapy* 58(Suppl 1): S1–S11.

Hamilton, Mark A. 2015. "Astrology as a Culturally Transmitted Heuristic Scheme for Understanding Seasonality Effects: A Response to Genovese (2014)." *Comprehensive Psychology* 4(7).

Hammerschlag, Richard, Michael Levin, Rollin McCraty, Namuun Bat, John A. Ives, Susan K. Lutgendorf, and James L. Oschman. 2015. "Biofield Physiology: A Framework for an Emerging Discipline." *Global Advances in Health Medicine* (November 4): 35–41.

Hand, Robert. 1987–1988. "The Emergence of an Astrological Discipline." *NCGR Journal:* 66–70.

Hand, Robert. 1982. *Essays in Astrology*. The Wave Theory of Astrology. Para Research.19–32.

Hand, Robert. 2004. Astrology as a Revolutionary Science. In *The Future of Astrology*, ed. A. T. Mann. Paraview. pp. 23–39.

Hand, Robert. 2014. *The Use of Military Astrology in Late Medieval Italy: The Textual Evidence*. Dissertation submitted to the Faculty of the Department of History School of Arts & Sciences of The Catholic University of America. P.26 ff.

Hanson, Kirby, George A. Maul, and William McLeish. 1987. "Precipitation and the Lunar Synodic Cycle: Phase Progression across the United States." *Journal of Applied Meteorology* 26(10);1358–62.

Hardin, Paul E., Jeffrey C. Hall, and Michael Rosbash. 1990. "Feedback of the Drosophila Period Gene Product on Circadian Cycling of its Messenger RNA Levels. *Nature* 343: 536–40.

Hart, Vlastimil, Petra Nováková, Erich Pascal Malkemper, Sabine Begall, Vladimír Hanzal, Miloš Ježek, Tomáš Kušta, Veronika Němcová, Jana Adámková, Kateřina Benediktová, Jaroslav Červený, and Hynek Burda. 2013. "Dogs are Sensitive to Small Variations of the Earth's Magnetic Field. *Frontiers in Zoology* 10, no. 1: 80.

Haskins, Charles Homer. 1927. *The Renaissance of the 12th Century*. Harvard University Press.

Havu, Sirkka. 2005. *Conrad Gesner: Father of Bibliography*. Finland: Helsinki University Library.

Hayes, Brian. 2005. "Life Cycles." *American Scientist* 93:299–303.

Hays, J. D., J. Imbrie, and N. J. Shackleton. 1976. "Variations in the Earth's Orbit: Pacemaker of the Ice Ages." *Science* 194(4270):1121–1132.

Heisenberg, Werner. 1958. *Physics and Philosophy: The Revolution in Modern Science*. New York: Harper and Row.

Henninger, S. K. Jr. 1960. *A Handbook of Renaissance Meteorology*. Durham, North Carolina: Duke University Press.

Hesiod. 1982. *Hesiod, The Homeric Hymns, and Homerica*. Translated by H. G. Evelyn-White. Cambridge, Mass: Harvard University Press.

Heusner, A. 1965. "Sources of Error in Study of Diurnal Rhythm in Energy Metabolism." In *Circadian Clocks,* edited by J. Aschoff. Amsterdam, Netherlands: North-Holland Publishing Company: 3–12.

Hill, Christopher. 1972. *The World Turned Upside Down*. New York: Viking Press.

Hill, Judith A. and Jacalyn Thompson. 1988–89. "The Mars-Redhead Link: A Scientific Test of Astrology." *NCGR Journal* (Winter).

Holden, James Herschel. 1996. *A History of Horoscopic Astrology*. Tempe, Arizona: American Federation of Astrologers.

Hoppman, Jürgen G. H. 1997. "The Lichtenberger Prophecy and Melanchthon's Horoscope for Luther." *Culture and Cosmos* 1, no. 2 (Autumn/Winter).

Howe, Ellic. 1967. *Urania's Children: The Strange World of the Astrologers.* Kimber London.

Hoyt, Douglas V. and Kenneth Schatten. 1997. *The Role of the Sun in Climate Change.* Oxford, United Kingdom: Oxford University Press.

Hunt, William. 1696. *Demonstration of Astrology.* George Sawbridge.

Imbrie, John, and Katherine Plamer Imbrie. 1979. *Ice Ages: Solving the Mystery.* Cambridge: Harvard University Press.

Jacobs, Jayj. "The Law and Astrology." *Association for Astrological Networking.* Beverly Hills, CA, November 6, 1994. AFAN website.

Jankovic, Vladimir. 2000. *Reading the Skies: A Cultural History of English Weather 1650–1820.* University of Chicago Press.

Jenks, Stuart. 1983. "Astrometeorology in the Middle Ages." *Isis* 74:185–210.

Johndro, L. Edward. 1933. "Tuning in Super-Heterodyne Man." *The National Astrological Journal.*

Johnson, C. H., and S. S. Golden. 1999. "Circadian programs in cyanobacteria: adaptiveness and mechanism." *Annual Review of Microbiology* 53:389–409.

Johnson, Francis R. 1937, 1968. *Astronomical Thought in Renaissance England.* Baltimore: Johns Hopkins Press. Reprinted London: Octagon Books.

Jose, Paul D. 1965. "Sun's Motion and Sunspots." *The Astronomical Journal* 10(1):193–200.

Kahneman, Daniel. 2011. *Thinking Fast and Slow.* New York: Farrar, Straus and Giroux.

Kastrup, Bernardo. 2015. *Brief Peeks Beyond: Critical essays on metaphysics, neuroscience, free will, skepticism and culture.* United Kingdom: Iff Books (imprint of John Hunt Publishing).

Keeling, Charles D. and Timothy P. Whorf. 2000. "The 1,800-year oceanic tidal cycle: A possible cause of rapid climate change." *Proceedings of the National Academy of Sciences* 97(8):3814-3819.

Kepler, Johannes. (1987). *Concerning the More Certain Fundamentals of Astrology.* 1602. Sequim, Washington: Holmes Publishing Group.

Kepler, Johannes. (1997). *The Harmony of the World.* 1619. Translated by E. J. Aiton, A. M. Duncan, and J. V. Field. Philadelphia: American Philosophical Society.

Kepler, Johannes. (2008). *Tertius Interveniens*. 1610. In *Kepler's Astrology*. Translated by Ken Negus. Kansas City, Missouri: Earth Heart Publications.

Kepler, Johannes. (1942). *De Fundamentis Astrologiae Certioribus or Concerning the More Certain Fundamentals of Astrology*. 1602. trans. by Edmund Meywald, Clancy Publications.

Kirschvink, Joseph L., Atsuko Kobayashi-Kirschvink, Juan C. DiazRicci, and Steven J. Kirschvink. 1992. "Magnetite in Human Tissues: A Mechanism for the Biological Effects of Weak ELF." *Magnetic Fields Bioelectromagnetics* 1: 101–13.

Kirschvink, J. L., Walker, and M. M. Deibel, C. 2001. "Magnetite-based Magnetoreception." *Current Opinion in Neurobiology* 11 :462–67.

Kitson, Annabella, ed. 1989. *History and Astrology*. New York: Mandala.

Klarsfeld, André. 2013. *At the Dawn of Chronobiology*. ESPCI ParisTech, Neurobiology laboratory (ESPCI/CNRS UMR 7637), "Genes, Circuits, Rhythms, Neuropathology" Team.

Klein Tank, A.M.G. and Coauthors. 2002. "Daily Dataset of 20th-century Surface Air Temperature and Precipitation Series for the European Climate Assessment." *International Journal of Climatology* 22:1441–53.

Knight, John Alden. 1972. *Moon Up, Moon Down*. Montoursville, Penn: Solunar Sales Co.

Koestler, Arthur. 1963. *The Sleepwalkers*. New York: Grosset & Dunlap.

Kolisko, Lilly. 1928. "Workings of the Stars in Earthly Substance" Experimental Studies from the Biological Institute of the Goetheanum. (Original publisher: Orient-Occident Verlag, Stuttgart.)

Kollerstrom, N. 1976. "The Correspondence of Metals and Planets— Experimental Studies." *The Astrological Journal* 18, no. 3: 65–72.

Kondo, Takao and Masahiro Ishiura. (1999) "The circadian clocks of plants and cyanobacteria." *Trends in Plant Science* 4(5):171-176.

Koukkari, Willard L. and Robert B. Sothern. 2006. *Introducing Biological Rhythms*. Switzerland: Springer.

Krajick, Kevin. 2019. "Scientists Track Deep History of Planets' Motions, and Effects on Earth's Climate: Newly Forming Map of Chaos in the Solar System." *State of the Planet* (News from the Columbia Climate School). March 4, 2019.

Krijgsman, W., Fortuin, A. R., Hilgen, F. J., and Sierro, F. J. 2001. "Astrochronology for the Messinian Sorbas basin (SE Spain) and orbital (precessional) forcing for evaporite cyclicity." *Sedimentary Geology* 140(1–2):43–60.

Kuhn, Thomas S. 1962. *The Structure of Scientific Revolutions*. Chicago: University of Chicago Press.

Kumar, Sunil. 2012. Adorno on Astrology. *Telos*, July 3, 2012.

Kusukawa, Sachiko. 1995. *The Transformation of Natural Philosophy*. Cambridge, England: Cambridge University Press.

Landscheidt, Theodor. 1973. *Cosmic Cybernetics: The foundations of a modern astrology*; Fribourg, Switzerland: Ebertin-Verlag.

Landscheidt, Theodor. 1987. "Long-Range Forecasts of Solar Cycles and Climate Change." In Rampino, Sanders, Newman and Konigsson, eds. *Climate: History, Periodicity, and Predictability*. Van Nostrand Reinhold.

Landscheidt, Theodor. 1989. *Sun, Earth, Man: A Mesh of Cosmic Oscillations —How Planets Regulate Solar Eruptions, Geomagnetic Storms, Conditions of Life and Economic Cycles*. Old Windsor, UK: Urania Trust.

Lang, Kenneth R. 2001. *The Cambridge Encyclopedia of the Sun*. Cambridge, England: Cambridge University Press.

Laszlo, Ervin. 1972. *The Systems View of the World*. New York: George Braziller, Inc.

Lawrence-Mathers, Anne. 2020. *Medieval Meteorology*. Cambridge, England: Cambridge University Press.

LeGrice, Keiron. 2009. "The Birth of a New Discipline: Archetypal Cosmology in Historical Perspective." *Archai: The Journal of Archetypal Cosmology*, vol. 1, number 1.

LeGrice, Keiron. 2011. *The Archetypal Cosmos*. Edinburgh: Floris Books.

Lehoux, Daryn. 2007. *Astronomy, Weather, and Calendars in the Ancient World: Parapegmata and Related Texts in Classical and Near-Eastern Societies*. Cambridge: Cambridge University Press.

Lemay, Richard. 1987. "The True Place of Astrology in Medieval Science and Philosophy: Towards a Definition." In *Astrology, Science and Society: Historical Essays*. Edited by Patrick Curry. Woodbridge, United Kingdom: Boydell & Brewer.

Leo, Alan. 1918. *Esoteric Astrology*. London: N. Fowler & Co.

Lethbridge, Mae DeVoe. 1970. "Relationship between Thunderstorm Frequency and Lunar Phase and Declination." *Journal of Geophysical Research* 75, no. 27 (September 20).

Lilly, William. 1644. *England's Prophetical Merlin*. London, John Partridge.

Lilly, William, ed. 1676. *Anima Astrologiae; or, A Guide for Astrologers*. B. Harris.

Lohmann, K. J. and A. O. Willows. 1987. "Lunar-modulated Geomagnetic Orientation by a Marine Mollusk." *Science* 16 (January).

Loring, Philip A. 2020. "Finding Our Niche: Toward a Restorative Human Ecology." Nova Scotia: Fernwood Publishing.

Mackey, Richard. 2007. "Rhodes Fairbridge and the Idea that the Solar System Regulates the Earth's Climate." *Journal of Coastal Research* 50:955–68.

Maclean, Ian. 1998. "Foucault's Renaissance Episteme Reassessed: An Aristotelian Counterblast." *Journal of the History of Ideas* 59(1):149–166.

Marchant, Jo. 2009. *Decoding the Heavens: A 2,000-year-old-computer and the Century Long Search to Discover Its Secrets*. Boston, Mass.: Da Capo Press.

Marsh, Nigel and Henrik Svensmark. 2000. "Low Cloud Properties influenced by Cosmic Rays." *Physical Review Letters* 85(23):5004–5007.

Marshack, A. 1972. *The Roots of Civilization: The Cognitive Beginning of Man's First Art, Symbol and Notation*. New York: McGraw-Hill.

Maturana, Humberto R. 1988. "Reality: The Search for Objectivity or the Quest for a Compelling Argument." *The Irish Journal of Psychology* 9(1):25–82

Maturana, Humberto R. and Francisco J. Varela. 1992. *The Tree of Knowledge: The Biological Roots of Human Understanding*. Berkeley: Shambhala.

Maxwell, Grant. 2017. "Archetype and Eternal Object: Jung, Whitehead, and the Return of Formal Causation." *Archai: The Journal of Archetypal Cosmology* 6.

Mayewski, Paul Andrew and Frank White. 2002. *The Ice Chronicles: The Quest to Understand Global Climate Change*. University Press of New England.

Mayo, J., O. White and H. J. Eysenck. 1978. "An Empirical Study of the

Relation between Astrological Factors and Personality." *The Journal of Social Psychology* 105(2).

McGrew, John H. and Richard M. McFall. 1990. "A Scientific Inquiry into the Validity of Astrology." *Journal for Scientific Exploration* 4(I):75–83.

McRitchie. Ken. 2011. "Support for Astrology from the Carlson Double-blind Experiment." *ISAR International Astrologer* 40, no. 2 (August): 34–39.

McRitchie, Ken. 2016. "Clearing the Logjam in Astrological Research: Commentary on Geoffrey Dean and Ivan Kelly's Article 'Is Astrology Relevant to Consciousness and Psi?'" *Journal of Consciousness Studies* 23(9–10):153–179.

Medvedev, Mikhail V. and Adrian L. Melott. 2007. "Do Extragalactic Cosmic Rays Induce Cycles in Fossil Diversity?" *The Astronomical Journal* 664:879–89.

Merchant, Carolyn. 1980. *The Death of Nature*. New York: Harper and Row.

Merton, Robert K. 1970. *Science, Technology and Society in Seventeenth-Century England*. New York: Harper and Row.

Methuen, Charlotte. 1996. "The Role of the Heavens in the Thought of Philip Melanchthon." *Journal of the History of Ideas* 57(3):385–403.

Mingers, John. 2004. "Can Social Systems be Autopoietic? Bhaskar's and Giddens' Social Theories." *Journal for the Theory of Social Behaviour* 34 (December): 4.

Moore-Ede, Martin C., Frank M. Sulzman and Charles A. Fuller. 1982. *The Clocks That Time Us: Physiology of the Circadian Timing System*. Harvard University Press.

Napier, W. M. and S. V. M. Clube. 1979. "A Theory of Terrestrial Catastrophism." *Nature* 282:455–59.

Nelson, J. H. 1968. "Forecasting Radio Weather." In Dewey, E. R. ed., *Cycles—Selected Writings*. Pittsburgh: Pittsburgh Foundation for the Study of Cycles.

Nelson, J.H. 1974. *Cosmic Patterns*. American Federation of Astrologers.

Nescolarde-Selva, Josué Antonio, José-Luis Usó-Doménech and Hugh Gash. 2017. "What Are Ideological Systems?" *Systems* 5(21).

Neugebauer O. and H.B. Van Hoesen. 1959. *Greek Horoscopes*. Philadelphia, Pennsylvania: American Philosophical Society.

Niroma, Timo. 2009. "Understanding Solar Behaviour and its Influence on Climate." *Energy & Environment* 20(1):145–59.

Nisbett, Richard E. 2004. *The Geography of Thought: How Asians and Westerners Think Differently . . . and Why.* New York: Free Press.

North, J. D. 1986. *Horoscopes and History.* London: Warburg Institute.

North, J. D. 1987. "Medieval Concepts of Celestial Influences: A Survey." In Curry, Patrick, ed. *Astrology, Science and Society: Historical Essays.* Suffolk: The Boydell Press.

Oll, Moonika. 2010. *Hellenistic Astrology as a Case Study of Cultural Translation.* A dissertation submitted to the University of Birmingham for the degree of MPhil(B) in Classics and Ancient History. Institute of Archaeology and Antiquity, College of Arts and Law, University of Birmingham, September 2010.

Olsen, Paul E. and Dennis V. Kent. 1996. "Milankovitch Climate Forcing in the Tropics of Pangaea During the Late Triassic." *Palaeogeography, Palaeoclimatology, Palaeoecology* 122 (1–4): 1–26.

Olsen, P. E. 1999 *Climate, Astronomical Forcing, and Chaos. Milankovitch Theory and Climate.* Report on the International Workshop for a Climatic, Biotic, and Tectonic, Pole-to-Pole Coring Transect of Triassic-Jurassic Pangea at Acadia University, Nova Scotia, Canada.

Olsen, P. E. and D. V. Kent. 1999. "Long-period Milankovitch Cycles from the Late Triassic and Early Jurassic of Eastern North America and their Implications for the Calibration of the Early Mesozoic Time Scale and the Long-term Behavior of the Planets." *Philosophical Transactions: Mathematical, Physical and Engineering Sciences* 357(1757): 1761–86.

Orey, Cal. 2006. *The Man Who Predicts Earthquakes: Jim Berkland, Maverick Geologist—How His Quake Warnings Can Save Lives.* Boulder, Colorado: Sentient Publications.

Osler, Margaret J. ed. 2000. *Rethinking the Scientific Revolution.* Cambridge: University Press.

Ouimet, Alan J., O.F.S. "The Condemnation of Astrology: The Secret Vatican Archives and Pope Sixtus V."

Ouyang, Yan, Carol R. Andersson, Takao Kondo, Susan S. Golden, and Carl Hirschie Johnson. 1998. "Resonating Circadian Clocks Enhance Fitness in Cyanobacteria. *Proceedings of the National Academy of Sciences* 95: 8660–64.

Palmer, John D. 1995. *Biological Rhythms and Clocks of Intertidal Animals.* Oxford, England: Oxford University Press.

Parker, D. E., T. P. Legg, and C. K. Folland. 1992. "A New Daily Central England Temperature Series, 1772–1991." *International Journal of Climatology* 12:317–42.

Parker, Derek. 1975. *Familiar to All: William Lilly and Astrology in the Seventeenth Century.* London: J. Cape.

Pawlik, K., and L. Buse. 1979. "Self-attribution as a Differential, Psychological Moderator Variable: Verification and Clarification of Eysenck's Astrology-personality Correlations." *Zeitschrift für Sozialpsychologie* 10(1):54–69.

Payne, Buryl. 1988. *The Body Magnetic & Getting Started in Magnetic Healing.* Santa Cruz: Psychophysics.

Pearce, Alfred John. 1970. *The Text-Book of Astrology.* 1895. Tempe, Arizona: American Federation of Astrologers.

Peart, Sanda J. 1991. "Sunspots and Expectations: W. S. Jevons's Theory of Economic Fluctuations." *Journal of the History of Economic Thought* 13(2).

Pellegrini, Robert J. 1973. "The Astrological 'Theory' of Personality: An Unbiased Test by a Biased Observer." *The Journal of Psychology: Interdisciplinary and Applied* 85(1).

Pelikan, Wilhelm. 1973. *The Secrets of Metals.* New York: Anthroposophic Press.

Pepper, Stephen. 1942. *World Hypotheses: A study of evidence.* Berkeley: University of California Press.

Phillips, Kenneth J. H. *Guide to the Sun.* 1992. Cambridge, England: Cambridge University Press.

Pingree, David. 1968. *The Thousands of Abu Ma'shar.* London: The Warburg Institute.

Pittendrigh, Colin S. 1960. "Circadian Rhythms and the Circadian Organization of Living Systems." *Cold Spring Harbor Symposium on Quantitative Biology* 25:159–84.

Plato. 1961. *Collected Dialogues of Plato.* Edited by Edith Hamilton and Huntington Cairns. Princeton, New Jersey: Princeton University Press.

Plotinus. 1952. "Second Ennead, Third Tractate: Are the Stars Causes?"

In *The Six Enneads*. Translated by Stephen Mackenna and B. S. Page. Chicago, London, Toronto: William Benton Publisher.

Pope Sixtus V. *Coeli et terrae*. January 1586. Sections of the text translated from the Latin by the author.

Ptolemy, Claudius. (c. 150). *Tetrabiblos*. Translated by J. M. Ashmand 1936. Eden, New York: The Aries Press.

Ptolemy, Claudius. (c. 150). *Tetrabiblos*. Translated by F. E. Robbins 1940. Cambridge, Massachusetts: Harvard University Press.

Ptolemy, Claudius. (c. 150). *Tetrabiblos*. Translated by R. Schmidt 1996. Cumberland, Maryland: Golden Hind Press.

Rabin, Sheila J. 1997. "Kepler's Attitude Toward Pico and the Anti-astrology Polemic." *Renaissance Quarterly* 50:750–70.

Rahmstorf, Stephan, David Archer, Denton S. Ebel, Otto Eugster, Jean Jouzel, Douglas Maraun, Urs Neu, Gavin A. Schmidt, Jeff Severinghaus, Andrew J. Weaver, and Jim Zachos. 2004. "Cosmic Rays, Carbon Dioxide, and Climate." *Eos Trans* 85(4):38–41.

Raisbeck, G. M., F. Yiou, J. Jouzel and J. R. Petit. 1990. "10Be and 2H in polar ice cores as a probe of the solar variability's influence on climate." *Philosophical Transactions of the Royal Society of London A*, 330(1615): 463–70.

Rampino, M. R. and R. B. Stothers. 1984. "Terrestrial Mass Extinctions, Cometary Impacts and the Sun's Motion Perpendicular to the Galactic Plane." *Nature* 308: 709–12.

Rampino, Michael R. and Bruce M. Haggerty. 1996. "The Shiva Hypothesis: Impacts, Mass Extinctions, and the Galaxy. *Earth, Moon, and Planets* 72 (1–3): 441–60.

Randall, Lisa and Matthew Reece. 2014. "Dark Matter as a Trigger for Periodic Comet Impacts." *Physical Review Letters* 112 (161301).

Raup, David and John Sepkoski Jr. 1982. "Mass Extinctions in the Marine Fossil Record." *Science* 19 (215), no. 4539: 1501–03.

Raup, D. M. and J. J. Sepkoski. 1984. "Periodicity of Extinction in the Geologic Past." *Proceeding of the National Academy of Sciences* 81:801–5.

Rawlins, Dennis. 1981. "sTARBABY" *Fate*, No. 34, October 1981, pp. 67–98.

Reid, G. C. 1987. "Influence of Solar Variability on Global Sea Surface Temperatures. *Nature* 329:142–43.

Rochberg, Francesca. 2007. *The Heavenly Writing: Divination, Horoscopy, and Astronomy in Mesopotamian Culture.* Cambridge, England: Cambridge University Press.

Rodriguez McRobbie, Linda. 2016. "How Are Horoscopes Still a Thing?" *Smithsonian.* January 5, 2016.

Rogers, J. H. 1998. "Origins of the Ancient Constellations: I. The Mesopotamian Traditions." *Journal of the British Astronomical Association* 108, no. 1: 9–28.

Rohde, Dirk. 2003. "Experiments at Moon-Saturn Conjunctions using the Capillary Dynamolysis Method of Lili Kolisko." *Elemente der Naturwissenschaft* 79(2):123–31.

Rosen, Edward. 1984. "Kepler's Attitude Toward Astrology and Mysticism." In *Occult and Scientific Mentalities of the Renaissance.* Edited by Brian Vickers. Cambridge, United Kingdom: Cambridge University Press.

Roy, Shouraseni Sen. 2006. "Impact of Lunar Cycle on the Precipitation in India." *Geophysical Research Letters* 33 (January).

Rudhyar, Dane. 1936. *The Astrology of Personality.* New York: Random House.

Rudhyar, Dane. 1967. *The Lunation Cycle.* Santa Fe, New Mexico: Aurora Press.

Rudhyar, Dane. 1969. *Astrology for New Minds.* Los Angeles: CSA Publishing.

Sagendorph, Robb. 1970. *America and her Almanacs.* Dublin, New Hampshire: Yankee Publishing, Inc.

Sandbach, F. H. 1975. *The Stoics.* Bristol, Conn.: The Bristol Press.

Scafetta, Nicola, Franco Milani, Antonio Bianchini, and Sergio Ortolani. 2016. "On the Astronomical Origin of the Hallstatt Oscillation Found in Radiocarbon and Climate Records Throughout the Holocene. *Earth-Science Reviews* 162 (November): 24–43.

Schaffer, Simon. 1987. Newton's Comets and the Transformation of Astrology. In Curry, Patrick, ed. *Astrology, Science and Society: Historical Essays.* Suffolk: The Boydell Press.

Schaffer, Simon. 1979. "Newton's Comets and the Transformation of Astrology." In *Astrology, Science and Society: Historical Essays.* Edited by Curry. Woodbridge, England: Boydell Press.

Schaffer, Simon. 2010. "The Astrological Roots of Mesmerism." *Studies in History and Philosophy of Science Part C: Studies in History and Philosophy of Biological and Biomedical Sciences* 41(2):158–68.

Schmid, Boris V., Ulf Büntgen, W. Ryan Easterday, Christian Ginzler, Lars Walløe, Barbara Bramanti, and Nils Chr. Stenseth. 2015. "Climate-driven introduction of the Black Death and successive plague reintroductions into Europe." *PNAS* March 10, 112 (10) 3020–3025.

Schoener, Johannes. (1994). *Opusculum Astrologicum*. 1539. Trans. Robert Hand. Cumberland, MD: Golden Hind Press.

Schulz, Michael. 2002. "On the 1470-year pacing of Dansgaard-Oeschger warm events." *Paleoceanography* 17(4)1–9.

Schulz, M., A. Paul, 2002. "Holocene climate variability on centennial-to-millennial time scales: 1. Climate records from the North-Atlantic realm." In: Wefer, G., Berger, W. H., Behre, K.-E. and Jansen, E., eds., *Climate development and history of the North Atlantic Realm*. Berlin: Springer Verlag.

Schwalb, A.; Dean, W. E.; Kromer, B. 2003. Centennial Drought Cyclicity in the Great Plains, USA: A Dominant Climate Pattern over the Past 4000 Years. EGS - AGU - EUG Joint Assembly, Abstracts from the meeting held in Nice, France, 6 - 11 April 2003.

Scofield, B. 1987–1988. "A Planetary Model of the Developing Self." *NCGR Journal*. pp. 62–65.

Scofield, B. 2000. *The Circuitry of the Self.* Amherst: One Reed Publications.

Scofield, B. 2013. "A Signal from Saturn in Daily Temperature Data." *Correlation* 29 (1).

Scofield, B. 2016. "Outer Planet Conjunctions and the 1500-Year Climate Cycle." *Journal of Research of the American Federation of Astrologers* 16 (2016): 151–80.

Scofield, B. 2019. "Astrology as an Analytical Technique for the Study of Self-organized Systems." *NCGR Syzygy* (Winter 2019–2020): 20–33.

Sextus Empiricus. (c. 200) 1949. *Against the Professors*. Trans. R. G. Bury. Loeb Classical Library. Harvard University Press.

Seymour, Percy. Adam Hilger. 1986. *Cosmic Magnetism*. Cambridge University Press.

Seymour, P. A. H., M. Willmott, and A. Turner. 1992. "Sunspots, Planetary Alignments and Solar Magnetism: A Progress Review." *Vistas in Astronomy* 35:39–71.

Seymour, Percy. 1998. "The Magus of Magnetism (An Interview With

Dr. Percy Seymour by Bronwyn Elko)." *The Mountain Astrologer.* 80 (August/September).

Seymour, Percy. 2008. *Dark Matters.* Franklin Lakes, New Jersey: Career Press.

Shackleton, Nicholas J. 2000. "The 100,000-Year Ice-Age Cycle Identified and Found to Lag Temperature, Carbon Dioxide, and Orbital Eccentricity." *Science* 289:1897–1902.

Shapin, Steven. 1994. *A Social History of Truth.* Chicago: University of Chicago Press.

Shapin, Steven. 1996. *The Scientific Revolution.* Chicago: University of Chicago Press.

Shaviv, N. J. 2002. "Cosmic Ray Diffusion from the Galactic Spiral Arms, Iron Meteorites and a possible Climatic Connection." *Physical Review Letters,* 89:051102.

Shaviv, N. J. 2003. "The Spiral Structure of the Milky Way, Cosmic-Rays, and Ice-Age Epochs on Earth." *New Astronomy* 8:39–77.

Shaviv, N. J & J. Veizer. 2003. "Celestial Driver of Phanerozoic Climate?" *GSA Today* 13:4–10.

Shaw, Sir Napier. 1926. *Manual of Meteorology. Vol. I, Meteorology in History.* Cambridge University Press.

Shindell, D., D. Rind, N. Balachandran, J. Lean and P. Lonergan. 1999. "Solar cycle Variability, Ozone, and Climate." *Science* 284:305-308.

Shumaker, Wayne. 1972. *The Occult Sciences in the Renaissance.* Berkeley: University of California Press.

Sibly, Ebenezer. 1798. *An Illustration of the Celestial Science of Astrology.* E. Sibley.

Simmonite, W. J. 1890. *The Celestial Philosopher, or the Complete Arcana of Astral Philosophy.* John Story.

Smith, George. 1921-1922. *The Dictionary of National Biography.* Oxford, United Kingdom: Oxford University Press.

Smuts, Jan Christian. 1926. *Holism and Evolution.* New York: MacMillan.

Solanki, S. K., I. G. Usoskin, B. Kromer, M. Schüssler, and J. Beer. 2004. "Unusual Activity of the Sun During Recent Decades Compared to the Previous 11,000 Years." *Nature* 431:1084–87.

Sprat, Thomas. 1667. *The History of the Royal-Society of London.* T. Sprat.

Stack S., Lester D. 1988. "Born under a bad sign? Astrological sign and suicide ideation." *Perceptual and Motor Skills*, Apr;66 (2):461–62.

Stefani, Frank, André Giesecke, and Tom Weier. 2018. "A Tayler-Spruit Model of a Tidally Synchronized Solar Dynamo." *Solar Physics*, 294(5).

Steirwalt, Sabrina. 2020. "Is Astrology Real? Here's What Science Says." *Scientific American Website* (June 20, 2020).

Strimer, Steve. 2000. "Synchronicity and Jung's Astrological Experiment." *NCGR Journal* (Winter 2000/2001): 45–53.

Sulzman F. M., D. Ellman, C. Fuller, M. C. Moore-Ede, and G. Wassmer. 1984. "Neurospora Circadian Rhythms in Space: A Reexamination of the Endogenous-Exogenous Question." *Science* 225: 232–34.

Svensmark, Henrik and Eigil Friis-Christensen. 1997. "Variations of Cosmic Ray Flux and Global Cloud Coverage—A Missing Link in Solar-Climate Relationships." *Journal of Atmospheric and Solar-Terrestrial Physics* 59(11):1225–32.

Svensmark, Henrik. 2007. "Cosmoclimatology: A New Theory Emerges." *Astronomy & Geophysics*, 48(1):18–24.

Sweeney, Beatrice M. 1960. "The Photosynthetic Rhythm in Single Cells of *Gonyaulax polyedra*." *Quantitative Biology* 25:145–148.

Sweeney, B. M. and Haxo, F. T. 1961. "Persistence of a Photosynthetic Rhythm in Enucleated *Acetabularia*." *Science* 134: 1361–63.

Sweeney, Beatrice M. 1987. *Rhythmic Phenomena in Plants*. Cambridge, Mass: Academic Press.

Symons, G. J. 1893. "English Meteorological Literature from the 15th to 17th Centuries." In *Report of the International Meteorological Congress*. Edited by Oliver L. Fassign. U.S. Department of Agriculture Weather Bureau. United States Weather Bureau. Bulletin No. 11. pp. 338–51.

Tamrazyan, Gurgen P. 1967. "Tide-forming Forces and Earthquakes. *Icarus*, 7(1–3):59–65.

Tamrazyan, Gurgen P. 1968. "Principal Regularities in the Distribution of Major Earthquakes Relative to Solar and Lunar Tides and Other Cosmic Forces. *Icarus* 9(1–3):574–92.

Tamrazyan, Gurgen P. 1988. "Prediction of the Destructive Armenian Earthquake of December 7, 1988." *Cycles,* 42 (March/April 1991): 93–99.

Tarnas, Richard. 2007. *Cosmos and Psyche*. New York: Plume.

Taylor, Bernie. 2004. *Biological Time*. Oregon: The Ea Press.

Tester, S. J. 1987. *A History of Western Astrology*. Woodbridge, United Kingdom: Boydell & Brewer Academic Press.

Thomas, Keith. 1971. *Religion and the Decline of Magic*. Oxford, United Kingdom: Oxford University Press.

Thompson, R. Campbell. 1900. *The Reports of the Magicians and Astrologers of Nineveh and Babylon in the British Museum*. London: Luzac and Co.

Thorndike, Lynn. 1923–1941. *A History of Magic and Experimental Science*. 8 Vols. New York: Columbia University Press.

Thorndike, Lynn. 1955. "The True Place of Astrology in the History of Science." *Isis* 46:273–278.

Tobyn, Graeme. 1997. *Culpeper's Medicine: A Practice of Western Holistic Medicine*. Ogden, Utah: Element Books.

Todd, Michael P., Patrice S. Salome, Hannah J. Yu, Taylor R. Spencer, Emily L. Sharp, Mark A. McPeek, Jose M. Alonso, Joseph R. Ecker, and C. Robertson McClung. 2003. "Enhanced Fitness Conferred by Naturally Occurring Variation in the Circadian Clock." *Science* 302:1049–53.

Tomaschek, R. 1959. "Great Earthquakes and the Astronomical Position of Uranus." *Nature,* no. 4681 (July 18, 1959): 177–78.

Ueno, Yusuke, Masayuki Hyodo, Tianshui Yang, and Shigehiro Katoh. 2019. "Intensified East Asian Winter Monsoon During the Last Geomagnetic Reversal Transition." *Scientific Reports* 9, no. 9389.

Ulansey, David. 1991. *The Origins of the Mithraic Mysteries: Cosmology and Salvation in the Ancient World*. Oxford, United Kingdom: Oxford University Press.

University of California—Los Angeles. 2004. "Strong Earth Tides Can Trigger Earthquakes, UCLA Scientists Report." *ScienceDaily* (October 22).

University of California—Berkeley. 2009. "Sun and Moon Trigger Deep Tremors on San Andreas Fault." *ScienceDaily* (December 25).

Vanden Broecke, Steven. 2003. *The Limits of Influence: Pico, Louvain, and the Crisis of Renaissance Astrology*. Leiden, Netherlands: Brill Academic Publishers.

Vanden Broecke, Steven. 2019. "Astrological Contingency: Between

Ontology and Epistemology (1300–1600)." In *Contingency and Natural Order in Early Modern Science* (Vol. 332). Edited by P. D. Omodeo and R. Garau. New York: Springer.

van Rooij, Jan J. F., Marianne A. Brak, and Jacques J. F. Commandeur. 1988. "Introversion-Extraversion and Sun-Sign." *The Journal of Psychology* 122(3):275–78.

Vaughan, Valerie. 2002. *Earth Cycles: The Scientific Evidence for Astrology.* Vol. 1, The Physical Sciences. Amherst, Massachusetts: One Reed Publications.

Veno, A. and P. Pamment. 1979. "Astrological Factors and Personality: A Southern Hemisphere Replication." *The Journal of Psychology* 101:73–77.

Verhulst, J. 2003. "World Cup Soccer Players Tend to be Born with Sun and Moon in Adjacent Zodiacal Signs." *British Journal of Sports Medicine* 2000(34):465–66.

Vidmar, Joseph E. 2008. "A Comprehensive Review of the Carlson Astrology Experiments." *Correlation* 26 (1).

Wake, Ryotaro, Takuya Misugi, Kenei Shimada, and Minoru Yoshiyama1. 2010. "The Effect of the Gravitation of the Moon on Frequency of Births." *Environmental Health Insights* 2010, no. 4:65–69.

Walker, D. P. 1958, 2000. *Spiritual and Demonic Magic from Ficino to Campanella.* Univeristy Park, Pennsylvania: Pennsylvania State University Press.

Ward, Richie R. 1971. *The Living Clocks.* New York: Alfred A. Knopf.

Wedel, Theodore Otto. 1920. *The Medieval Attitude Toward Astrology.* New Haven, Connecticut: Yale University Press.

Weigelius, Valentine. 1649. *Astrology Theologized.* George Whittington.

West, Anthony. 1991. *The Case for Astrology.* London: Arkana.

Westman, Robert S. and J. E. McGuire. 1977. *Hermeticism and the Scientific Revolution.* Los Angeles: William Andrews Clark Memorial Library.

Whitehouse, Dr. David. 1999. "Medieval Astromer's Horscope Discovered." *BBC News Website* (March 3, 1999).

Whittaker, Robert H. 1969. "New Concepts of Kingdoms or Organisms." *Science* 163: 150–94.

Wilkie, R. 1974. "The Process Method Versus the Hypothesis Method: A Nonlinear Example of Peasant Spatial Perception and Behavior." In

Proceedings of the 1972 Meeting of the International Geographical Union Commission on Quantitative Geography. Edited by Maurice Yeates. Montreal & London: McGill University Press.

Wilsford, Thomas. 1642. *Nature's Secrets.* N. Brooke.

Wilson, I. R. G., B. D. Carter, and I. A. Waite. 2008. "Does a Spin-Orbit Coupling Between the Sun and the Jovian Planets Govern the Solar Cycle?" *Publications of the Astronomical Society of Australia* 25:85–93.

Winkless III, Nels and Iben Browning. 1980. *Climate and the Affairs of Men.* Flint Hill, Virginia: Fraser Publishing Co.

Woolum, John C. 1991. "A Re-Examination of the Role of the Nucleus in Generating the Circadian Rhythm in *Acetabularia.*" *Journal of Biological Rhythms* 6(2):129–36.

Wright, Peter. 1975. "Astrology and Science in Seventeenth Century England" *Social Studies of Science* 5(4).

Yates, Frances A. 1964. *Giordano Bruno and the Hermetic Tradition.* New York: Vintage Books.

Zeilik, Michael. 2002. *Astronomy: The Evolving Universe.* Cambridge, England: Cambridge University Press.

Zhou, Jiansong and Ka-Kit Tung. 2010. "Solar Cycles in 150 Years of Global Sea Surface Temperature Data." *Journal of Climate* 23(12):3234–48.

Zivkovic, Bora. 2011. "Circadian Clock Without DNA—History and the Power of Metaphor." *Scientific American* (blog). February 11, 2011.

INDEX

Note: Page numbers in *italics* indicate illustrations.

512

BOOKS OF RELATED INTEREST

Astrology for Mystics
Exploring the Occult Depths of the Water Houses in Your Natal Chart
by Tayannah Lee McQuillar

Moon Phase Astrology
The Lunar Key to Your Destiny
by Raven Kaldera

360 Degrees of Your Star Destiny
A Zodiac Oracle
by Ellias Lonsdale

The Chiron Effect
Healing Our Core Wounds through Astrology, Empathy, and
Self-Forgiveness
by Lisa Tahir, LCSW

Aspects in Astrology
A Guide to Understanding Planetary Relationships in the Horoscope
by Sue Tompkins

The Twilight of Pluto
Astrology and the Rise and Fall of Planetary Influences
by John Michael Greer

Alchemical Tantric Astrology
The Hidden Order of Seven Metals, Seven Planets, and Seven Chakras
by Frederick Hamilton Baker

Astrology and the Rising of Kundalini
The Transformative Power of Saturn, Chiron, and Uranus
by Barbara Hand Clow

INNER TRADITIONS • BEAR & COMPANY
P.O. Box 388
Rochester, VT 05767
1-800-246-8648
www.InnerTraditions.com

Or contact your local bookseller